This publication forms part of an Open University module. Details of this and other Open University modules can be obtained from the Student Registration and Enquiry Service, The Open University, PO Box 197, Milton Keynes MK7 6BJ, United Kingdom (tel. +44 (0)845 300 6090; email general-enquiries@open.ac.uk).

Alternatively, you may visit the Open University website at www.open.ac.uk where you can learn more about the wide range of modules and packs offered at all levels by The Open University.

To purchase a selection of Open University materials visit www.ouw.co.uk, or contact Open University Worldwide, Walton Hall, Milton Keynes MK7 6AA, United Kingdom for a brochure (tel. +44 (0)1908 858779; fax +44 (0)1908 858787; email ouw-customer-services@open.ac.uk).

The Open University, Walton Hall, Milton Keynes, MK7 6AA.

First published 2015.

Edited, designed and typeset by The Open University, using the Open University TeX System.

Printed in the United Kingdom by Hobbs the Printers Ltd, Totton, Hampshire.

The Open University has had Woodland Carbon Code Pending Issuance Units assigned from Doddington North forest creation project (IHS ID103/26819) that will, as the trees grow, compensate for the greenhouse gas emissions from the manufacture of the paper in MST210 Block F. More information can be found at https://www.woodlandcarboncode.org.uk/

ISBN 978 1 7800 7872 4

1.1

Contents

Contents

Reviewing the model

Introduction

In Unit 8, mathematical modelling was introduced via a set of key stages as follows.

1 **Specify the purpose of the model:**
   define the problem;
   decide which aspects of the problem to investigate.

2 **Create the model:**
   state assumptions;
   choose variables and parameters;
   formulate mathematical relationships.

3 **Do the mathematics:**
   solve equations;
   draw graphs;
   derive results.

4 **Interpret the results:**
   collect relevant data;
   describe the mathematical solution in words;
   decide what results to compare with reality.

5 **Evaluate the model:**
   test the model by comparing its predictions with reality;
   criticise the model.

The diagram in Figure 1 was introduced to emphasise these key stages.

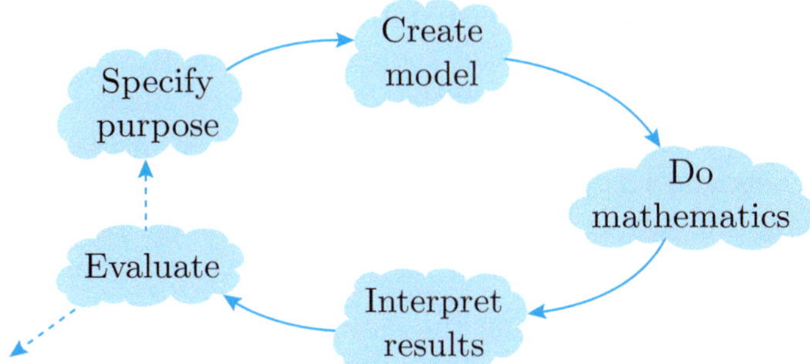

Figure 1   Mathematical modelling process

In Unit 8 the emphasis was on creating a first model, by making sufficient simplifying assumptions so that the essential features of the model are considered, without getting lost in a myriad of practical details. In this unit you will have the opportunity to improve your modelling skills and to see how models might be improved by assessing their deficiencies, before going round the modelling cycle again, taking steps to rectify some of those deficiencies. It should be noted that most complex models start out as very simple models, with more features added stage by stage to overcome the various shortcomings of the current model.

It should be stated at the outset that in modelling there is no right model. Any model that follows on from the assumptions and simplifications made is a useful model. Other models may give better predictions, but a simple model that includes the essential details may be much better than a complex model that tries to cover too much. When you start writing modelling reports, you should produce a fairly simple model and suggest how this might be revised. In writing a report, the key thing is to follow the modelling process, for example, listing the assumptions and simplifications, and showing how this establishes the model. In this module, this is more important than the model that you derive.

In Section 1 we introduce the method of dimensional analysis as a way of creating relationships between variables that are dimensionally consistent. In Section 2 we look at the sensitivity of the solution to variations in the parameters as a method of estimating the accuracy of the solution. We also revisit the tin can model created in Unit 8, to discuss its limitations with a view to trying to improve the model. In Section 3 we return to the problem of pollution in the Great Lakes to give examples of how the very simple model developed in Unit 8 can be gradually made more sophisticated by revising some of the simplifying assumptions that were made. Finally, Section 4 presents a new case study as a demonstration of the way in which a modelling investigation can be written up.

1 Method of dimensional analysis

In the final section of Unit 8, we used the base dimensions of physical quantities to check the consistency of equations and to find the dimensions of parameters in equations. However, the fact that an equation must be dimensionally consistent can also be used to *suggest* or check for a possible form of the relationship between the physical quantities involved in modelling a situation, once the key variables and parameters have been identified. This process is called the method of **dimensional analysis**.

We will illustrate the method by considering the example of a projectile, which we met in Unit 3, Section 5. Our aim is to find the form of the formula for the range of a projectile thrown on horizontal ground in terms of the parameters involved in the situation. If we assume that the projectile is thrown from ground level and neglect any effects due to air resistance, then the principal quantities involved in the situation are those listed in Table 1.

In this case we already know the formula, so this example illustrates the general method in a situation with which we are already familiar.

Table 1

Physical quantity	Symbol	Dimensions
Range of projectile	R	L
Mass of projectile	m	M
Launch speed of projectile	u	$\mathrm{L\,T^{-1}}$
Launch angle of projectile	θ	1
Magnitude of acceleration due to gravity	g	$\mathrm{L\,T^{-2}}$

Our aim is to find the range R as a function of the four parameters. In other words, we are looking for a function f of four variables such that

$$R = f(m, u, \theta, g).$$

We begin by assuming that

$$R = k\, m^\alpha\, u^\beta\, \theta^\gamma\, g^\delta, \tag{1}$$

where k is a (dimensionless) real number, and the powers α, β, γ and δ are to be found.

For this equation to be dimensionally consistent, we must have

$$[R] = [k\, m^\alpha\, u^\beta\, \theta^\gamma\, g^\delta] = [k]\,[m]^\alpha\,[u]^\beta\,[\theta]^\gamma\,[g]^\delta.$$

As in Unit 8, square brackets denote 'the dimensions of'.

In terms of the base dimensions, we have

$$\mathrm{L} = \mathrm{M}^\alpha\,(\mathrm{L\,T^{-1}})^\beta\,(\mathrm{L\,T^{-2}})^\delta = \mathrm{M}^\alpha\,\mathrm{L}^{\beta+\delta}\,\mathrm{T}^{-\beta-2\delta},$$

as k and θ are dimensionless. Equating powers of M, L and T on both sides of this equation, we have

$$0 = \alpha, \quad 1 = \beta + \delta, \quad 0 = -\beta - 2\delta.$$

Solving these equations for α, β, γ and δ gives

$$\alpha = 0, \quad \beta = 2, \quad \delta = -1.$$

So the requirement that the equation must be dimensionally consistent has given us values of α, β and δ. Because the parameter θ is dimensionless, the method cannot give us a value for γ.

We have shown that

$$R = k\, u^2\, g^{-1}\, \theta^\gamma$$

is dimensionally consistent for any value of the power γ. We can take an additive combination of such expressions with different values of γ and k, which will still be dimensionally consistent. Hence the best we can deduce is that

$$R = k(\theta)\, u^2\, g^{-1} = k(\theta)\, \frac{u^2}{g}, \tag{2}$$

where k is now a function of θ. As the launch angle θ is dimensionless, the function $k(\theta)$ is also dimensionless, and equation (2) is dimensionally consistent.

Dimensional analysis can tell us nothing further about the form of the function $k(\theta)$. In order to find this function, we would have to use Newton's second law to determine it analytically, as we did in Unit 3, finding that

$$k(\theta) = \sin 2\theta,$$

or we would have to use a suitable experiment to find it numerically.

In the following exercise, you will find the form of the relationship for the period of a pendulum. From Unit 11, we already know this relationship for *small* oscillations of the pendulum, where we can use the approximation $\sin\theta \simeq \theta$ throughout the oscillations, but here we will consider the case of *large* oscillations.

Exercise 1

In this exercise we neglect air resistance.

A pendulum is oscillating in a vertical plane. Use dimensional analysis to find a possible form for the expression for the period τ of the pendulum in terms of the mass m of the bob, the length l of the pendulum stem, the angular amplitude Φ of the oscillations, and g, the magnitude of the acceleration due to gravity.

In Exercise 1, we derived the result

$$\tau = k(\Phi)\sqrt{\frac{l}{g}} \tag{3}$$

for the period of a pendulum. You may have been disappointed in this result, in that we have not been able to determine the function $k(\Phi)$. But in fact the result is very powerful. First, it shows that the period of oscillation of the pendulum is independent of the mass of the bob − a fact that we can easily verify by experiment in situations where air resistance is small. Second, it implies that for a given angular amplitude, the period is proportional to $\sqrt{l/g}$. So if we can determine the period of oscillation for one known length of pendulum stem for a particular angular amplitude, say $\Phi = \pi/4$, then we can deduce the period of oscillation of any pendulum, no matter what its length is, for this amplitude, even if the pendulum is on Mars!

Let us now return to our initial assumption that the relationship between the parameters involves their powers, such as in equation (1). You might think that the relationship could involve transcendental functions such as exp, ln, sin, cos, tan, and so on. However, as we remarked in Unit 8, the argument of any such function must be dimensionless. In our previous example of the simple pendulum, if we ignore air resistance, there is no dimensionless combination of the parameters m, l and g, so the only possible transcendental function involved is in the term $k(\Phi)$.

So far we have discussed situations where there is only one possible combination of the parameters with the required dimensions. We will now discuss a situation where this is not the case.

We consider again our investigation of the range of a projectile, but include air resistance. Now the important parameters are R, m, u, θ, g (as before) and air resistance F. (Of course, air resistance is a function of speed at time t, which is not a parameter, but we can define F to be the magnitude of the air resistance at a typical speed, such as the launch speed u. The key thing here is that F will have the dimensions of force.) These parameters are listed in Table 2, together with their dimensions.

We have been rather vague about defining F. However, the second-order differential equation derived from Newton's second law can be solved numerically to determine the range, and this will depend uniquely on the initial conditions such as the launch speed and launch angle.

Table 2

Physical quantity	Symbol	Dimensions
Range of projectile	R	L
Mass of projectile	m	M
Launch speed	u	$\mathrm{L\,T^{-1}}$
Launch angle	θ	1
Acceleration due to gravity	g	$\mathrm{L\,T^{-2}}$
Typical air resistance	F	$\mathrm{M\,L\,T^{-2}}$

Again, we will assume a relationship between these parameters of the form

$$R = k\, m^{\alpha}\, u^{\beta}\, \theta^{\gamma}\, g^{\delta}\, F^{\varepsilon}, \tag{4}$$

where k is a dimensionless number. For this equation to be dimensionally consistent, we must have

$$[R] = [k\, m^{\alpha}\, u^{\beta}\, \theta^{\gamma}\, g^{\delta}\, F^{\varepsilon}]$$
$$= [k]\,[m]^{\alpha}\,[u]^{\beta}\,[\theta]^{\gamma}\,[g]^{\delta}\,[F]^{\varepsilon}.$$

In terms of the base dimensions M, L and T, we have

$$\mathrm{L} = \mathrm{M}^{\alpha}\,(\mathrm{L\,T^{-1}})^{\beta}\,(\mathrm{L\,T^{-2}})^{\delta}\,(\mathrm{M\,L\,T^{-2}})^{\varepsilon}$$
$$= \mathrm{M}^{\alpha+\varepsilon}\,\mathrm{L}^{\beta+\delta+\varepsilon}\,\mathrm{T}^{-\beta-2\delta-2\varepsilon},$$

as k and θ are dimensionless. Equating powers of M, L and T on both sides of this equation, we have

$$\alpha + \varepsilon = 0, \quad \beta + \delta + \varepsilon = 1, \quad -\beta - 2\delta - 2\varepsilon = 0.$$

We have three equations involving the four unknowns α, β, δ and ε. As before, the constant γ remains indeterminate. By adding twice the second of these three equations to the third, we deduce that $\beta = 2$. But we still have two equations, namely

$$\alpha + \varepsilon = 0, \quad \delta + \varepsilon = -1,$$

in the three unknowns α, δ and ε.

To proceed, we choose to write two of these unknowns in terms of the third. If we let $\varepsilon = a$, where a is an unknown parameter, then we have $\alpha = -a$ and $\delta = -1 - a$. Equation (4) becomes

$$R = k\, m^{-a}\, u^2\, \theta^{\gamma}\, g^{-1-a}\, F^{a}$$
$$= k\, \frac{u^2}{g}\, \theta^{\gamma} \left(\frac{F}{mg}\right)^{a}.$$

In the absence of air resistance, the range is given by

$$R = \frac{u^2 \sin 2\theta}{g},$$

which is of this form, as we saw earlier.

Each choice of values of the parameters γ and a gives a possible expression for the range R, as does a linear combination of these possibilities. More generally, we can write

$$R = \frac{u^2}{g} f\left(\theta, \frac{F}{mg}\right), \tag{5}$$

where f is a function of two variables. The two quantities θ and F/mg are dimensionless.

You may be wondering whether we would have arrived at a different result if, when solving the equations $\alpha + \varepsilon = 0$, $\delta + \varepsilon = -1$, we had made a choice other than $\varepsilon = a$ for the unknown parameter. Suppose that we let $\delta = b$ instead. Then the solution of the equations is $\alpha = 1 + b$ and $\varepsilon = -1 - b$. So equation (4) becomes

$$R = k\, m^{1+b}\, u^2\, \theta^\gamma\, g^b\, F^{-1-b}$$
$$= k\, \frac{mu^2}{F}\, \theta^\gamma \left(\frac{mg}{F}\right)^b,$$

and we deduce that the general expression for the range is

$$R = \frac{mu^2}{F}\, h\left(\theta, \frac{mg}{F}\right),$$

where h is a function of two variables.

At first sight this appears different from equation (5). However,

$$\frac{mu^2}{F}\, h\left(\theta, \frac{mg}{F}\right) = \frac{u^2}{g}\, \frac{mg}{F}\, h\left(\theta, \frac{mg}{F}\right)$$
$$= \frac{u^2}{g}\, f\left(\theta, \frac{F}{mg}\right),$$

where

$$f(x, y) = \frac{1}{y}\, h\left(x, \frac{1}{y}\right).$$

(Similarly, if we had made a third possible choice, and let $\alpha = c$, we would again have arrived at an expression for the range R that is equivalent to equation (5).)

Dimensionless combinations of physical quantities, such as F/mg in equation (5), are called **dimensionless groups**. These are particularly important in situations where we cannot find an analytical solution to our model. For example, in the case of the range of a projectile with air resistance, equation (5) means that we have reduced the problem to finding, either numerically or by experiment, an unknown function of two dimensionless groups, namely θ and F/mg, so we do not have to consider the effect of varying every individual parameter involved in the model.

The above discussion of the range of a projectile, taking into account the effects of air resistance, illustrates the method of dimensional analysis, which is summarised in the following box.

Method of dimensional analysis

To find dimensionally consistent relationships, carry out the following steps.

1. List the parameters $y, x_1, x_2, \ldots, x_n$ that represent the important features involved in the situation being modelled, and determine their dimensions.

2. Assume a relationship involving the powers of these parameters, namely

    $$y = k\, x_1^{\alpha}\, x_2^{\beta}\, x_3^{\gamma} \cdots x_n^{\nu},$$

    where k is a dimensionless constant.

3. Use the principle of dimensional consistency to write

    $$[y] = [x_1]^{\alpha}\, [x_2]^{\beta}\, [x_3]^{\gamma} \cdots [x_n]^{\nu},$$

    and equate the powers of M, L and T on both sides of this equation.

4. Solve the four simultaneous equations obtained in Step 3 for the powers $\alpha, \beta, \ldots, \nu$. This solution will usually involve unknown parameters.

5. Substitute for the powers in the expression in Step 2, and rewrite it in terms of dimensionless groups. Hence write down a general expression for y, which will usually involve an unknown function of these dimensionless groups.

In practice, if you omit some of the key parameters, you may end up with a set of equations for the powers where there are, for example, more equations than variables, and these equations may well have no solution.

Exercise 2

Use dimensional analysis to find an expression for the frequency f of vibration of a guitar string, assuming that it depends on only the mass m of the string, the length l of the string, and the tension T_{eq} of the string.

This problem was considered in Unit 14.

2 Evaluating the model

Having used the techniques outlined in Unit 8, and in Section 1, to formulate mathematical models, in this section we will focus on their evaluation. We will describe three ways in which a model can be evaluated. The first, which you saw in Unit 8, is to use the model to predict the outcome of several events, which can then be compared with real data, if they are available. For example, in Unit 8, the tin can problem was based on the assumption that a manufacturer of cans would wish to minimise the amount of material used to construct a cylindrical can to hold a specified volume of food. After making a number of additional simplifying assumptions, we constructed a model that predicted that the height of the can would be equal to its diameter. Checking with a range of tin cans used to store food, you will find that rarely does the height of the can equal its diameter (so that, sideways on, the can would be square-shaped), so something has gone wrong.

If the predictions are accurate, then the model can be used with confidence. If they are accurate only for specific ranges of the parameters, then these limitations can be noted. If they are inaccurate for a range of the parameters in which the model is required to make predictions, then we need to look at the assumptions that underpin the model to see if one or two of them can be changed to try to improve the accuracy of the model. This second process is called *criticising the model*.

There is a third check that we need to carry out on our model, to see how sensitive it is to variations in the parameters of the model. We may have made many assumptions and simplifications in order to arrive at a mathematical model. If a small change in one of the parameters causes a significantly larger change in the prediction, then we have a problem, since it is likely that neither the model nor the data used to interpret the results will be particularly accurate, and such sensitivities may undermine the usefulness of the model to make predictions. We start this section by looking at the sensitivity of models.

2.1 Sensitivity analysis

The aim of this subsection is to determine the sensitivity of the output value, obtained from a calculation, to the values of the parameters used in that calculation. You will need an understanding of partial derivatives, as described in Unit 7, to follow the analysis in this subsection. We will also look at empirical methods of assessing sensitivity, by carrying out numerical experiments on the model.

This approach is very similar to the approach to ill-conditioning in Unit 4.

A modelling process usually results in an equation that gives the value of the variable, or variables, of interest in terms of the model parameters. Figure 2 shows this in diagrammatic form.

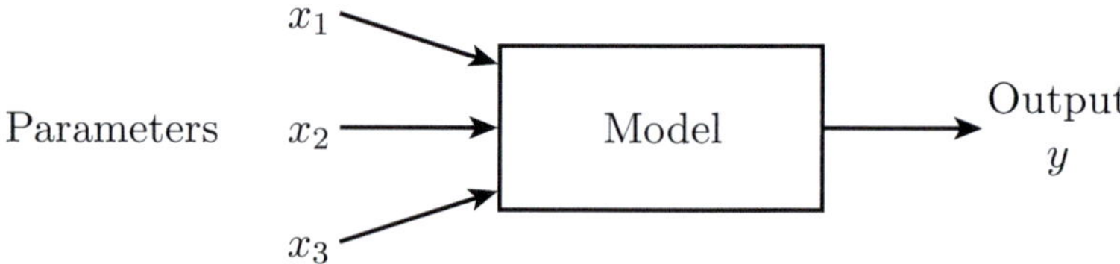

Figure 2 Diagram showing the inputs and output for a model

In most situations, the values of the parameters are not known exactly, and they will have some error margin associated with them, as shown in Figure 3.

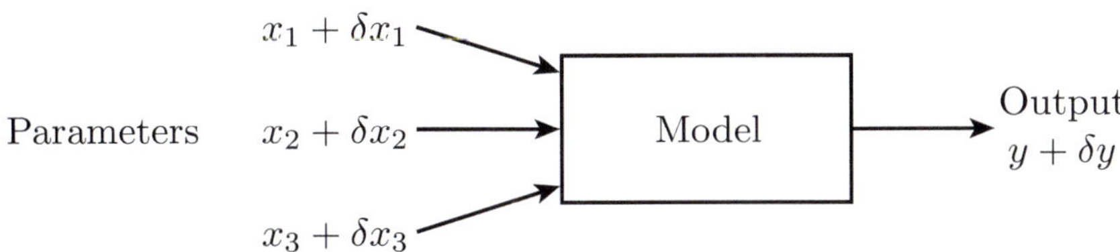

Figure 3   Diagram showing the inputs and output, with errors, for a model

Sensitivity analysis is used to quantify the effect that parameter values have on the output of the model. Errors in some of the parameters may have more effect on the final value than errors in other parameters. The output is sensitive to a particular parameter if a small change in that parameter has a significant effect on the output. To examine this, we need some definitions.

The **absolute sensitivity** of a solution value y to a parameter value x is defined as $\delta y/\delta x$, where δy is the change in y caused by a small change δx in the value of the parameter x.

The **relative sensitivity** of a solution value y to a parameter value x is defined as

$$\frac{\delta y/y}{\delta x/x} = \frac{x}{y}\frac{\delta y}{\delta x}, \tag{6}$$

where $\delta y/y$ is the relative (or proportionate) change in y for a small relative change $\delta x/x$ in the value of the parameter x.

Where the dimensions, or units, of x and y are different, it is better to consider the relative sensitivity, since a change in the unit of measurement will affect the value of the absolute sensitivity, but not the relative sensitivity, which is dimensionless.

There are two ways of determining the sensitivity of the final answer to changes in the parameter values. One way is an empirical approach, achieved by changing the value of a parameter by a small amount, then considering its effect on the final solution. The other method is analytical, based on finding the rate of change of the variable of interest with respect to each of the parameters.

In the empirical approach, only one parameter value at a time is changed; if more than one parameter value is changed, then the effects of one change may be masked by another.

For the analytical approach, where $y = f(x_1, x_2, \ldots, x_n)$, the absolute sensitivity of y with respect to x_i for infinitesimally small changes in x_i is $\partial y/\partial x_i$; the relative sensitivity is

$$\frac{x_i}{y}\frac{\partial y}{\partial x_i}. \tag{7}$$

We consider both approaches in the following example.

Example 1

We want to estimate the depth of a well by measuring the time that a pebble takes to hit the surface of the water when dropped into the well. A simple model, neglecting air resistance, gives

$$h = ut + \tfrac{1}{2}gt^2,$$

where h is the depth of the well, u is the initial speed of the pebble, g is the value of the acceleration due to gravity, and t is the time for the pebble to drop.

In one measurement, the pebble was dropped from rest and a splashing sound was heard 2.8 s later. The estimate for the depth of the well is

$$h = \tfrac{1}{2} \times 9.81 \times 2.8^2 = 38.455.$$

So the well is about 38.5 m deep. The question is, how accurate is this estimate of the depth? Is it unduly sensitive to the measurement of the time taken? How accurate should the value of g be?

Solution

Since the values that are used to estimate the depth are in units different to those for the depth, it is better to consider the *relative* change in the values; otherwise a change of units would affect the calculation of the sensitivity.

- First, try the empirical approach. Change the value of the time taken to fall, then consider the effect that this has on the estimate of the depth. On increasing the measured time from 2.8 to 2.83, a relative change of about 1%, the estimate of the depth changes from 38.455 to 39.284, an increase of approximately 0.8 m. This is equivalent to a relative change of $(39.284 - 38.455)/38.455 \simeq 0.02$ or 2%. So the small relative change in the time for the pebble to drop, when $t = 2.8$, has magnified the relative change in the depth by a factor of 2.

  Now, instead, change the value of g to 9.80, a relative change of $|(9.80 - 9.81)/9.81| \simeq 0.001$, which is 0.1%. The resulting value for the depth is now 38.416, a decrease of approximately 0.3 m. The relative change in the value for the depth of the well is $|(38.416 - 38.455)/38.455| \simeq 0.001$ or 0.1%. The relative change in depth is approximately the same as the relative change in the value of g when $g = 9.81$.

- Now consider the analytical approach. Think of the expression for h as a function of the two variables t and g:

  $$h(t, g) = ut + \tfrac{1}{2}gt^2.$$

  The first-order Taylor approximation around the nominal values $t = t_0$ and $g = g_0$ is

  $$h(t_0 + \delta t, g_0 + \delta g) \simeq h(t_0, g_0) + \delta t \, \frac{\partial h}{\partial t}(t_0, g_0) + \delta g \, \frac{\partial h}{\partial g}(t_0, g_0),$$

  where δt is the small change in the value of t, and δg is the small change in the value of g.

This model was derived in Unit 3.

Although the pebble may be dropped from rest, so that $u = 0$, u is included so that the effect of a non-zero value can be considered.

Although the accuracy quoted is far in excess of what is needed, when performing a sensitivity analysis it is necessary to use greater precision, since the changes to the parameter values will be small.

Taylor approximations for functions of two variables were developed in Unit 7.

Thus the change in the depth δh is given by

$$\delta h = h(t_0 + \delta t, g_0 + \delta g) - h(t_0, g_0) \simeq \delta t \, \frac{\partial h}{\partial t}(t_0, g_0) + \delta g \, \frac{\partial h}{\partial g}(t_0, g_0).$$

For this problem, the partial derivatives are

$$\frac{\partial h}{\partial t}(t, g) = u + gt \quad \text{and} \quad \frac{\partial h}{\partial g}(t, g) = \tfrac{1}{2}t^2,$$

so we have, at $t = t_0$ and $g = g_0$,

$$\delta h \simeq (u + g_0 t_0)\, \delta t + \tfrac{1}{2}t_0^2 \, \delta g.$$

When $u = 0$, $g_0 = 9.81$ and $t_0 = 2.8$, we have $\partial h/\partial t = u + g_0 t_0 = 27.5$ and $\partial h/\partial g = \tfrac{1}{2}t_0^2 = 3.92$. Thus the absolute change in the depth of the well for small changes in t and g is

$$\delta h \simeq 27.5\, \delta t + 3.92\, \delta g.$$

As we said earlier, these results are not particularly useful since h, t and g have different units, and a change in the units of measurement will affect the results. We can, however, say that with the units that we have chosen, the model is absolutely sensitive to changes in t.

We describe the module convention for absolute sensitivity at the end of this example.

A more important measure is the effect of relative changes. The relative error in the value of the depth of the well is $\delta h/h_0$, where $h_0 = u t_0 + \tfrac{1}{2}g_0 t_0^2$. The Taylor approximation can be rearranged to give the relative error in h as

$$\frac{\delta h}{h_0} \simeq \frac{\frac{\partial h}{\partial t}(t_0, g_0)}{h_0}\, \delta t + \frac{\frac{\partial h}{\partial g}(t_0, g_0)}{h_0}\, \delta g$$

$$= \frac{t_0}{h_0}\, \frac{\partial h}{\partial t}(t_0, g_0) \left(\frac{\delta t}{t_0}\right) + \frac{g_0}{h_0}\, \frac{\partial h}{\partial g}(t_0, g_0) \left(\frac{\delta g}{g_0}\right),$$

which gives the relative error in the depth of the well h in terms of the relative errors in t and g.

For our model, where

$$h = ut + \tfrac{1}{2}gt^2, \quad \frac{\partial h}{\partial t}(t, g) = u + gt \quad \text{and} \quad \frac{\partial h}{\partial g}(t, g) = \tfrac{1}{2}t^2,$$

we have, at $t = t_0$ and $g = g_0$,

$$\frac{\delta h}{h_0} \simeq \frac{t_0(u + g_0 t_0)}{u t_0 + \tfrac{1}{2}g_0 t_0^2}\left(\frac{\delta t}{t_0}\right) + \frac{g_0 \tfrac{1}{2}t_0^2}{u t_0 + \tfrac{1}{2}g_0 t_0^2}\left(\frac{\delta g}{g_0}\right)$$

$$= \frac{(u + g_0 t_0)}{u + \tfrac{1}{2}g_0 t_0}\left(\frac{\delta t}{t_0}\right) + \frac{\tfrac{1}{2}g_0 t_0}{u + \tfrac{1}{2}g_0 t_0}\left(\frac{\delta g}{g_0}\right).$$

If we assume that the pebble is dropped from rest, so that $u = 0$, then considerable further simplification occurs and we have

$$\frac{\delta h}{h_0} \simeq 2\left(\frac{\delta t}{t_0}\right) + \left(\frac{\delta g}{g_0}\right)$$

as we predicted in the earlier empirical approach. However, the analytical approach tells us that these results hold for a range of values for t_0 and g_0, provided that the pebble is dropped from rest.

We conclude that a small relative change in the time is magnified by a factor of 2 in the relative change in the depth of the well, while a small relative change in the value of g gives a similar relative change in the depth of the well. Hence the predicted depth is less sensitive to relative changes in g than to relative changes in t.

Example 1 illustrates that for absolute and relative sensitivity, the empirical approach to the effect of a change in the parameters $x_1, x_2, \ldots, x_n$ on the value of $y(x_1, x_2, \ldots, x_n)$ can be done numerically by simply changing one of the parameters by a small amount and assessing its effect on the output, while the analytic approach is based on a first-order Taylor series approximation

$$\delta y = y(x_1 + \delta x_1, x_2 + \delta x_2, \ldots, x_n + \delta x_n) - y(x_1, x_2, \ldots, x_n)$$
$$\simeq \delta x_1 \frac{\partial y}{\partial x_1} + \delta x_2 \frac{\partial y}{\partial x_2} + \cdots + \delta x_n \frac{\partial y}{\partial x_n}.$$

We could use this to consider a change in one variable at a time as

$$\delta y_i = y(x_1, x_2, \ldots, x_i + \delta x_i, \ldots, x_n) - y(x_1, x_2, \ldots, x_i, \ldots, x_n)$$
$$\simeq \delta x_i \frac{\partial y}{\partial x_i},$$

or to consider the effect of changing several of the variables, if we so wished. Similarly, the corresponding relative changes are given by

$$\frac{\delta y}{y} \simeq \frac{\delta x_1}{x_1} \left(\frac{x_1}{y} \frac{\partial y}{\partial x_1} \right) + \frac{\delta x_1}{x_2} \left(\frac{x_2}{y} \frac{\partial y}{\partial x_2} \right) + \cdots + \frac{\delta x_n}{x_n} \left(\frac{x_n}{y} \frac{\partial y}{\partial x_n} \right).$$

This equation relates the relative change in y, namely $\delta y/y$, to the relative change in each variable x_j, namely $\delta x_j/x_j$. So the coefficient of $\delta x_j/x_j$ must be the relative sensitivity, as given in equation (7).

If we simply wish to consider the effect of a relative change in the parameter x_j, with all the other parameters fixed, then we have

$$\frac{\delta y}{y} \simeq \frac{\delta x_j}{x_j} \left(\frac{x_j}{y} \frac{\partial y}{\partial x_j} \right),$$

and this formula will enable us to analyse the sensitivity of the output y for a range of values for the parameters, not just a particular value. Hence this approach is more thorough than the empirical approach, which gives results only for particular values of the parameters.

Conditions for absolute and relative sensitivity

Suppose that small absolute or relative changes are made in the parameters of a model. The model is absolutely (relatively) sensitive if it is possible for the absolute (relative) change in the solution to be significantly larger than the absolute (relative) change in one or more of the parameters.

Usually the interpretation of *significantly larger* is dependent on the context. However, for the sake of clarity and certainty, we adopt the module convention. A model is judged to be:

- absolutely (relatively) insensitive with respect to a particular parameter if a small absolute (relative) change in the parameter is magnified by a factor less than or equal to 5

- neither absolutely (relatively) sensitive nor absolutely (relatively) insensitive with respect to a particular parameter if a small absolute (relative) change in the parameter is magnified by a factor of between 5 and 10

- absolutely (relatively) sensitive with respect to a particular parameter if a small absolute (relative) change in the parameter is magnified by a factor greater than or equal to 10.

Exercise 3

Consider the following problem of mortgage repayment. Let C_n be the capital sum owing in year n, let i be the annual interest rate (assumed constant), and let R be the constant sum repaid each year, where C_n and R are measured in pounds. The capital sum owing the next year is the capital sum owing in the previous year plus the interest on this capital sum less the repayment. Expressed mathematically, this becomes

$$C_{n+1} = C_n + iC_n - R.$$

This is an iterative formula whose explicit solution is

$$C_n = \left(C_0 - \frac{R}{i} \right)(1 + i)^n + \frac{R}{i},$$

The solution can be checked by substituting values of $0, 1, 2, \ldots$ for n.

where C_0 is the capital sum originally owed.

Take an example where the initial mortgage is £100 000, and £8000 is paid every year, with an interest rate of 5%. Using the defined symbols, this gives $R = 8000$, $i = 0.05$ and $C_0 = 100\,000$, and the capital sum owing is given by

$$C_n = \left(100\,000 - \frac{8000}{0.05} \right)(1.05)^n + \frac{8000}{0.05} = 160\,000 - 60\,000(1.05)^n.$$

So the sum still outstanding 15 years after the start of the mortgage will be

$$C_{15} = 160\,000 - 60\,000 \times 1.05^{15} = 160\,000 - 60\,000 \times 2.078\,928$$
$$= 35\,264.31.$$

How sensitive is the amount owing after 15 years to changes in the values of the following parameters? Use the empirical approach to determine the sensitivity of the solution to changes in the values of both parameters.

(a) The annual payment

(b) The interest rate

In Exercise 3 we considered the effect of increasing the annual repayment. As the number of years increases, the amount owing decreases more rapidly as more is paid off each year, which also reduces the amount of interest on the remaining mortgage. Consequently the extra payment can have a significant effect on the time taken to repay the mortgage. The model becomes increasingly sensitive to the values of the parameters as n increases and the amount owing decreases. For example, suppose that the model is used to predict how long it will take to pay off the mortgage. For the given values of the parameters, the model suggests that the mortgage will be repaid some time in the second month of the 21st year. If the repayment is increased by £100 per year, then the mortgage will be repaid in the ninth month of the 20th year, some five months earlier. Similarly, if the interest rate is reduced by 0.1% then the mortgage will be repaid in the tenth month of the 20th year, four months earlier, assuming that the repayments are continued at the same rate.

Exercise 4

Consider the trajectory of a particle launched from a height h, with speed u, at an angle θ to the horizontal. If we assume that the only force acting on the particle is the weight of the particle, then the trajectory is given by

See Unit 3, Subsection 5.2 for a derivation of this model.

$$y = h + x \tan \theta - x^2 \frac{g}{2u^2} \sec^2 \theta.$$

We are interested in how far the particle travels (i.e. the value of x) before it hits the ground (i.e. when $y = 0$). The analytical approach is to be used to investigate the relative sensitivity of the model.

(a) Derive a formula that can be used to determine the relative sensitivity of this model to changes in the launch speed u.

(b) If $h = 2$, $u = 20$, $\theta = \pi/4$ and $g = 9.81$, then it can be shown that the particle hits the ground when $x = 42.68$. Use the formula found in part (a) to estimate how much further the particle travels before it hits the ground if u is increased to 21.

2.2 Comparing predictions with reality and criticising the model

In Unit 8 we developed a model for the manufacture of tin cans in which we wished to determine the height h and radius r of a cylindrical can that would minimise the amount of metal used for a given volume V of food. The following simplifying assumptions were made.

We have modified the wording slightly here.

Assumption 1: The cans are manufactured from sheet metal of a uniform thickness.

Assumption 2: Any overlapping metal needed to form joins is ignored.

Assumption 3: Any leftover metal can be recycled.

Assumption 4: The costs of manufacture can be ignored.

We thus had to minimise the surface area A given by the cylindrical portion of the can plus two circular pieces, that is,

$$A = 2\pi r h + 2\pi r^2,$$

such that the volume $V = \pi r^2 h$ is fixed. Substituting $h = V/(\pi r^2)$ into the equation for A leads to the problem of determining the value of r that minimises

$$A = \frac{2V}{r} + 2\pi r^2.$$

Differentiating with respect to r to find the maximum and minimum values of A gives

$$\frac{dA}{dr} = -\frac{2V}{r^2} + 4\pi r = 0,$$

from which we deduced that $r = (V/(2\pi))^{1/3}$ and $h = 2r$, and the minimum area of metal needed is

$$A = 2\pi r h + 2\pi r^2 = 4\pi r^2 + 2\pi r^2 = 6\pi r^2 = 6\pi \left(\frac{V}{2\pi}\right)^{2/3} = 3\sqrt[3]{2V^2\pi}.$$

This is a minimum since

$$\frac{d^2 A}{dr^2} = \frac{4V}{r^3} + 4\pi > 0.$$

We concluded that the height of the can should be the same as its diameter.

It was commented in Unit 8 that this rarely matches reality, with most cans being taller than they are wide. So what has gone wrong? It is tempting to conclude that manufacturers have no knowledge of mathematics and hence are not finding the optimal solution, but this might well be wrong. Having found a solution that does not match reality, we need to revisit our modelling process and re-examine what has been done. For example, is the real problem to minimise the amount of metal used? The sides of the can are often used to advertise the product. Is this important? Studies have suggested that certain shapes are more pleasing to the eye, and such products may fly off the shelves more rapidly than cans with a square-shaped profile. These may well be issues that the manufacturer takes into consideration.

Before we look at the assumptions, it is helpful to consider the sensitivity of the area A of metal used to changes in r while keeping the volume fixed. At the minimum, $dA/dr = 0$, so the model is relatively insensitive to changes in r.

We can also express A in terms of h as

$$A = 2\sqrt{\pi V h} + \frac{2V}{h},$$

and minimise the area by solving $dA/dh = 0$ for h. Hence we can deduce that at the minimum, the model is also insensitive to changes in h.

Away from the optimal value, suppose that we set $h_1 = 4r_1$ so the height of the can is twice the width. The volume $V = \pi r_1^2 h_1 = 4\pi r_1^3$ is still fixed, so $r_1 = (V/(4\pi))^{1/3}$. The output from the model is the area

$$A_1 = 2\pi r_1 h_1 + 2\pi r_1^2$$

$$= 8\pi r_1^2 + 2\pi r_1^2 = 10\pi r_1^2 = 10\pi \left(\frac{V}{4\pi}\right)^{2/3} = 5\sqrt[3]{V^2\pi/2}.$$

The relative change in the area of metal needed is

$$\frac{A_1 - A}{A} = \frac{\left(5\sqrt[3]{1/2} - 3\sqrt[3]{2}\right)\sqrt[3]{V^2\pi}}{3\sqrt[3]{2}\sqrt[3]{V^2\pi}}$$

$$= \frac{5\sqrt[3]{1/2} - 3\sqrt[3]{2}}{3\sqrt[3]{2}} = \frac{5}{3\sqrt[3]{4}} - 1 = 0.050.$$

This is a rather remarkable result since it tells us that having a can that is twice as high as it is wide requires only an extra 5% of material to manufacture. This might well explain why manufacturers may use different criteria than minimising the amount of material.

Exercise 5

(a) How much extra material is needed for the tin can if $h_2 = r_2$ so that the height is half the width of the can? What can you conclude about the sensitivity of the amount of material A_2 needed for $r \leq h \leq 4r$?

(b) For the cases $h_1 = 4r_1$ and $h_2 = r_2$, what is the change in the amount of advertising space $W = 2\pi rh$ on the wall of the can?

However, let us stay with the idea that we are trying in some way to minimise the amount of material, or perhaps the total cost of manufacture, and have a look at the simplifying assumptions that we made.

Assumption 1 looks fairly reasonable, although it could be noted that some cans are sometimes *ribbed* to give added strength to the sides. It is also possible to manufacture cans where the top and bottom are made of a different thickness than the sides. Would this enable savings to be made?

Assumption 2 could certainly be modified to take into account the overlap for the circular pieces and in the vertical wall of the can.

Assumption 3 could be modified. Maybe all components of the can are made from rectangular pieces of metal, with leftover pieces discarded or scrapped.

Assumption 4 could be reviewed. The cost of manufacture includes the cost of the materials plus the cost of soldering the joins. Should the soldering costs be included?

The examination of the simplifying assumptions has led to at least four different ways of modifying the first model. You are asked to consider two of these in the following exercises.

Exercise 6

For the tin can problem, suppose that Assumption 3 is modified to

Assumption 3: Any leftover metal is discarded.

In this case the parts of the can are cut from rectangular pieces of metal.

In particular, the top and bottom of the can are cut from squares of side $2r$, with the leftover bits discarded. We will assume that these are the only bits that will be discarded.

(a) What is the new model for the area of material used?

(b) How much material is needed for the optimal-sized can, and what is the ratio of the height to the diameter of the can?

Exercise 7

For the tin can problem, consider a modification to Assumption 2:

Assumption 2: An overlap of width b is included in the revised model for the circular pieces and the vertical seam.

(a) What is the new model for the area of material used?

(b) What is the equation to be solved to find the radius that minimises (or maximises) the area of material?

Exercise 8

Consider the problem of estimating the depth of a well in Example 1. The model for estimating the height h of the well is

$$h = ut + \tfrac{1}{2}gt^2,$$

where u is the initial velocity of a pebble, and t is the time that it takes to drop to the bottom of the well. It is based on the following assumptions.

Assumption 1: The pebble is treated as a particle.

Assumption 2: All forces other than the gravitational force are ignored.

Assumption 3: The sound of the pebble hitting the water is heard instantaneously.

For a particular well, a pebble of diameter $1\,\mathrm{cm}$ was dropped from rest and the sound was heard $2.8\,\mathrm{s}$ later. This suggested that the well was $38.5\,\mathrm{m}$ deep. A subsequent measurement of the well gave the depth as $34\,\mathrm{m}$.

(a) Discuss each of the assumptions in terms of the discrepancy in the depth of the well.

(b) How would the first model change if a linear model for air resistance was assumed?

(c) How would the first model change if the time for the sound to travel up the well was incorporated into the model? What difference would this make to the estimated depth of the well?

The speed of sound is $343\,\mathrm{m\,s^{-1}}$.

In the next section we return to the Great Lakes model, developed in Unit 8, to see how the assumptions that underpin it can be reassessed in order to develop an improved model.

3 A return visit to the Great Lakes

The information for this revised case study comes from two main sources:

The Great Lakes: An Environmental Atlas and Resource Book, available online at www.epa.gov/grtlakes/ atlas/intro.html (accessed 23 November 2014);

R.V. Thomann and J.A. Mueller (1987) *Principles of Surface Water Quality Modeling and Control*, Harper & Row.

This section evaluates and revises the model developed in Section 1 of Unit 8, for pollution in the Great Lakes of North America. It is worth revisiting that section to remind yourself of the basic model and of the assumptions and simplifications that led to it. These are fundamental to a critical evaluation.

Subsection 3.1 evaluates the success of the first model by comparing the model's predictions against data on contamination levels in the Great Lakes. Subsection 3.2 proposes some revisions to the first model in an attempt to overcome some of its deficiencies. In Subsection 3.3 we briefly discuss the techniques that were used, in order to highlight the key activities involved.

3.1 Evaluation of the model

◀ Evaluate ▶

To see if the model is realistic, consider its purpose. The problem is to predict how long it will take for the level of pollution in a lake to reduce to a target level if all sources of pollution are eliminated. It is intended to use the model to investigate pollution levels in any one of the Great Lakes, although the model could be used for any polluted lake.

The usual way of evaluating a model is to compare the predictions of the model with real data. These data can come either from your own experiments or from some published source. In the current case, collecting data from your own experiments is not feasible. However, there are useful published data, based on research studies conducted for the US Environmental Protection Agency and the US National Oceanographic and Atmospheric Administration by the University of Minnesota and the University of Wisconsin. One such set of data is shown in Table 3.

PCB stands for polychlorinated biphenyl. PCBs have been used in the manufacture of transformers, capacitors and other electrical equipment.

Table 3   Total PCB concentrations in Lake Superior

Year	PCB concentration $(10^{-9}\,\mathrm{kg\,m^{-3}})$
1978	1.73
1979	4.04
1980	1.13
1983	0.80
1986	0.56
1988	0.33
1990	0.32
1992	0.18

The model considered pollution in general, but applies equally well to any single pollutant. The set of data in Table 3 is based on one particular type of pollutant, PCBs, which were a major source of pollution in the Great Lakes in the 1960s. The manufacture and importation of PCBs was prohibited in the USA in 1975. It is useful to consider this pollutant in the model since it is fairly stable, that is, it does not react with other chemicals, is not broken down by sunlight, and so on.

Exercise 9

Consider the discussion above.

(a) Why do you think we need to consider a stable pollutant?

(b) Is it relevant that the manufacture and importation of PCBs stopped in 1975?

(c) What are the advantages of using Lake Superior for the evaluation of the model?

A plot of the data in Table 3 reveals that there is a curious and unexpectedly high concentration in 1979. All of the other data points seem to lie close to a smooth curve. For the moment, put this anomaly aside, and concentrate on the remaining data points. From the data it looks as if it takes 14 years (from 1978 to 1992, the actual time range of the data) for the concentration to fall to approximately a tenth of its former value, which relates to the specific question posed in the original model. However, the simple model predicts that it will take 440 years for the pollution level in Lake Superior to drop by a factor of ten. Clearly there is an enormous quantitative discrepancy between the model and these data.

Are the data at fault, or is the model at fault? One might question the data, particularly on the basis of the unexpectedly high concentration given for 1979. Was this due to measurement error? Does this particular value indicate that the data may be unreliable? Does it really reflect the level of pollution in that year? Do PCBs actually satisfy the assumptions of the model, or is there some process, relevant to PCBs, that has been overlooked? These are the kinds of question that you should ask when using data to evaluate the success of a model.

Without going back over the data and checking how they were obtained, it is difficult to explain the anomalous value for 1979. It has been suggested that this may have been caused by unusual disturbances of the sediment (for example, by storms, earthquakes or high winds). The sediment, which may contain high doses of the pollutant, would be drawn into the water and raise the measured value, while the reversion back to a much lower level the following year could have been caused by the PCBs settling back into the sediment from which they were disturbed. Regardless of whether this anomaly can be explained, these data will not validate the quantitative predictions of the model derived in Unit 8.

The model is not supported by the data. However, it is difficult to decide how to revise the model on the basis of a mismatch over a single value, namely, the time that it takes for the pollution concentration to reach a given level. A set of values or an equation to validate might be a better way to assess the success of the model and to give ideas on how best to revise it.

The main equation in the simple model proposes a relationship between the pollutant concentration, at any time after all pollution has ceased, and the time:

$$c(t) = c(0)\, e^{-kt},$$

where $c(0)$ is the initial concentration of pollutant. Do the data bear any qualitative resemblance to this equation? Since it involves a negative exponential, taking the logarithm of both sides gives

$$\ln(c(t)) = \ln(c(0)) - kt,$$

which predicts a straight-line graph when $\ln(c(t))$ is plotted against t (in years).

As Figure 4 illustrates, except for the anomalous value for 1979, the data are reasonably well approximated by a straight line. This confirms that the simple model does have some good qualitative agreement with the data, in the sense that a negative exponential model approximates reasonably well the data for PCBs. The value of k is given by the negative of the slope in Figure 4, that is, by $4.8 \times 10^{-9}\,\mathrm{s}^{-1}$. The value of k derived from the model is $0.166 \times 10^{-9}\,\mathrm{s}^{-1}$, which differs substantially from the experimental value, by a factor of approximately 30.

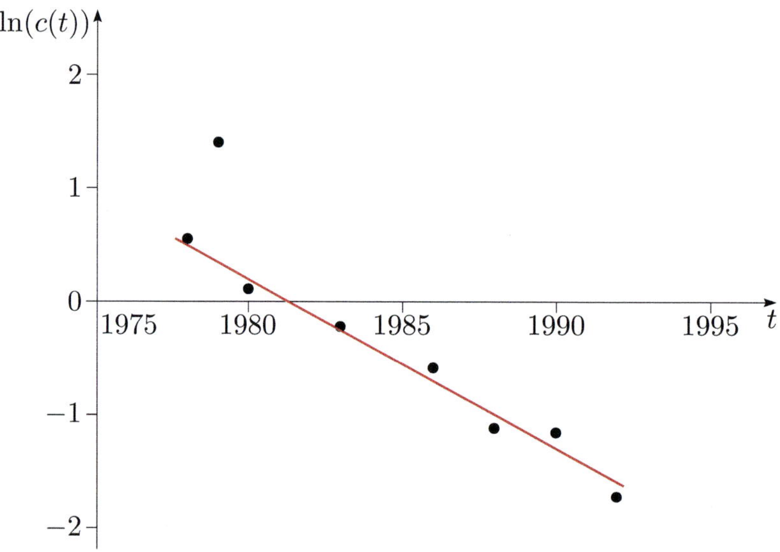

Figure 4   Plot of the data for Lake Superior

This is only one set of data, and more sets of data ought to be tried. Unfortunately, there are few published data on the actual concentrations of pollutants in the lakes (possibly because their measurement is time-consuming and costly). However, there are extensive data available on the contaminant concentration in fish and birds inhabiting the Great Lakes. It has been shown that over the long term, trends of contaminants (including PCBs) in such creatures have followed those in the water and thus provide a measure of the trends in the ecosystem of the Great Lakes. So it is reasonable to assume that if the simple model applies to the pollutant concentrations in the lakes, then it will apply to those in its inhabitants too.

Source: D. De Vault and R. Hesselberg (1994) 'Contaminant trends in lake trout and walleye from the St Lawrence Great Lakes', *Journal of Great Lakes Research*.

Table 4 gives the concentrations of PCBs in herring gull eggs collected around Lake Superior over a number of years. The concentrations are given in micrograms per gram (or equivalently, parts per million), rather than in kilograms per cubic metre, since this is a more common method of measuring concentrations in solids.

Table 4   PCB concentrations in herring gull eggs from Lake Superior

Year	PCB concentration $(\mu\mathrm{g\,g}^{-1})$	Year	PCB concentration $(\mu\mathrm{g\,g}^{-1})$
1974	65	1985	21
1975	81	1986	17
1977	59	1987	14
1978	44	1988	15
1979	60	1989	17
1980	28	1990	13
1981	36	1991	15
1982	37	1992	15
1983	23	1993	16
1984	19		

To compare the concentration in herring gull eggs with the concentration in water, note that $1\,\mathrm{m}^3$ of water weighs $1000\,\mathrm{kg}$. Hence in 1978 the concentration of PCBs in the water (from Table 3) was 1.73×10^{-12} grams of PCB per gram of water. In the herring gull eggs that year, the concentration was 44×10^{-6} grams of PCB per gram of egg, which is a magnification of about 25 million.

The high concentrations here as compared with those in the lake are an example of *biomagnification*, in which small concentrations of pollutant at one point cause higher and higher concentrations when moving up the food chain.

Exercise 10

When evaluating the model, why should you disregard at least the first data point in Table 4?

Eliminating the first data point from Table 4 gives a value for k of 3.1×10^{-9}, which is similar to the value obtained with the first set of data. This confirms that while the negative exponential form of the relationship is approximately correct, the actual numerical values predicted by the model are in error.

While the quantitative comparison (between the data and the model) of the time for the concentration of pollutant to drop to a tenth of its initial value gives no clue how to revise the model, the qualitative comparison gives some information on where to focus revision of the model – in the assumptions concerning the rate at which the pollutant dissipates.

The comparison being made here is valid provided that the magnification factor (from concentration in the water to concentration in gulls) and the time delay (referred to in the solution to Exercise 10) are both constant.

3.2 Revisions to the model

This subsection investigates two possible revisions to the model as applied to PCBs. The first takes account of a process that affects the rate at which PCB concentration diminishes. The second revision tries to take into account the variation in concentration across the lake.

In considering revisions to the model, first review the modelling assumptions.

Exercise 11

Consider the assumptions made for the first model. Which of these assumptions do you think should be re-examined?

First revision: adding a process

◀ Create model ▶

To explain the discrepancy between the prediction of the model and evidence of the data, the physics and chemistry of PCBs need examination, to determine the likeliest ways in which PCB concentrations can diminish other than by being flushed out of the lake by the flow of water. For any chemical in solution, there are four principal mechanisms that can cause its concentration to diminish:

The details of these processes would be important to a mathematical modeller investigating this problem, but you only need to understand that there are various additional processes that may affect the model.

- *volatilisation*, where the chemical is absorbed into and from the atmosphere at the surface of the lake

- *photolysis*, where the action of sunlight on water near the surface of the lake causes certain chemicals to degrade

- *hydrolysis*, where the chemical reacts with water, causing it to break down

- *biodegradation*, where the chemical is degraded by bacteria and other living organisms.

Now, PCBs are stable. This means that they are difficult to break down, which tends to rule out photolysis, hydrolysis and biodegradation as relevant processes. However, there is evidence that PCBs are volatile, so we consider this mechanism further. Inclusion of volatilisation amends Assumption (b) as follows:

(b) The pollutant does not biodegrade in the lake or decay through any other biological, chemical or physical process *except through volatilisation*.

An additional assumption is also needed, about the process of volatilisation.

Exercise 12

On what physical properties does the rate of volatilisation depend?

For a simple revision of the model, consider only the first two factors mentioned in the solution to Exercise 12. This results in the following additional assumption:

(h) The pollutant is lost from the lake to the atmosphere at a rate that is proportional to the total surface area of the lake and to the concentration of pollutant in the lake.

Suppose that the surface area of the lake is A (measured in square metres). The assumption is that the mass of the pollutant is lost at a rate that is proportional to both A and $c(t)$, where $c(t)$ is the pollutant concentration in the lake. Hence pollutant is lost at a rate proportional to $A\,c(t)$. Thus the differential equation that models the rate of loss of the mass of pollutant is

$$\frac{dm}{dt} = -k\,m(t) - pA\,c(t), \tag{8}$$

A detailed derivation of this differential equation would involve use of the input–output principle.

where p is the constant of proportionality, known as the *volatilisation rate*. Now, the model includes the equation $c(t) = m(t)/V$, which is still valid. So, by eliminating $c(t)$, equation (8) can be expressed entirely in terms of the mass m of pollutant, as

$$\frac{dm}{dt} = -k\,m(t) - \frac{pA}{V}\,m(t) = -\left(k + \frac{pA}{V}\right) m(t) = -\kappa\,m(t), \tag{9}$$

κ is the Greek letter kappa.

where $\kappa = k + pA/V$. This is essentially the same model as the first one, except that the *proportionate flow rate k* has now become a *proportionate decay rate κ*, and κ should be greater than k.

The solution of equation (9) is

◀ Do mathematics ▶

$$m(t) = m(0)\,e^{-\kappa t},$$

where $m(0)$ is the initial mass of pollutant. The corresponding equation for the concentration is

$$c(t) = c(0)\,e^{-\kappa t},$$

where $c(0)$ is the initial concentration of pollutant. The time T taken for the pollutant concentration to reduce to the target level $c(T) = c_{\text{target}}$ is given by

$$T = -\frac{1}{\kappa}\ln\left(\frac{c_{\text{target}}}{c(0)}\right).$$

The proportionate decay rate κ determines the value of T.

Now we use this model to predict what will happen to PCB concentrations in Lake Superior. It is a negative exponential decay model, therefore it will have the same desirable qualitative behaviour that was apparent in the first model. Since the proportionate decay rate κ is larger than the proportionate flow rate k, it is also likely to give better quantitative agreement than the first model.

◀ Interpret results ▶

R.V. Thomann and J.A. Mueller (1987) *Principles of Surface Water Quality Modeling and Control*, Harper & Row.

Note that pA/V, which was assumed to be zero in the first model, is about 50 times as big as k, and is clearly the more dominant process.

Thomann and Mueller suggest that volatilisation probably occurs at a rate of about 0.1 m (that is, $0.1\,\mathrm{m}^3$ per m^2) per day. So in SI units ($\mathrm{m\,s}^{-1}$), we have $p \simeq 0.1/(24 \times 60 \times 60) \simeq 1.16 \times 10^{-6}$. The surface area of Lake Superior is given as $A \simeq 8.21 \times 10^{10}\,\mathrm{m}^2$. Hence in SI units,

$$\kappa = k + pA/V$$
$$\simeq 0.166 \times 10^{-9} + 1.16 \times 10^{-6} \times 8.21 \times 10^{10}/(12.10 \times 10^{12})$$
$$\simeq 8.03 \times 10^{-9}.$$

The estimated proportionate decay rate is about 50 times larger than the proportionate flow rate. It is clear that the volatilisation of PCBs has a much greater influence on the rate at which their concentration is reduced than does the flow of pollutant out of the lake. The revised time, in seconds, for the pollutant concentration to fall by a factor of ten is

$$T = \frac{\ln 10}{\kappa} \simeq 2.87 \times 10^8,$$

which is approximately 9.1 years.

◀ Evaluate ▶

The results of this revised model are much more encouraging. The values in Table 3 indicate that PCB concentrations should take about 14 years to reduce by a factor of ten, while the revised model predicts about 9 years (compared with the 440 years predicted by the original model). So from a model that grossly overestimates the time taken, the revised model underestimates it. It may be that for some purposes, the revised model is sufficient. If it is not, then look again for possible reasons for the discrepancy between the revised model's prediction and reality.

You saw how values of A, V and k for Lake Superior led to values for κ and for the decay time T (the time for pollutant concentration to fall by a factor of ten). Table 5 gives the corresponding results for κ and decay time for all of the lakes.

The values of k in Table 5 were obtained, from the water flow rate and volume of each lake, in Unit 8. The decay time is given (in seconds) by
$$T = (\ln 10)/k$$
in the original model, and by
$$T = (\ln 10)/\kappa$$
in the revised model.

Table 5   Proportionate flow and decay rates, and times to fall by a factor of ten

| Lake | Volume V $(10^{12}\,\mathrm{m}^3)$ | Surface area A $(10^{10}\,\mathrm{m}^2)$ | Original model | | Revised model | |
			Prop. flow rate k $(10^{-9}\,\mathrm{s}^{-1})$	Decay time (years)	Prop. decay rate κ $(10^{-9}\,\mathrm{s}^{-1})$	Decay time (years)
Superior	12.10	8.21	0.166	440	8.02	9.1
Michigan	4.92	5.78	0.319	229	13.92	5.2
Huron	3.54	5.96	1.441	51	20.93	3.5
Erie	0.48	2.57	12.292	6	74.26	1.0
Ontario	1.64	1.90	4.134	18	17.54	4.2

Possible reasons for the discrepancy between the data in Table 3 and the prediction of the revised model in Table 5, for Lake Superior, include the following.

- The volatilisation rate is given to only one significant figure, and this may cause a sizeable numerical error. A more accurate estimate for this rate is required to ensure reliable predictions.

- The volatilisation of PCBs from the lake to the atmosphere has been included, but the volatilisation of PCBs from the atmosphere to the lake has been ignored.

- Although PCBs were banned in 1975, there will be a residual amount in the catchment area that will continue to wash into the lake over time, so Assumption (a) may need to be revised.

- Some of the PCBs settle into the sediment at the bottom of the lake, and will be flushed out of this sediment at a much slower rate, so the early samples of water will have lower PCB concentrations than expected.

- The volatilisation rate is based on the assumption that the pollutant is uniformly distributed over the whole lake. However, it is known that the pollutant concentration varies considerably within the lake, so this assumption may have to be reconsidered.

The first item in this list is not a criticism of the model but a requirement for more accurate data. Each of the other criticisms could form the basis of a second revised model.

The process of improving a model is one of evaluating the model at each stage and then addressing the assumption that seems most likely to improve the model; it is an iterative process. It is best to change one feature at a time, even though that may affect more than one assumption.

In the second revision of the model below, the assumption of a uniform concentration of pollutant in the whole lake is revised, to take some account of the last criticism.

Second revision: segmenting the lake

To account for the variation in pollutant concentration across a lake, consider the technique of breaking down the region considered by the model into a number of smaller subregions or cells, where each cell has its own distinct properties. A model for the whole region is built up by considering what happens in each cell, and how it interacts with the other cells.

◄ Create model ►

This technique is known as *segmentation* or *compartmentalization*.

The mathematical model for pollution in the Great Lakes used by the US Environmental Protection Agency considers what happens in each lake by segmenting it into a large number of smaller volumes. This segmentation is done by first constructing a horizontal grid on the lake. The horizontal segmentation takes into account, for example, regions of the lake where the pollutant concentration is much higher or lower than average. Then for each cell created by the horizontal segmentation, there is also vertical segmentation. This allows for different behaviour between the upper regions of the lake, where photolysis and volatilisation take place, and the lower regions, where the presence of sediment may have an effect.

In this way, each lake is divided up into hundreds of small cells, on which more precise data are available and for which more precise information may be required. Each cell has its own variables representing, for example, its pollutant concentration and the rates of flow to and from adjacent cells.

Models that use segmentation can be very complicated, and this is true of the models used to analyse levels of pollution in the Great Lakes. So to keep the analysis relatively simple, the revised model here is a simple segmentation of Lake Huron. In the south-west of Lake Huron is a large bay, called Saginaw Bay (see Figure 5). The Saginaw River flows into this bay, and was a major source of pollution for the whole lake. Suppose that a model of the concentration of PCBs in Lake Huron is required. It would be reasonable to consider Saginaw Bay separately from the rest of the lake, as shown in Figure 5, since the concentration of pollutant in this bay is likely to be significantly higher than that in the rest of the lake.

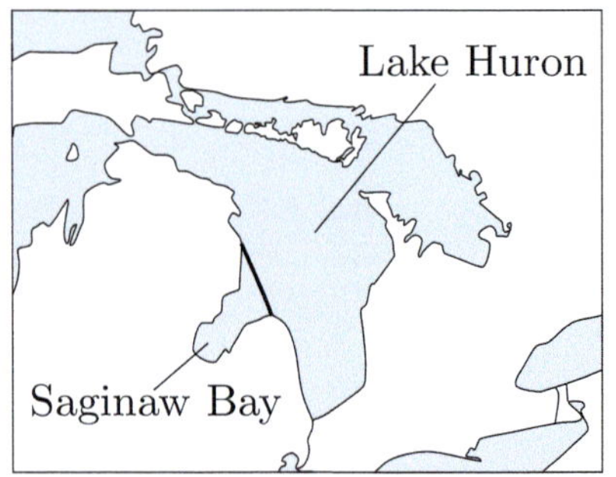

Figure 5   Saginaw Bay in Lake Huron

Two further assumptions are now needed, in addition to those made for the first revised model:

(i)   Saginaw Bay and the rest of Lake Huron can be considered as two separate lakes.

Diffusion of pollutant is a transportation process that involves the movement of molecules of the pollutant relative to the water, rather than bodily movement of the water itself.

(j)   The rate of diffusion of pollutant from Saginaw Bay into the rest of Lake Huron is proportional to the difference in concentrations between the two regions of the lake.

The first assumption here implies that the pollutant in Saginaw Bay is evenly dispersed throughout the bay, so that the concentration of pollutant in the bay varies only with time and not with position. The second assumption models the interaction between the two regions of the lake.

There will be approximately twice as many variables and parameters as before, with one set for each region of the lake.

The variables are:

t the time, in seconds, since all sources of pollution cease;

$m_B(t)$ the mass, in kilograms, of pollutant in the bay at time t;

$m_L(t)$ the mass, in kilograms, of pollutant in the rest of the lake at time t;

$c_B(t)$ the concentration, in kilograms per cubic metre, of pollutant in the bay at time t;

$c_L(t)$ the concentration, in kilograms per cubic metre, of pollutant in the rest of the lake at time t.

The parameters are:

V_B the volume, in cubic metres, of water in the bay;

V_L the volume, in cubic metres, of water in the rest of the lake;

r_B the water flow rate, in cubic metres per second, into and out of the bay;

r_L the water flow rate, in cubic metres per second, into and out of the rest of the lake;

A_{B} the surface area, in square metres, of the bay;

A_{L} the surface area, in square metres, of the rest of the lake;

p the volatilisation rate, in metres per second, for the pollutant;

E the bulk exchange rate, in cubic metres per second.

The meaning of E will become apparent shortly.

Starting with Saginaw Bay, consider what happens to the mass of pollutant in the bay over a time interval $[t, t + \delta t]$. According to Assumptions (a) and (g), there is no new pollution entering the lake, so there is no input of pollutant to the bay from external sources. The output (in kilograms per second) of pollutant from the bay takes three main forms.

- Pollutant flows from Saginaw Bay into the rest of Lake Huron at a rate $r_{\mathrm{B}} \, c_{\mathrm{B}}(t) = r_{\mathrm{B}} \, m_{\mathrm{B}}(t)/V_{\mathrm{B}}$.

- Pollutant is lost to the atmosphere, through volatilisation, at a rate $p A_{\mathrm{B}} \, c_{\mathrm{B}}(t) = p A_{\mathrm{B}} \, m_{\mathrm{B}}(t)/V_{\mathrm{B}}$.

- Pollutant diffuses from a region of higher concentration (the bay) to a region of lower concentration (the rest of the lake) at a rate $E(c_{\mathrm{B}}(t) - c_{\mathrm{L}}(t)) = E(m_{\mathrm{B}}(t)/V_{\mathrm{B}} - m_{\mathrm{L}}(t)/V_{\mathrm{L}})$, where the parameter E depends, for example, on the vertical cross-sectional area separating Saginaw Bay from the rest of Lake Huron.

This follows from Assumption (j). The parameter E is known as the *bulk exchange rate* for diffusion of pollutant across the boundary between the bay and the rest of the lake.

Thus the total output of pollutant from the bay during the time interval $[t, t + \delta t]$ is

$$\left(\frac{r_{\mathrm{B}}}{V_{\mathrm{B}}} \, m_{\mathrm{B}}(t) + \frac{p A_{\mathrm{B}}}{V_{\mathrm{B}}} \, m_{\mathrm{B}}(t) + \frac{E}{V_{\mathrm{B}}} \, m_{\mathrm{B}}(t) - \frac{E}{V_{\mathrm{L}}} \, m_{\mathrm{L}}(t) \right) \delta t.$$

The accumulation of the mass of pollutant within the bay is

$$m_{\mathrm{B}}(t + \delta t) - m_{\mathrm{B}}(t).$$

Using the input–output principle then gives

$$m_{\mathrm{B}}(t + \delta t) - m_{\mathrm{B}}(t)$$
$$= -\left(\frac{r_{\mathrm{B}}}{V_{\mathrm{B}}} \, m_{\mathrm{B}}(t) + \frac{p A_{\mathrm{B}}}{V_{\mathrm{B}}} \, m_{\mathrm{B}}(t) + \frac{E}{V_{\mathrm{B}}} \, m_{\mathrm{B}}(t) - \frac{E}{V_{\mathrm{L}}} \, m_{\mathrm{L}}(t) \right) \delta t.$$

Dividing through by δt, then letting $\delta t \to 0$, leads to the differential equation

$$\frac{dm_{\mathrm{B}}}{dt} = -k_1 \, m_{\mathrm{B}}(t) + k_2 \, m_{\mathrm{L}}(t), \tag{10}$$

where $k_1 = r_{\mathrm{B}}/V_{\mathrm{B}} + p A_{\mathrm{B}}/V_{\mathrm{B}} + E/V_{\mathrm{B}}$ and $k_2 = E/V_{\mathrm{L}}$.

Consider now the rest of Lake Huron. The input (in kilograms per second) of pollutant to this part of the lake comes from two main sources:

- the flow of polluted water from Saginaw Bay, at a rate $r_{\mathrm{B}} \, m_{\mathrm{B}}(t)/V_{\mathrm{B}}$

- diffusion across the boundary between the two parts of the lake, at a rate $E(m_{\mathrm{B}}(t)/V_{\mathrm{B}} - m_{\mathrm{L}}(t)/V_{\mathrm{L}})$.

This ignores any input of pollutant to Lake Huron from the upstream lakes, in keeping with Assumption (g). This assumption may not be justified for Lake Huron, but it keeps the model simple.

Hence the input of pollutant to the rest of Lake Huron in the time interval $[t, t + \delta t]$ is

$$\left(\frac{r_{\mathrm{B}}}{V_{\mathrm{B}}} \, m_{\mathrm{B}}(t) + \frac{E}{V_{\mathrm{B}}} \, m_{\mathrm{B}}(t) - \frac{E}{V_{\mathrm{L}}} \, m_{\mathrm{L}}(t) \right) \delta t.$$

During this same time interval, the output of pollutant from this part of the lake takes two main forms.

- Pollutant flows from Lake Huron to Lake Erie at a rate $r_L\, c_L(t) = r_L\, m_L(t)/V_L$.

- Pollutant is lost to the atmosphere, through volatilisation, at a rate $pA_L\, c_L(t) = pA_L\, m_L(t)/V_L$.

Thus the total output of pollutant from the rest of the lake in the time interval $[t, t + \delta t]$ is

$$\left(\frac{r_L}{V_L}\, m_L(t) + \frac{pA_L}{V_L}\, m_L(t) \right) \delta t.$$

The accumulation of the mass of pollutant within the rest of the lake is

$$m_L(t + \delta t) - m_L(t).$$

Using the input–output principle applied to the rest of Lake Huron then leads to the differential equation

$$\frac{dm_L}{dt} = \frac{r_B}{V_B}\, m_B(t) + \frac{E}{V_B}\, m_B(t) - \frac{E}{V_L}\, m_L(t) - \frac{r_L}{V_L}\, m_L(t) - \frac{pA_L}{V_L}\, m_L(t)$$

$$= k_3\, m_B(t) - k_4\, m_L(t), \tag{11}$$

where $k_3 = r_B/V_B + E/V_B$ and $k_4 = E/V_L + r_L/V_L + pA_L/V_L$.

Techniques for solving systems of linear differential equations can be found in Unit 6.

The revised model has led to equations (10) and (11). These are a pair of linear differential equations that describe the behaviour of pollutant in the two regions of the lake.

Exercise 13

It is worthwhile estimating the contribution from each term that arose in applying the input–output principle. Table 6 gives the relevant data on parameter values for the two regions of the lake.

Table 6   Parameter values for the second revised model

Parameter	Value
V_B	$25 \times 10^9 \, \mathrm{m}^3$
A_B	$4.2 \times 10^9 \, \mathrm{m}^2$
r_B	$153 \, \mathrm{m}^3 \, \mathrm{s}^{-1}$
V_L	$3.24 \times 10^{12} \, \mathrm{m}^3$
A_L	$57 \times 10^9 \, \mathrm{m}^2$
r_L	$4967 \, \mathrm{m}^3 \, \mathrm{s}^{-1}$
p	$1.16 \times 10^{-6} \, \mathrm{m}\,\mathrm{s}^{-1}$
E	$11\,000 \, \mathrm{m}^3 \, \mathrm{s}^{-1}$

(a) Suppose that the concentration of pollutant in the rest of the lake is half that in Saginaw Bay. Compare the rates of pollutant output from the bay for each of the three forms of pollutant transfer: water flow, volatilisation and diffusion. Hence decide which of these mechanisms is the most significant in removing pollutant from the bay.

(b) How will the rate of pollutant output for each of these three forms change if the concentration of pollutant in the rest of the lake is less than half that in the bay?

(c) What is the main mechanism for the removal of pollutant from the rest of the lake?

3.3 Review

In this section you have seen how data were used to evaluate a simple model for the behaviour of pollution in the Great Lakes. The data were obtained from published sources, but it was not possible to check their reliability. There was good qualitative agreement with the model, but rather poor quantitative agreement.

Prior to making revisions, the assumptions made for the first model were evaluated. The first revision was based on the inclusion of an additional process, to try to explain the reason for the substantial quantitative differences between the published data and the predictions of the model. This revision was very effective, since it demonstrated that volatilisation is likely to have a much more substantial effect on the removal of PCBs than does the flow of water out of the lake.

The second revision was based on a segmentation of the lake, or division into compartments, to take into account the non-uniformity of the concentration of pollutant. This technique is widely applicable in mathematical modelling. Exercise 13 shows that diffusion, as well as volatilisation, is more significant than water flow in removing pollutant from Saginaw Bay. In fact, the only form of pollutant output considered in the original model turns out to be the least important in the revised models.

The original model, based on reasonable simplifying assumptions, made a start in the problem-solving process. The evaluation of this model revealed its deficiencies, and as a result other forms of output were considered to improve the model. The development that was undertaken here emphasises the role of the first simple model in initialising the modelling process.

4 Sample modelling report

The marginal notes give more information about what is required in each section.

We round off this unit by looking at and commenting on a report on a modelling activity based on bungee jumping, starting with the statement of the problem that is to be investigated. Remember that it is the modelling process rather than the model that is important in writing up the report. The report given here is intended to be not a model report, but simply an attempt to outline the approach to writing a modelling report.

Try not to worry about the *details* of the sample report. It is provided to show you the sort of thing that you might produce, not to teach you any new mathematics.

In order to follow this report, you need one extra piece of physics that is not taught in this module. From Unit 9, you know that the magnitude of the force exerted by a spring is proportional to the extension, and that this constant of proportionality k is called the stiffness of the spring. Here we need to include something that links in the cross-sectional area of the elastic material used in making the bungee, and the required model is that the magnitude of the force F is given by

$$F = \frac{EAe}{L},$$

The difference between a spring and a bungee is that the cross-sectional area of the bungee changes when it is stretched.

where A is the original (unstretched) cross-sectional area of the elastic material, L is the natural length of the bungee, e is the extension, and E is the *Young's modulus* (sometimes called the modulus of elasticity) for the elastic material. We can deduce that the stiffness is $k = AE/L$.

Problem statement: Bungee jumping

The extreme sport of bungee jumping involves a participant diving off a high platform, below which is clear space, while attached by a safety harness to an elastic cord. The cord eventually brings the jumper momentarily to rest before he begins to move upwards. It is paramount that those who organise such jumps should ensure the safety of the jumpers, which means among other things that if there is a solid surface such as the ground beneath the jump, then the jumper should not collide with it. (With water beneath, as in several famous jump locations, this safety issue is less crucial, but typically the jumper will wish to remain above the water at all times.) However, within this concern for safety, the jumper will wish to experience the greatest possible thrill.

Develop a mathematical model that will enable you to advise the organisers of a bungee jump on how to ensure the safety or dryness of participants, while not diminishing the element of excitement more than is necessary.

Title: Bungee jumping

Author: Model Student, 30 September 2014

Specify the purpose of the model

Definition of the problem

In the sport of bungee jumping, each participant (jumper) drops into space from the top of a high tower or platform. The jumper is tied to one end of an elastic cord (called a bungee), the other end of which is secured to the top of the tower. The safety and wellbeing of the jumpers are most important. For a jump over water, the jumper may not wish to be soaked in the process. Another important aim is that the force magnitudes experienced by the jumper should not be too high, since this also can cause injury or significant discomfort.

This report considers what advice should be provided to those who organise bungee jumps over water, to ensure that each jumper remains dry during the jump and is not subject at any stage to excessive forces. In particular, the aim is to specify the characteristics of the bungee so as to maximise the distance that a jumper falls without entering the water beneath, while keeping the maximum force encountered to an acceptable limit.

In many cases, the problem specified may be rather vague. This gives you scope to choose which aspects of the problem you think are important. You should outline, in a couple of sentences, the approach that you have adopted in your report.

Aspects of the problem to be investigated

The most important elements of a bungee jump are the mass of the jumper, and the length, cross-sectional area and elastic properties of the bungee. It will be shown that these factors determine both the forces encountered and the distance fallen. Consequently, to achieve a given distance of fall in safety, the length and elastic properties of the bungee can be selected for a jumper of given mass. See also Appendix 3.

Create the model

Outline of the approach in the first model

The situation will be simplified by ignoring dissipative effects (like friction or air resistance). This enables the principle of conservation of mechanical energy to be applied. The constituent quantities that will feature in applying energy conservation are the kinetic energy and gravitational potential energy of the jumper, and the elastic potential energy of the bungee. The sum of these quantities will be constant throughout the jump. For a chosen total distance of fall, the energy conservation equation provides a relationship between the mass of the jumper on the one hand and the unstretched length and stiffness of the bungee on the other. The stiffness is in turn related to other physical properties of the bungee.

You need to specify the approach that you will take in the first model. This should be an outline of how you will tackle the model, giving the reader some idea of what you are attempting to do. It should not be mathematical, but you may need to use some mathematical jargon occasionally.

Assumptions

It should be possible to justify each major step in the formulation of mathematical relationships in your model, by appealing to one or more of the assumptions. On the other hand, an assumption that is unnecessary for the derivation of your model is superfluous, and should be omitted.

In your first model, it is best to simplify the problem as much as you can, without losing any of its most essential features.

1. The jumper is a particle (so his size and shape and any rotations are ignored).

2. The motion is entirely in the vertical direction and hence in one dimension.

3. The initial speed of the jumper (on leaving the platform) is zero.

4. The mass of the bungee is ignored throughout.

5. Until the bungee is extended (beyond its unstretched length), it exerts no force on the jumper.

6. Once the bungee is extended, it acts as a model spring (satisfying Hooke's law) with natural length equal to its unstretched length.

7. The bungee is made of uniform material, which behaves according to a standard model for elastic materials.

8. The physical properties of the bungee are unaffected by the local air temperature, exposure to sunshine or other weather conditions.

9. The distance between the take-off platform and the water surface beneath is constant in time.

10. Air resistance is ignored.

It is helpful to number the assumptions for reference later in the report.

11. The force exerted on the jumper by the bungee should not be too great.

12. The bungee will at no point be in danger of breaking.

Definition of variables and parameters

You may decide to list all the variables and parameters together in this section, or you may decide to give just the main ones here and to introduce others where they are needed in developing the model. In either case, make sure that each variable or parameter is clearly and unambiguously defined, and that the units or dimensions are given if appropriate.

The report is much clearer if all the variables and parameters are defined in a table.

Symbol	Description	Units
m	Mass of the jumper	kg
x	Distance of jumper below platform	m
H	Distance from jumping platform to lowest point of motion, at surface of water	m
L	Unstretched length of bungee	m
e	Extension of bungee when stretched	m
A	Cross-sectional area of unstretched bungee	m^2
F	Magnitude of force applied to end of bungee	N
E	Modulus of elasticity (Young's modulus) for the bungee material	$N\,m^{-2}$
k	Stiffness of bungee	$N\,m^{-1}$
g	Magnitude of acceleration due to gravity	$m\,s^{-2}$
λg	Maximum acceleration permitted for the jumper	$m\,s^{-2}$
p	Constant ($p = mg/(AE)$)	–

Formulation of mathematical relationships

By Assumptions 5, 6 and 10, there is no energy lost during the motion, hence the law of conservation of mechanical energy can be applied. By Assumptions 1 and 4, the system to be considered is a single particle of mass m, together with a model spring when in extension. By Assumption 2, the motion is in one (vertical) dimension, starting at $x = 0$ and reaching its lowest point at $x = H$. A diagram of the situation is given in Figure 1.

In formulating each of the mathematical relationships in your model, you need to explain to the reader, in words, what the relationship is and how it arises from your assumptions. This is probably the most difficult part of the report, so make sure that you re-read what you have written with a critical eye to ensure that it is clear. If possible, ask someone else to read this part of the report, and ask him or her whether your description of the formulation is easy to follow. You may wish to check the dimensional consistency of any mathematical relationships.

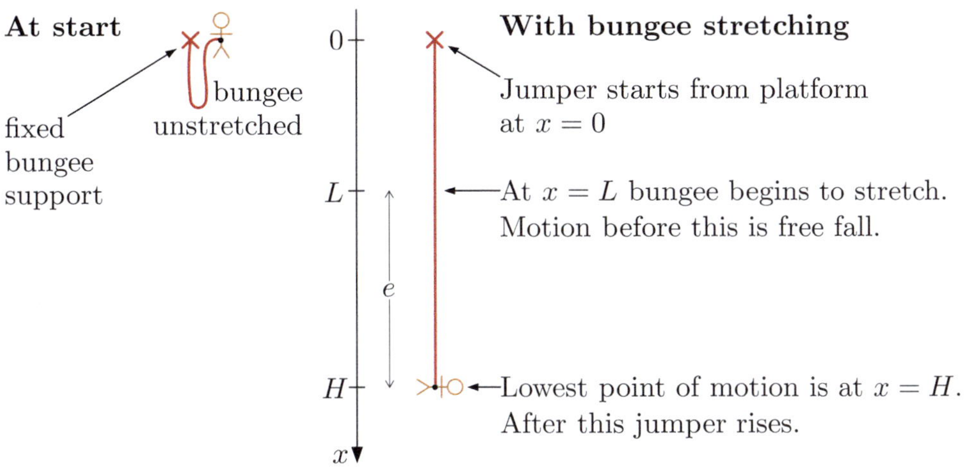

Figure 1   Positions and distances in bungee jumping

By Assumption 9, H is constant. Let $x = H$ be the datum for gravitational potential energy.

At $x = 0$ the jumper has zero kinetic energy (since he starts with speed zero, by Assumption 3) and gravitational potential energy mgH (by the choice of datum). The bungee, being massless (by Assumption 4) and unstretched, has zero potential energy.

At $x = H$ the jumper has zero kinetic energy (since he comes momentarily to rest at this point) and gravitational potential energy zero (by the choice of datum). The bungee (by Assumptions 6 and 8) has spring potential energy $\frac{1}{2}ke^2$, where the stiffness k is constant and $e = H - L$.

Equating the sums of potential and kinetic energies at $x = 0$ and at $x = H$, we have

$$mgH = \tfrac{1}{2}ke^2 = \tfrac{1}{2}k(H - L)^2. \tag{1}$$

It is helpful to number your equations and relations as you develop the solution.

By Assumptions 6 and 8, we can apply Hooke's law to obtain the magnitude of the total upwards force at the bottom of the jump as

$$F = ke - mg. \tag{2}$$

Using the standard elasticity model referred to in Assumption 7, we note that the stiffness of the spring k is inversely proportional to the length of the bungee. We have

$$k = \frac{AE}{L}, \tag{3}$$

where A is the cross-sectional area of the bungee, and E is the Young's modulus for the material of the bungee.

By Assumption 11, the acceleration exerted on the jumper should be no more than the maximum multiple of g permitted. From equation (2) this occurs at the bottom of the jump, so

$$\frac{F}{m} \le \lambda g. \tag{4}$$

Equations (1)–(3) and inequality (4) constitute the first model. In addition, by Assumption 12, F will not exceed (or even come close to) the breaking strength of the bungee.

Do the mathematics

Solution of the equations

Explain to the reader, in words, what mathematical methods you use and how this will solve the problem. Write out the solution. A computer algebra package may be useful here.

With $p = mg/(AE)$, we have, from equations (1) and (3),

$$\frac{mgH}{k} = \frac{mgHL}{AE} = pHL = \tfrac{1}{2}(H - L)^2,$$

so

$$L^2 - 2(1 + p)LH + H^2 = 0.$$

We can solve this for L (with $H > L$, since the stretched length of the bungee is greater than the unstretched length), in terms of the mass of the jumper m and the height H, as

$$L = \left((1 + p) - \sqrt{p^2 + 2p} \right) H. \tag{5}$$

The negative square root is taken here because $L < H$. This is dimensionally consistent provided that p is dimensionless. We have

$$[p] = [mg/(AE)] = \frac{[m][g]}{[A][E]} = \frac{\mathrm{M}\,(\mathrm{L\,T^{-2}})}{(\mathrm{L^2})\,(\mathrm{M\,L^{-1}\,T^{-2}})} = 1,$$

as required.

The alternative of solving equation (1) for the extension e in terms of L (choosing the positive square root to ensure that $e > 0$) gives

$$e = pL \left(1 + \sqrt{1 + \frac{2}{p}} \right). \tag{6}$$

From equations (2), (3) and (6), the maximum upwards acceleration experienced by the jumper due to the bungee is

$$\frac{F}{m} = \frac{ke}{m} - g = \left(\frac{e}{Lp} - 1 \right) g = \left(\sqrt{1 + \frac{2}{p}} \right) g.$$

From inequality (4), the condition for maximum acceleration magnitude can therefore be written as

$$\sqrt{1 + \frac{2}{p}} \leq \lambda, \quad \text{where } \lambda > 1,$$

and this can be interpreted as a limit on the weight of the jumper, since using $p = mg/(AE)$ we can deduce that

$$mg \geq \frac{2AE}{\lambda^2 - 1}. \tag{7}$$

This means that to avoid excessive acceleration at the bottom of the jump, the weight of the jumper must be sufficiently large. We can also deduce that for a given mass of jumper, the cross-sectional area of the bungee must satisfy

$$A \leq \frac{mg(\lambda^2 - 1)}{2E}. \tag{8}$$

Graphs

Various graphs may be drawn based on the relationships above – two examples are shown below.

A computer algebra package (as used here) may help in presenting graphs of the solution for sample values of the parameters.

Figure 2 is a graph of the unstretched bungee length L (as a proportion of H) against the parameter $p = mg/(AE)$ ($0.2 < p < 0.4$), which can be interpreted as a graph of the unstretched length of the bungee against the mass of the jumper, when A is fixed, using equation (5).

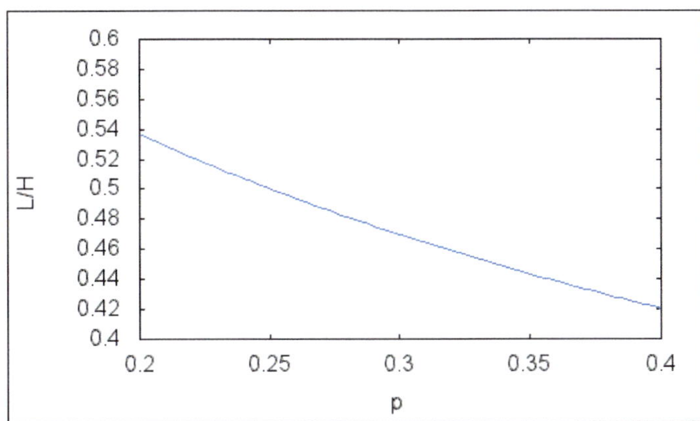

Figure 2   Graph of the unstretched length L of the bungee as a proportion of the height H, against p

An expected feature here is that the unstretched length L of the bungee decreases as the mass increases.

Figure 3 gives the maximum extension e of the bungee (as a proportion of L) against the parameter $p = mg/(AE)$ ($0.2 < p < 0.4$), using equation (6).

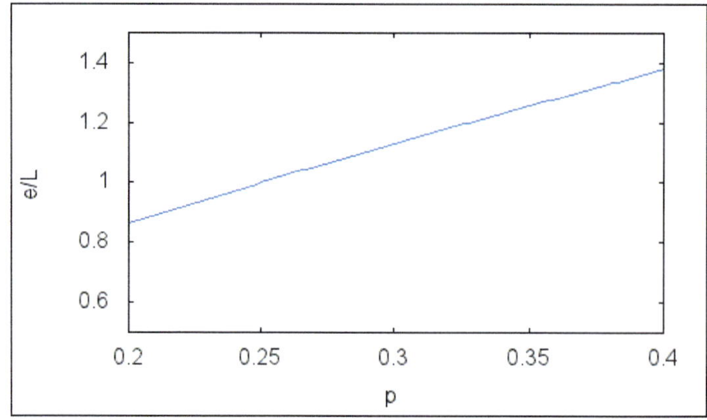

Figure 3   Graph of the maximum extension e of the bungee as a proportion of the unstretched length L, against p

Results

Give the mathematical solution to the problem in algebraic form.

As the maximum permitted acceleration λg to be exerted by the bungee occurs at the lowest point of motion, to obtain the maximum thrill of the jump, the inequality (8) becomes an equation. In this case, we have

$$A = \frac{mg(\lambda^2 - 1)}{2E} \quad (\lambda > 1). \tag{9}$$

Also, using equation (5) we have an expression for the length of the bungee, given the weight of the jumper, as

$$L = \left(\left(1 + \frac{mg}{AE}\right) - \sqrt{\left(\frac{mg}{AE}\right)^2 + 2\frac{mg}{AE}}\right) H. \tag{10}$$

Equations (9) and (10) constitute the main results from the model: for a jumper of mass m, a drop of total distance H and a maximum acceleration exerted of λg, these two equations give the appropriate values to be selected for the unstretched bungee cross-sectional area A and length L. Values are needed for the parameters g and E.

Interpret the results

Data

You should present the data that you have collected, giving references to any sources of data that you have used (though we do not do so here). `Internet sources` have been suppressed in this sample report, but you should give any internet sources that you have used, together with the date when you referenced each source.

Quantity	Value	Source or justification
g	$9.8\,\mathrm{m\,s^{-2}}$	Standard value
E	$1\text{–}5 \times 10^6\,\mathrm{N\,m^{-2}}$	For natural rubber, `internet source`

There is a wide variation in the values for Young's modulus of natural rubber to be found in various sources, and some are well outside the range quoted above. We will take $E = 1 \times 10^6 \, \mathrm{N\,m^{-2}}$.

The masses of jumpers are assumed typically to vary between 40 kg and 120 kg, though lighter and heavier jumpers are also accommodated on some jumps.

The allowable maximum acceleration of the bungee jumper is often laid down in legislation. For example, **internet source** specifies that the maximum G-force allowable on a jumper should be $4\frac{1}{2}g$ for a waist-and-chest harness, and $3\frac{1}{2}g$ for an ankle harness. Other jurisdictions lay down restrictions in terms of cord extension. The Hong Kong Code of Practice for Bungee Jumping states a condition that is equivalent to $2.7 \le \lambda \le 3.3$.

Values of H also vary widely. A famous example is the first permanent commercial bungee jump site at Kawarau Bridge, Queenstown, New Zealand, where the drop distance is 43 m. However, the model should also apply to miniature bungee jumps, such as that considered later.

Note that the extension of the bungee, at which it is likely to break, is of the order of 750–850% (according to **internet source**). The extensions estimated by the current model are well outside this range, so should be completely safe, as required by Assumption 12.

Interpretation of results

Using the given data for g and E, equation (9) becomes

$$A = 4.9 \times 10^{-6} m(\lambda^2 - 1) \quad (\lambda > 1),$$

whereas

$$p = \frac{mg}{AE} = \frac{2}{\lambda^2 - 1}.$$

Selecting as input the reasonable values $H = 43$ and $\lambda = 3$ gives $p = 0.25$, so

$$L = \left((1 + p) - \sqrt{p^2 + 2p}\right) H = 0.5H = 21.5$$

and

$$A = 39.2 \times 10^{-6} m.$$

Hence for a person of mass 80 kg jumping from Kawarau Bridge, the bungee should have length 21.5 m and cross-sectional area $A = 3.14 \times 10^{-3}$ or 31.4 cm^2 (i.e. a diameter of about 6 cm). For these values of H and λ, the graph of A against m is shown in Figure 4.

If you have done your own experiments to collect data, then the descriptions of those experiments should appear in an appendix at the end of the report. (See Appendices 1 and 2 here.)

You should, if possible, give both a qualitative and quantitative interpretation of your solution. The qualitative interpretation might include how the solution varies with different parameters and whether this agrees with common sense, or it might examine whether the solution behaves as expected in limiting cases, or in simpler situations. Choose an appropriate level of accuracy for your numerical results.

Be careful to present your solution in such a way that it addresses the purpose of the mathematical model.

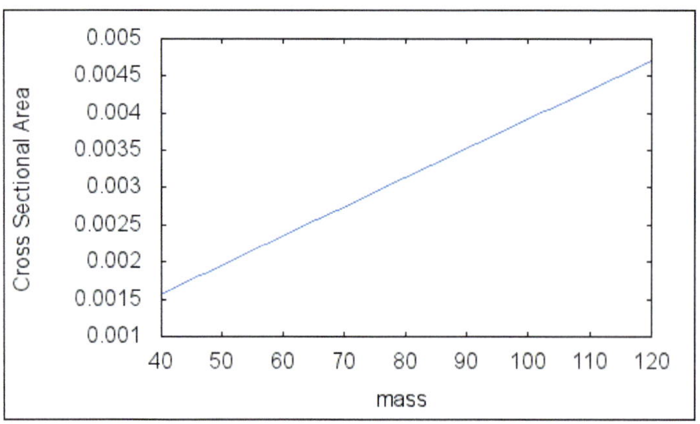

Figure 4   Graph of bungee cross-sectional area against jumper mass (for $H = 43$, $\lambda = 3$)

Clearly (from either this graph or equation (9)), the cross-sectional area should be chosen in proportion to the mass of the jumper. This conclusion applies for any values of H and λ. Provided that this rule is adhered to, each jumper, whatever their mass, should fall the same distance and experience the same maximum acceleration at the bottom of the jump. The length L of the bungee required to achieve this is given by equation (5), and once λ has been selected, L should be a fixed proportion of the drop distance H.

To specify a precise cross-sectional area for an unstretched bungee does not appear to be very practical advice. However, bungee cords are typically made up of a large number of individual rubber threads whose cross-sectional area is very small (for example, '22 gauge' at 0.643 mm diameter is common). Achieving a particular cross-sectional area is then a matter of aggregating the correct number of rubber threads. In practice there are four or five set numbers of threads that may be selected, each appropriate for a certain range of jumper mass, so there will be some (acceptable) variation in jump performance, for a given bungee, between jumpers at either end of the corresponding mass range. The assessment of 'worst case' performance must be carried out with respect to the heaviest mass that each type of bungee will be expected to carry.

As regards the sensitivity of the results to small changes in the parameters, it is clear from the proportionality of A and m that a 10% change in the value of m will lead to a 10% change in the value of A. Similarly, a 10% change in the value of H will lead to a 10% change in the value of L. The relative sensitivity is therefore 1 in each case.

It can be shown that the unstretched length L of the bungee can be expressed in terms of λ as

$$L = \frac{(\lambda - 1)H}{\lambda + 1}.$$

Thus the relative sensitivities of A and L to changes in λ are given respectively by

$$\frac{\lambda}{A}\frac{\partial A}{\partial \lambda} = \frac{2\lambda^2}{\lambda^2 - 1} \quad \text{and} \quad \frac{\lambda}{L}\frac{\partial L}{\partial \lambda} = -\frac{2\lambda}{\lambda^2 - 1}.$$

Both variables are relatively insensitive provided that λ is not close to 1. The typical data for λ quoted earlier show that values close to 1 do not occur in practice.

The conclusion is that neither the unstretched bungee cross-sectional area nor the length is unduly sensitive to changes in the parameter values of the model.

Choice of results to compare with reality

There is no shortage of relevant data available on internet sites, but it is difficult to find any site that provides a complete and consistent set of data to validate this model. The premises that underlie the data provided are often not made explicit. In what follows, some data from the internet are supplied, for comparison with certain aspects of the model, but the chief validation is based on an experiment. The experiment seeks to check the validity of equation (6), which predicts the maximum extension to be expected for given bungee length, bungee stiffness and jumper mass. The details are given in Appendix 1.

Which of your results can be tested against reality? These tests do not have to be directly related to the purpose of the model, and there may be ways of testing the reliability of your model by using it to make predictions in simple situations, or for only parts of the model.

Evaluate the model

Comparison with reality

The experimental results reported in Appendix 1 (for which the data and graphs are shown in Appendix 2) indicate some deficiencies in the model. One notable feature is that Hooke's law does not seem to apply at all closely over the range of masses considered. A comparison of the predicted and measured values for e as m is varied is shown in Figure 5. This indicates that the measured values are quite close numerically to the values predicted by the model. However, the shapes of the two graphs are noticeably different.

You should try to compare some of the predictions of your model with real data, in order to test the model's accuracy and reliability. Again, you should give references to any sources of data, and if you have carried out any experiments, then you should present your results here. The descriptions of the experiments can be put in an appendix at the end of the report. How reliable are your results? Are any of the simplifications that you made likely to have led to an answer that is too large or too small?

If you feel that there are no appropriate tests for your model, then you should explain why.

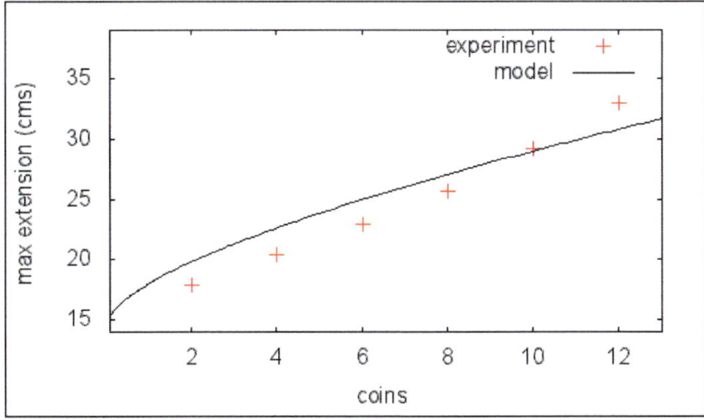

Figure 5   Comparison of predicted and experimental values for maximum extension

The data from internet sites given in Appendix 3 confirm that Hooke's law does not hold at all closely for bungee cords. However, there is considerable support here for the prediction that the cross-sectional area of a bungee (or equivalently, the number of threads from which it is made) should be approximately proportional to the mass of the jumper.

Criticism of the model

Criticising your work is a healthy practice, provided that it is done in moderation and constructively. You should praise the positive aspects of your model as well as criticising the negative ones. If the model is deficient, either qualitatively or quantitatively, then you should reconsider the assumptions that you made earlier. Decide whether each assumption is valid, based on the outcome of evaluation. Try to identify the key assumption(s) that, if changed, would lead to improvement in the predictions of the model. Make sure that the discussion is relevant.

In some respects the predictions of the model match both common sense and available data. This applies in particular to the outcome that the cross-sectional area of the bungee should be proportional to the mass of the jumper. However, the experimental results and data from the internet both indicate that the extension of a bungee cord is not proportional to the force applied at its end, leading to significant inaccuracies in the model's predictions.

Consider in turn each of the assumptions at the start of the report.

1.   Since the jumper could be up to 2 m in length, the assumption that he is a particle might need revision to avoid some part of the jumper entering the water.

2.   There will be minor departures from the vertical in the motion, but these should hardly affect the predictions of the model.

3.   The jumper might project himself with a non-zero velocity, but this need not be a major factor for revision. It could be incorporated if desired, however.

4.   The mass of the bungee is, in fact, quite significant (as one of the sources quoted in Appendix 3 indicates), and its weight provides an additional downward force on the jumper in the initial stage of the jump. This could be taken into account in a revised model.

5.   Assuming that the bungee exerts no force on the jumper is reasonable as regards the elastic properties of the bungee, but its weight has an effect, as discussed above.

6.   The experiment, and other available data, indicate that the elastic material does *not* behave like a model spring, and a revised model might attempt to capture more closely the behaviour actually observed. In fact, there are 'viscoelastic' energy losses that cause the bungee to heat up.

7.   It is reasonable to assume that the bungee material is uniform, but the standard model referred to appears not to apply at all closely, except for very small suspended masses.

8.   The assumption that the physical properties of the bungee are unaffected by weather etc. is not too bad, though the behaviour of rubber does alter significantly with temperature, and over-exposure to sunlight causes a deterioration in performance.

9.   It is reasonable to assume a constant height between platform and water. If there were slight seasonal variations, then the model could take these into account.

10. Air resistance has an effect, but in practice the principal energy losses are within the bungee itself, as remarked above.

11. It is sensible to limit G-forces for safety, and this is in accordance with regulations for bungee jumping.

12. It is essential that the bungee should not break, but this hardly needed to feature in the model.

The chief candidates for revision, in the extent to which they affect the model's predictions, seem to be Assumption 4, on the one hand, and Assumptions 6 and 7 on the other; the latter pair are linked to each other. We will consider here the effect of removing Assumption 4, and hence taking the mass of the bungee into account. This can be a significant proportion of the mass of the jumper.

Revise the model

Description of the revision

Of the original assumptions, Assumption 4 is now ignored, while the other assumptions remain in force. By Assumption 7, the mass of the bungee (denoted by M, in kg) is distributed uniformly. The law of conservation of mechanical energy can still be applied, since there are no dissipative forces. Taking into account the mass of the bungee increases the loss of gravitational potential energy that takes place during the jump. Since this is translated into spring potential energy, the jumper can be expected to fall further than in the first model, unless the unstretched bungee length is reduced to take this into account. At both $x = 0$ and $x = H$ the system is momentarily at rest, so the kinetic energy of the bungee need not be considered.

The revised model

At $x = 0$ the jumper has gravitational potential energy mgH as before. The bungee hangs down from the attachment at $x = 0$ to approximately $x = \frac{1}{2}L$, then rises back up to the jumper at $x = 0$ again, so its centre of mass can be taken to be at $x = \frac{1}{4}L$. Hence the potential energy of jumper and bungee together at the start is

$$mgH + Mg\left(H - \tfrac{1}{4}L\right).$$

When the jumper reaches $x = H$ (the bottom of the jump), the centre of mass of the bungee has moved to $x = \frac{1}{2}H$, hence its gravitational potential energy is $\frac{1}{2}MgH$. As before, the bungee then has spring potential energy $\frac{1}{2}ke^2$. Hence conservation of mechanical energy leads to the equation

$$mgH + Mg\left(H - \tfrac{1}{4}L\right) = \tfrac{1}{2}MgH + \tfrac{1}{2}ke^2,$$

or

$$\left(mH + \tfrac{1}{4}M(2H - L)\right)g = \tfrac{1}{2}ke^2 = \tfrac{1}{2}k(H - L)^2.$$

Your best policy for revising and improving your model is to relax one (and no more) of the simplifying assumptions that you made and subsequently evaluated. In writing this section, you need to make clear which assumption has been changed. You do not need to solve the revised model, but you do need to follow through the changes to the formulation of your first model that occur as a consequence of the new assumption.

You should make a qualitative assessment of the effects of the change to the model that result.

Note that using your first model again with a new set of values for the parameters does not constitute a revision of the model. The revision needs to be based on a change in the assumptions that underpin the first model.

After changing an assumption, you will probably need to modify the work that you did in creating the first model.

The other equations from the first model remain unchanged. The consequence of having $M > 0$ is that for a given drop distance H, the extension e will be greater than with $M = 0$. So either the cross-sectional area A of the bungee must increase or its unstretched length L must decrease, to avoid accidents.

Conclusions

This is your opportunity to sum up the work that you have done, to say whether you have obtained a satisfactory solution to the problem that you set out to solve, and to indicate how you might further improve your model.

One prediction from the model – that the cross-sectional area or number of threads in the bungee should be proportional to the mass of the jumper – is supported both by common sense and by the available data.

The model also predicts that as the mass increases, for a fixed cross-sectional area of bungee, the length of the unstretched bungee should decrease. The formula (10) for bungee length in terms of drop distance should be treated with suspicion, since it is based on a linear model for the deformation of rubber that does not apply at all closely. The mass of the bungee is also a significant factor, but as indicated above, the model could be revised to take this into account.

Any general advice provided on the basis of the model should be accompanied by appropriate warnings as to its limitations. A practical approach at the actual jump site would appear to be most important. This could involve the dropping of a sandbag of known mass, to gauge the appropriate length of bungee. The result could then be scaled (in terms of cross-sectional area) to match the mass of each jumper.

Appendix 1: The experiment and summary of outcomes

You should use appendices to describe any experiments that you have conducted to collect data for the interpretation or evaluation of your model. You might also use an appendix to present data from referenced sources, if you need to adapt them for use with your model.

An experiment was conducted to check the validity of equation (6) from the model, expressed in the form

$$e = \frac{mg}{k}\left(1 + \sqrt{1 + \frac{2Lk}{mg}}\right). \tag{11}$$

In the first part of the experiment, masses were hung from a rubber band, to provide data from which the stiffness k could be estimated. Then the maximum extensions obtained by dropping various masses were measured.

Part 1 of the experiment

Note that it made sense to give measurements in centimetres here, rather than in the SI units of metres.

The experimental set-up is shown in Figure 6. The masses used were £1 coins, each of mass 9.5 g. The range of masses was extended using a bag of ground almonds weighing 11 times as much as a £1 coin. The mass of the plastic bag containing the coins was about $1\frac{1}{2}$ g, and this mass was neglected.

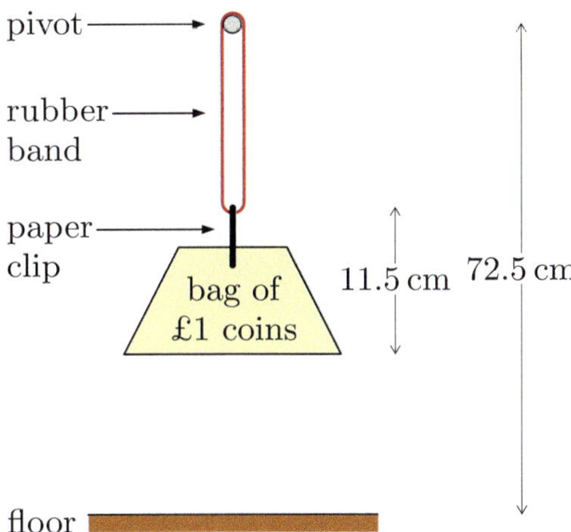

Figure 6   Set-up for both parts of the experiment

While it was not possible to obtain a direct measurement for the unstretched length of the rubber band, the data obtained for small suspended masses indicated that the appropriate value to take was $L = 14.6$ (in cm). The detailed data obtained for extensions are shown in Appendix 2.

From the data on extension e against suspended mass m, an average value for e/m was obtained, leading to an estimate for stiffness of $k = 22.2$. (Other approaches to fitting the data would lead to different values for k.)

However, the graphs indicate that except for small masses within the range considered, Hooke's law does not hold for this situation. Another notable feature was that the extensions measured when coins were being added progressively were smaller than the extensions when coins were being removed one by one. (This behaviour of natural rubber, known as 'hysteresis', is confirmed by various internet sources; see, for example, 'Hysteresis and rubber bands' at **internet source**.) To match the initial circumstances of a bungee jump, it seems sensible to consider only the data obtained when masses are being added and the rubber band is being initially stretched, and only those values for e/m were used in evaluating k.

Part 2 of the experiment

With values established for L and k from Part 1, equation (11) provides a prediction of what maximum extension e is to be expected for a given mass m attached via the rubber band and dropped from the pivot point. In this part of the experiment, the raw data obtained were measurements upwards from floor level, obtained using a pile of paperback books and judging whether the bag containing the dropped mass did or did not touch the top book. Hence the accuracy of these measurements was not as good as in Part 1. These data were then transformed appropriately into data for e. Again, the detailed outcomes are shown in Appendix 2.

It can be seen from the graph of maximum extension against dropped mass that while the quantitative results are in the right general area, the functional behaviour predicted seems to be at some odds with that observed. In particular, the second derivatives of the theoretical and empirical functions have opposite signs. This may well be due to the approximations involved in using a single value for the stiffness k instead of modelling the static extension behaviour more accurately.

Appendix 2: Details of experimental results

The results of the experiments outlined in Appendix 1 will be analysed using a computer algebra system. We begin by inputting the data for the length of the elastic band as coins are added to the bag, as `load1`, and as the coins are removed from the bag as `unload1`. The experiment is repeated to obtain `load2` and `unload2`. The average values for two sets of data are used to specify `load` and `unload`.

```
(%i1)   load1:  [[1,14.9],[2,15.2],[3,15.5],[4,15.8],[5,16.1],
[6,16.4],[7,16.9],[8,17.3],[9,17.8],[10,18.3],[11,18.9],
[12,19.4]]$
```

```
(%i2)   unload1:  [[1,15.1],[2,15.5],[3,15.8],[4,16.1],[5,16.5],
[6,16.9],[7,17.3],[8,17.7],[9,18.2],[10,18.6],[11,19.2],
[12,19.7]]$
```

```
(%i3)   load2:  [[1,14.9],[2,15.2],[3,15.5],[4,15.8],[5,16.1],
[6,16.4],[7,16.8],[8,17.2],[9,17.7],[10,18.2],[11,18.8],
[12,19.4]]$
```

```
(%i4)   unload2:  [[1,15.1],[2,15.4],[3,15.8],[4,16.2],[5,16.5],
[6,16.9],[7,17.3],[8,17.8],[9,18.2],[10,18.8],[11,19.2],
[12,19.7]]$
```

```
(%i5)   load: (load1+load2)/2$
```

```
(%i6)   unload: (unload1+unload2)/2$
```

The data in `load` can be used to predict the stiffness k for the elastic band by finding a least squares fit (simple linear regression) to the data.

```
(%i7)   res: simple_linear_regression(load);
```

$$(\%o7) \quad \begin{pmatrix} \text{SIMPLE LINEAR REGRESSION} \\ \text{model} = 0.40297202797203\,x + 14.23484848484848 \\ \text{correlation} = 0.99095108077158 \\ \text{v\_estimation} = 0.042602855477855 \\ \text{b\_conf\_int} = [0.36451341780587, 0.44143063813818] \\ \text{hypotheses} = \text{H0}:\text{b}=0, \text{H1}:\text{b}\neq 0 \\ \text{statistic} = 23.34659584269099 \\ \text{distribution} = [\text{student\_t}, 10] \\ \text{p\_value} = 4.7062287400478908\,10^{-10} \end{pmatrix}$$

(%i8)   bestfit: take_inference(model,res);

(%o8) $0.40297202797203\,x + 14.23484848484848$

The package calculates the best fit as

$$y = 0.40297202797203x + 14.23484848484848,$$

where y is the extension in centimetres and x is the number of coins. Now the stiffness k measured in newtons per metre satisfies

$$k \times \text{extension in metres} = \text{mass in kilograms} \times g,$$

so

$$k = \frac{\text{mass in kilograms} \times g}{\text{extension in metres}}.$$

The slope of the graph, suitably converted to the correct units, gives an estimate for k as

$$k = \frac{0.0095 \times 9.81 \times 100}{0.40297202797203} = 23.13,$$

so the estimate for the stiffness is $23.13\,\mathrm{N\,m^{-1}}$.

The experimental data and the least squares fit are compared in the following figure.

```
(%i9)  wxplot2d([[discrete,load], [discrete,unload], bestfit],
[x, 0,13], [xlabel, "coins"], [y, 14,22],
[ylabel, "length (cms)"], [style, points,points,lines],
[point_type, plus,times,minus], [color, red,blue,black],
[legend,"loading","unloading","fit"])$
```

(%t9)

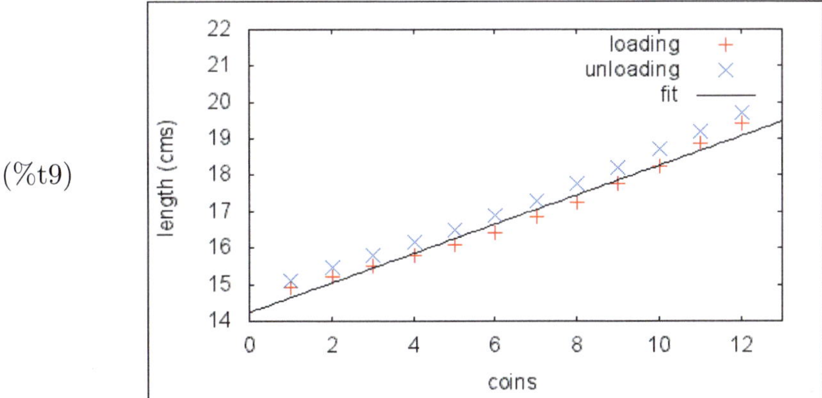

The fit looks pretty reasonable, but we need to see if the relationship still holds for large extensions. We thus continue the experiment by adding more coins in load3, unload3, load4 and unload4.

```
(%i10)  load3: [[11,18.8],[12,19.5],[13,20.0],[14,20.7],
[15,21.4],[16,22.1],[17,22.9],[18,23.9],[19,24.7],[20,25.7],
[21,26.7],[22,27.8],[23,28.8]]$
```

```
(%i11)   unload3: [[11,20.5],[12,21.1],[13,21.8],[14,22.5],
[15,23.2],[16,23.8],[17,24.5],[18,25.4],[19,26.2],[20,27.0],
[21,27.8],[22,28.6],[23,29.4]]$
```

```
(%i12)   load4: [[11,18.8],[12,19.6],[13,20.1],[14,20.8],
[15,21.4],[16,22.2],[17,23.0],[18,24.1],[19,25.0],[20,26.0],
[21,27.0],[22,28.0],[23,29.0]]$
```

```
(%i13)   unload4: [[11,19.9],[12,20.3],[13,21.6],[14,22.3],
[15,23.1],[16,24.0],[17,24.8],[18,25.5],[19,26.3],[20,27.0],
[21,27.9],[22,28.6],[23,29.5]]$
```

The data are again averaged and used with the original data to draw a
graph, which shows that the linear fit is very unsatisfactory for the rubber
band, where the cross-sectional area of the rubber band reduces as the
band extends.

```
(%i14)   allload: append(load,(load3+load4)/2)$
```

```
(%i15)   allunload: append(unload,(unload3+unload4)/2)$
```

```
(%i16)   wxplot2d([[discrete,allload],[discrete,allunload],bestfit],
[x, 0,24],[xlabel, "coins"],[y, 14,32], [ylabel, "length (cms)"],
[style, points,points,lines],[point_type, plus,times],
[color, red,blue,black],[legend, "loading","unloading","fit"])$
```

(%t16)

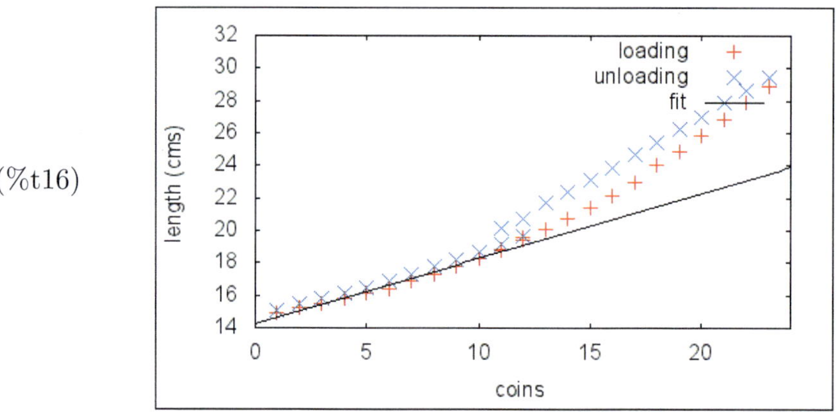

Part 2 of the experiment is to estimate the extension when the bag of coins
is dropped. The results of two runs of the experiment are recorded in
`drop1` and `drop2`, and averaged in `drop`. This is used to compare the
results with the predicted results, using the model established in the
following equation for the extension:

$$e = \frac{mg}{k}\left(1 + \sqrt{1 + \frac{2kL}{mg}}\right).$$

```
(%i17)   drop1: [[2,18.0],[4,20.5],[6,23.0],[8,25.5],
[10,28.0],[12,32.5]]$
```

```
(%i18)  drop2: [[2,17.8],[4,20.2],[6,22.7],[8,25.8],
[10,30.4],[12,33.5]]$

(%i19)  drop: (drop1+drop2)/2$

(%i20)  extension: 14.235+100*0.0095*x*9.81/23.13
*(1+sqrt(2*0.14235*23.13/(0.0095*x*9.81)))$

(%i21)  wxplot2d([[discrete,drop],extension], [x, 0.1,13],
[xlabel, "coins"], [y, 14,39], [ylabel, "max extension (cms)"],
[style, points,lines], [point_type, plus,times],
[color, red,black], [legend, "experiment","model"])$
```

(%t21)

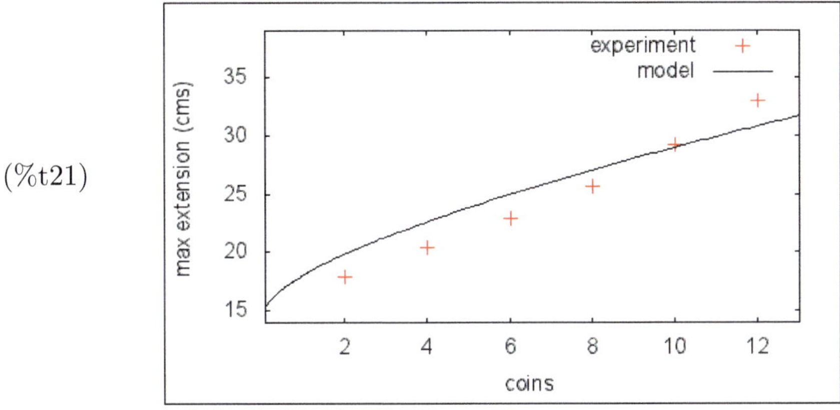

This graph is also reproduced in the main body of the report (Figure 5).

Appendix 3: Further data obtained from the internet

From one internet website

This site is at: `internet source`

Bungee type	Number of threads	Mass range (kg)
Green	960	40–61
Red	1200	62–84
Blue	1440	85–98
Black	1600	99–118

The threads are 22 gauge. Extension achieved is 325–360%. Cord retirement is recommended after 250 jumps. Tensile break occurs at an extension of 650–700%.

From a second internet website

This site is at: `internet source`

Bungee type	Diameter (mm)	Number of threads	Mass range (kg)	Mass of cord ($\mathrm{kg\,m^{-1}}$)
Yellow	37.5	920	45–62	1.10
Blue	41.0	1120	62–79	1.22
Red	46.0	1320	79–96	1.47
Black	48.0	1520	96–113	1.55

Extension of 350% is allowed in each case. Direct ultraviolet rays and ozone are the main culprits for shortening the lifespan of rubbers. It is recommended to use and discard cords within 150 days of the purchase date, and the number of jumps should not exceed 800.

Learning outcomes

After studying this unit, you should be able to:

- appreciate the role of dimensional analysis in creating relationships between the variables and parameters of a model

- understand the need to examine the sensitivity of a model to absolute and relative changes in its parameters, and be able to test the sensitivity of a model by analytic means in simple cases, and empirically in others

- appreciate the role of the assumptions of a model in thinking about possible revisions

- appreciate that there are a number of possible revisions to a model

- use a variety of techniques to revise a simple model

- appreciate that a complex model may be built up from a simple model by gradually including more and more features

- write a short report outlining the key stages of the modelling process associated with a particular problem.

Solutions to exercises

Solution to Exercise 1

The dimensions of the quantities assumed to be involved are listed in the
table below.

Physical quantity	Symbol	Dimensions
Period	τ	T
Mass of bob	m	M
Length of pendulum stem	l	L
Angular amplitude of oscillations	Φ	1
Magnitude of acceleration due to gravity	g	$\mathrm{L\,T^{-2}}$

We wish to find an expression for the period τ in terms of the other
quantities. We assume that this expression takes the form

$$\tau = k\,m^{\alpha}\,l^{\beta}\,\Phi^{\gamma}\,g^{\delta},$$

where k is a dimensionless constant. For this equation to be dimensionally
consistent, we must have

$$[\tau] = [k]\,[m]^{\alpha}\,[l]^{\beta}\,[\Phi]^{\gamma}\,[g]^{\delta}.$$

Hence

$$\mathrm{T} = \mathrm{M}^{\alpha}\,\mathrm{L}^{\beta}\,(\mathrm{L\,T^{-2}})^{\delta} = \mathrm{M}^{\alpha}\,\mathrm{L}^{\beta+\delta}\,\mathrm{T}^{-2\delta}.$$

Equating powers of M, L and T on both sides of this equation leads to

$$0 = \alpha, \quad 0 = \beta + \delta, \quad 1 = -2\delta.$$

Solving these equations gives

$$\alpha = 0, \quad \beta = \tfrac{1}{2}, \quad \delta = -\tfrac{1}{2}.$$

Hence

$$\tau = k\,l^{1/2}\,g^{-1/2}\,\Phi^{\gamma}.$$

Since additive combinations of this relationship with different values of k
and γ are also dimensionally consistent, we conclude that

$$\tau = k(\Phi)\,l^{1/2}\,g^{-1/2} = k(\Phi)\sqrt{\frac{l}{g}},$$

where $k(\Phi)$ is an (undetermined) dimensionless function of Φ.

Since $\alpha = 0$, any formula based only on the physical quantities in the table will not involve the mass of the bob.

Solution to Exercise 2

The dimensions of the quantities involved in the situation are listed below.

Physical quantity	Symbol	Dimensions
Frequency of vibration of string	f	T^{-1}
Mass of string	m	M
Length of string	l	L
Tension of string	T_{eq}	$\mathrm{M\,L\,T}^{-2}$

We wish to find an expression for the frequency f in terms of the other quantities. We assume that this expression takes the form

$$f = k\, m^{\alpha}\, l^{\beta}\, T_{\mathrm{eq}}^{\gamma},$$

where k is a dimensionless constant. For this equation to be dimensionally consistent, we must have

$$[f] = [k]\, [m]^{\alpha}\, [l]^{\beta}\, [T_{\mathrm{eq}}]^{\gamma},$$

or

$$\mathrm{T}^{-1} = \mathrm{M}^{\alpha}\,\mathrm{L}^{\beta}\,(\mathrm{M\,L\,T}^{-2})^{\gamma} = \mathrm{M}^{\alpha+\gamma}\,\mathrm{L}^{\beta+\gamma}\,\mathrm{T}^{-2\gamma}.$$

Equating powers of M, L and T on both sides of this equation leads to

$$0 = \alpha + \gamma, \quad 0 = \beta + \gamma, \quad -1 = -2\gamma.$$

The solution of these equations is

$$\alpha = -\tfrac{1}{2}, \quad \beta = -\tfrac{1}{2}, \quad \gamma = \tfrac{1}{2}.$$

The relationship derived in Unit 14 was
$$f = \frac{1}{2}\sqrt{\frac{T_{\mathrm{eq}}}{ml}}.$$

Hence

$$f = k\, m^{-1/2}\, l^{-1/2}\, T_{\mathrm{eq}}^{1/2}$$

$$= k\sqrt{\frac{T_{\mathrm{eq}}}{ml}},$$

where k is a dimensionless constant.

Solution to Exercise 3

(a) Suppose that we increase the annual payment from £8000 to £8100. The capital sum owing becomes

$$C_n = \left(100\,000 - \frac{8100}{0.05}\right)(1.05)^n + \frac{8100}{0.05}$$
$$= 162\,000 - 62\,000(1.05)^n.$$

After 15 years, the sum outstanding is $C_{15} = 33\,106.45$, which is a reduction $\delta C_{15} = -2157.86$. This is significantly larger than the £100 increase in the annual repayment. However, this comparison makes no sense, not least because the annual payment is made 15 times in the course of 15 years. The relative sensitivity is more useful, where a change of $(100/8000) \times 100$, that is, 1.25%, results in a change of

$(2157.86/35\,264.31) \times 100$, that is, a 6.12% reduction in the sum owing after 15 years. By the module definition of relative sensitivity, when $C_0 = 100\,000$, $R = 8000$ and $i = 0.05$, the sum owing after 15 years is relatively insensitive to changes in the annual payment, since the magnification is $6.12/1.25 = 4.90$ to two decimal places, and this is less than 5.

(b) Suppose that we change the interest rate from 5% to 4.9%, a decrease of $(0.001/0.05) \times 100$, that is, 2%. The capital sum owing becomes

$$C_n = \left(100\,000 - \frac{8000}{0.049} \right) (1.049)^n + \frac{8000}{0.049}$$
$$= 163\,265.31 - 63\,265.31(1.049)^n.$$

After 15 years, the sum outstanding is $C_{15} = 33\,607.71$, which is a reduction $\delta C_{15} = -1656.60$, that is, a 4.70% reduction in the sum owing after 15 years. Again, by the module definition of relative sensitivity, when $C_0 = 100\,000$, $R = 8000$ and $i = 0.05$, the sum owing after 15 years is not relatively sensitive to changes in the interest rate, since the magnification is $4.70/2 = 2.35$ to two decimal places, and this is less than 5.

Solution to Exercise 4

(a) We are interested in the positive value of x when $y = 0$. The change in x for a change in u is given by $\partial x/\partial u$. Differentiating partially with respect to u keeping all the other parameters fixed gives

$$0 = 0 + \frac{\partial x}{\partial u} \tan \theta - 2x \frac{\partial x}{\partial u} \frac{g}{2u^2} \sec^2 \theta + x^2 \frac{g}{u^3} \sec^2 \theta.$$

Collecting up the terms in $\partial x/\partial u$, we have

$$\frac{\partial x}{\partial u} \left(x \frac{g}{u^2} \sec^2 \theta - \tan \theta \right) = x^2 \frac{g}{u^3} \sec^2 \theta.$$

For relative sensitivity, we calculate expression (7):

$$\frac{u}{x} \frac{\partial x}{\partial u} = \frac{xg \sec^2 \theta}{xg \sec^2 \theta - u^2 \tan \theta} = \frac{1}{1 - u^2 \sin \theta \cos \theta/(xg)}.$$

For a given set of data, we can use this formula to assess the relative sensitivity of the output x to changes in the launch speed u.

(b) Using the given values we have

$$\frac{u}{x} \frac{\partial x}{\partial u} = \frac{1}{1 - 200/(42.68 \times 9.81)} \simeq 1.9145.$$

$$u^2 \sin \theta \cos \theta = 20^2 \times \tfrac{1}{\sqrt{2}} \times \tfrac{1}{\sqrt{2}}$$
$$= 200.$$

Thus a 5% increase in the launch speed results in a 1.9% increase in the distance, which is approximately 0.81 m.

Since the ratio of the relative change in u to the relative change in x is approximately $1.91/5 = 0.38$, which is small, we can conclude that the model, for this set of data, is relatively insensitive to changes in the launch speed.

Solution to Exercise 5

(a) If $h_2 = r_2$, then $V = \pi r_2^2 h_2 = \pi r_2^3$, giving $r_2 = (V/\pi)^{1/3}$, and $A_2 = 2\pi r_2 h_2 + 2\pi r_2^2 = 4\pi r_2^2 = 4\pi(V/\pi)^{2/3} = 4\sqrt[3]{V^2 \pi}$. The relative increase in the amount of material needed is

$$\frac{A_2 - A}{A} = \frac{(4 - 3\sqrt[3]{2})\sqrt[3]{V^2 \pi}}{3\sqrt[3]{2}\sqrt[3]{V^2 \pi}}$$

$$= \frac{4 - 3\sqrt[3]{2}}{3\sqrt[3]{2}} = \frac{4}{3\sqrt[3]{2}} - 1 = 0.058.$$

Thus an extra 5.8% of material is needed for a can where the diameter is twice the height. The amount of material required appears to be pretty insensitive to changes in the height, for a given volume, for $r \le h \le 4r$.

(b) For the optimal solution, when $h = 2r$ and $r = (V/(2\pi))^{1/3}$, the area of the wall of the can is

$$W = 2\pi r h = 4\pi r^2 = 4\pi \left(\frac{V}{2\pi}\right)^{2/3} = 2\sqrt[3]{2V^2 \pi}.$$

When the height is twice the diameter, $h_1 = 4r_1$ and $r_1 = (V/(4\pi))^{1/3}$, so the area of the wall of the can is

$$W_1 = 2\pi r_1 h_1 = 8\pi r_1^2 = 8\pi \left(\frac{V}{4\pi}\right)^{2/3} = 2\sqrt[3]{4V^2 \pi}.$$

The relative change in this area is

$$\frac{W_1 - W}{W} = \frac{2\sqrt[3]{4} - 2\sqrt[3]{2}}{2\sqrt[3]{2}} = \sqrt[3]{2} - 1 = 0.260.$$

Thus the increase in advertising area when the height is twice the width is 26%.

When the height is half the diameter, $h_2 = r_2$ and $r_2 = (V/\pi)^{1/3}$, so the area of the wall of the can is

$$W_2 = 2\pi r_2 h_2 = 2\pi r_2^2 = 2\pi \left(\frac{V}{\pi}\right)^{2/3} = 2\sqrt[3]{V^2 \pi}.$$

The relative change in this area is

$$\frac{W_2 - W}{W} = \frac{2 - 2\sqrt[3]{2}}{2\sqrt[3]{2}} = \frac{1}{\sqrt[3]{2}} - 1 = -0.206.$$

Thus the decrease in advertising area when the height is half the width is 21%.

These results support the idea that increasing the height gives more advertising space, which may explain why cans tend to have height greater than width.

Solution to Exercise 6

(a) The area of material used is

$$A_3 = 2\pi r h + 8r^2,$$

where the term $8r^2$ is the area of two squares of side $2r$ of material used to make the circular end pieces, with the excess discarded. The volume is still $V = \pi r^2 h$. Thus $h = V/(\pi r^2)$ and the area of material to be minimised is

$$A_3 = \frac{2V}{r} + 8r^2.$$

(b) Differentiating with respect to r to find the maximum and minimum values of A_3 gives

$$\frac{dA_3}{dr} = -\frac{2V}{r^2} + 16r = 0,$$

from which we can deduce that $r = V^{1/3}/2$. This gives a minimum (rather than a maximum) since $d^2 A/dr^2 > 0$. We can also deduce that $h = V/(\pi r^2) = 4V^{1/3}/\pi$, so $h/r = 8/\pi = 2.55$, which is larger than the ratio $h/r = 2$ for our first model.

The area of material used is

$$A_3 = \frac{2V}{V^{1/3}/2} + 8(V^{2/3}/4) = 6\sqrt[3]{V^2}.$$

The relative increase in the amount of material used is

$$\begin{aligned}
\frac{A_3 - A}{A} &= \frac{(6 - 3\sqrt[3]{2\pi})\sqrt[3]{V^2}}{3\sqrt[3]{2\pi}\sqrt[3]{V^2}} \\
&= \frac{6 - 3\sqrt[3]{2\pi}}{3\sqrt[3]{2\pi}} = 2/\sqrt[3]{2\pi} - 1 = 0.084.
\end{aligned}$$

Thus discarding the leftover pieces requires an extra 8.4% of material to manufacture the cans. It is clearly more important to recycle the leftover pieces than to worry about the precise ratio of h to r.

Solution to Exercise 7

(a) The radius of the circular pieces is now $r + b$, and the length of material for the walls of the can is now $2\pi r + b$. The area of material required is

$$A_4 = (2\pi r + b)h + 2\pi(r + b)^2,$$

where $h = V/(\pi r^2)$.

(b) Replacing h in the equation for A_4 gives

$$A_4 = \frac{(2\pi r + b)V}{\pi r^2} + 2\pi(r + b)^2.$$

Differentiating with respect to r to find the maximum and/or minimum values of A_4 gives

$$\frac{dA_4}{dr} = \frac{2\pi V}{\pi r^2} - \frac{2V(2\pi r + b)}{\pi r^3} + 4\pi(r + b) = 0.$$

Multiplying through by $r^3/(2\pi)$ and collecting up the terms in powers of r gives

$$2r^4 + 2r^3 b - \frac{rV}{\pi} - \frac{Vb}{\pi^2} = 0.$$

This is a quartic equation in r with potentially four roots, which will depend on the size of the overlap b. Only one of these roots will be the value of r for a practical solution that minimises the area.

Solution to Exercise 8

(a) Assumption 1 is reasonable, since the pebble will be very small. Assumption 2 could be changed to incorporate air resistance. This would slow the pebble down and hence reduce the estimate for the depth of the well. Assumption 3 could be varied to include the time taken for the sound to travel back up the well. This would also reduce the estimate for the depth of the well.

(b) Starting with Newton's second law, with the x-axis pointing down the well, the equation of motion is

$$m\ddot{x}\mathbf{i} = mg\mathbf{i} - c_1 D\dot{x}\mathbf{i},$$

where D is the diameter of the pebble, and $c_1 = 1.7 \times 10^{-4}$ for $D|\dot{x}\mathbf{i}| < 10^{-5}$ is a constant, given in Unit 3. Writing $k = c_1 D/m$, the equation to be solved is

$$\ddot{x} + k\dot{x} = g.$$

This linear equation can be solved using the techniques for second-order differential equations given in Unit 1, to give the general solution

This exact problem was solved in Example 7(a) in Unit 3 and in Example 1 in Unit 10.

$$x = A + Be^{-kt} + gt/k.$$

The particular solution with $x(0) = \dot{x}(0) = 0$ is

$$x = \frac{g}{k^2}\left(e^{-kt} - 1\right) + \frac{gt}{k}.$$

For a pebble of diameter $1\,\text{cm}$ ($D = 0.01$) and mass $0.01\,\text{kg}$ we have $k = 1.7 \times 10^{-4}$. For a time of $2.8\,\text{s}$ for the pebble to reach the bottom of the well, the model estimates the depth of the well as $38.45\,\text{m}$, which is almost identical to the estimate without air resistance.

It should be noted that the linear model of air resistance is not suitable for this problem since $D|\dot{x}\mathbf{i}| > 10^{-5}$ for a considerable part of the motion. The quadratic model applied to this problem gives the estimate $37.5\,\text{m}$.

(c) The time for the pebble to drop down the well, from rest, is

$$T_1 = \sqrt{2h/g}.$$

The time taken for the sound to travel up again is $T_2 = h/c$, where c is the speed of sound. Thus the total time T is

$$T = T_1 + T_2 = \sqrt{2h/g} + h/c.$$

Taking the h/c over to the other side and squaring both sides gives

$$\left(T - \frac{h}{c}\right)^2 = \frac{2h}{g}.$$

Hence to find h given T we need to solve the quadratic equation

$$\frac{1}{c^2} h^2 - \left(\frac{2T}{c} + \frac{2}{g}\right) h + T^2 = 0.$$

With $c = 343$, $T = 2.8$ and $g = 9.81$, the quadratic is

$$8.5 \times 10^{-6} h^2 - 0.22020 h + 7.84 = 0,$$

and this has roots

$$h = \frac{0.22020 \pm \sqrt{0.22020^2 - 4 \times 8.5 \times 10^{-6} \times 7.84}}{1.7 \times 10^{-5}}.$$

Thus the two estimates for the depth of the well are $h = 35.65$ and the spurious solution $h = 25\,870$ (which was introduced when $T - h/c$ was negative when we squared both sides). Taking into account the time for the sound to travel back up the well has reduced the estimate of the depth from $38.5\,\text{m}$ to $35.7\,\text{m}$. So there is evidence that this has made a significant improvement to the model, at least for wells around $35\,\text{m}$ deep.

Solution to Exercise 9

(a) The model is based on the assumption that the pollutant does not biodegrade or decay. The model would not be valid if the pollutant were unstable, in that it might be broken down by other elements in the water or by sunlight, for example.

(b) A significant reduction in the input of PCBs after this time would be expected, although there may be some residual pollutants in the catchment area of the lake that may take a number of years before they are washed into the lake. The model is based on the assumption that all sources of new pollution cease. This assumption may be reasonable if the input of PCBs to the lake has, as a result of the ban, reduced to insignificant levels.

(c) As one of the two lakes (Lake Michigan is the other) that is not downstream of any other lake in the system, there would be no input of pollution to Lake Superior from any of the other lakes.

Solution to Exercise 10

The model assumes no new pollution, and PCBs were banned in 1975. Any data from before this are therefore inappropriate for comparison purposes. Since there is a time delay in the ecosystem as pollution works its way up the food chain, it is likely that the level of pollution in herring gull eggs reflects the pollution level in the lake at an earlier time. It may thus be wise to also disregard the data point for 1975.

Solution to Exercise 11

Any of the assumptions, listed in Unit 8, could be challenged.

Assumption (a) is that all sources of pollution have been removed. In reality there may continue to be new pollution for several years.

Assumption (b) is that the pollutant does not biodegrade or decay. In reality all substances biodegrade or decay, although some do so more quickly than others.

Assumption (c) is that the pollutant is evenly dispersed. In reality some areas of the lake, close to the sources of pollution, will have a higher concentration of pollutant than the rest of the lake.

Assumption (d) is that water flows into and out of the lake at a constant rate. In reality there will be seasonal effects, with more water entering the lake during rainy periods, and less water entering during dry periods.

Assumption (e) is that all other water gains and losses can be ignored. In reality drinking water is extracted, water leaves the lake by evaporation, and so on.

Assumption (f) is that the volume of water in the lake is constant. In reality there will be seasonal effects, with more water in the lake during rainy periods, and less water during dry periods.

Assumption (g) is that negligible pollution flows into downstream lakes from upstream lakes (this is not relevant for Lake Superior and Lake Michigan). This appears to be unlikely.

Since the rate at which PCB concentration diminishes in Lake Superior is much faster than that predicted by the model, the likeliest candidate for re-examination is Assumption (b).

Solution to Exercise 12

The rate of volatilisation will depend on the surface area of the lake and on the concentration of pollutant near to the surface of the lake. It will also depend on the strength of the wind and on the concentration of the pollutant already in the atmosphere. It may also depend on temperature.

Solution to Exercise 13

(a) Let $c_B = c$ be the concentration of pollutant in the bay, so $c_L = \frac{1}{2}c$ is the concentration of pollutant in the rest of the lake. The rate of output of pollutant from the bay (in $\mathrm{kg\,s^{-1}}$) by each of the three mechanisms is as follows.

Water flow: $r_B c_B = 153c$.
Volatilisation: $pA_B c_B = 4872c$.
Diffusion: $E(c_B - c_L) = 5500c$.

Hence volatilisation and diffusion are the most important mechanisms for removing pollutant from the bay. The amount of pollutant removal by water flow is very small by comparison.

(b) The relative importance of the three mechanisms will not change greatly if the concentration of pollutant in the rest of the lake is less than half that in the bay (assuming that the concentration in the bay is still given by $c_B = c$). Even if c_L/c_B is very small, the loss of pollutant from the bay by diffusion will at most double. (If c_B and c_L are nearly equal, then the loss of pollutant by diffusion may be small, but since the bay is in practice significantly more polluted than the rest of the lake, this scenario is unlikely to occur.)

(c) Two mechanisms, water flow and volatilisation, are involved here. The rate of output of pollutant from the rest of the lake (in $\mathrm{kg\,s^{-1}}$) by each mechanism is as follows, for a pollutant concentration c_L.

Water flow: $r_L c_L = 4967c_L$.
Volatilisation: $pA_L c_L = 66\,120c_L$.

Hence volatilisation is the most important mechanism for pollutant output from the lake. (This is in line with a conclusion drawn from the first revised model.)

Systems of particles

Introduction

Figure 1 shows three positions of a diver performing a forward somersault, and the curves trace out the paths of the diver's head and feet. The motion of the diver is very complicated, which is reflected in the shapes of these curves; but in spite of this complexity, it is possible to obtain some simple information that adds considerably to our understanding of the dive. Associated with each position of the diver, there is a point known as the diver's *centre of mass*, and this point moves along a simpler path. (You met the idea of centre of mass in Unit 2.)

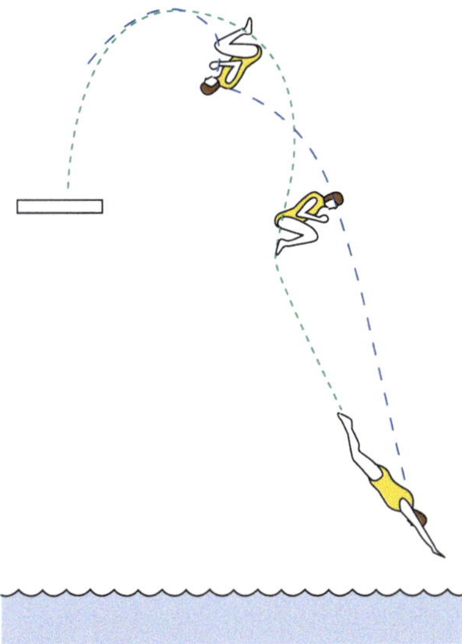

Figure 1   The paths of the head and feet of a diver

Our first objective in this unit is to obtain some useful information on the complicated motion of an object or system, and the concept of centre of mass is crucial to this process. It transpires that in a number of circumstances, we can justify modelling a complicated system, possibly composed of very many objects, by a single particle (of the same mass as the system) placed at the centre of mass of the system. This process has the dual benefit of making the behaviour of quite complicated systems much easier to understand and of considerably reducing the work involved when analysing them. It will form a central theme of the unit.

We start, in Section 1, by examining some specific examples of two-particle systems, from which we will be able to extract some general principles that apply to all two-particle systems and that can be extended to systems involving any number of particles. We move on, in Section 2, to discuss general systems of particles and the important concept of centre of mass. Then, in Section 3, we discuss the behaviour of objects during a collision, and here we introduce the principle of conservation of linear momentum. Section 4 concerns Newton's law of restitution, which models the effects of inelasticity in collisions.

1 Two-particle systems

The three mechanical systems illustrated in Figure 2 appear to have little in common. However, they all illustrate an important principle of mechanics – a principle that we intend to develop in this unit.

Figure 2   Three mechanical systems

In Subsection 1.1 we will use various techniques from previous units to examine each system in turn. Then, in Subsection 1.2, we will begin to draw together the ideas common to all three examples. This drawing together will lead to the statement of a general principle for two-particle systems, based on Newton's second law. First, we give a definition.

You have met two-particle systems before. For example, in Unit 2, Example 11, you saw how a scarf draped over the edge of a table can be modelled as a two-particle system. In Unit 11, many of the spring–mass systems are two-particle systems.

A **two-particle system** is a system modelled so that the total mass of the system is divided between two particles.

1.1 Examples of two-particle systems

A simple design for a lift

In this example we investigate a (rather impractical) design for a lift, as shown in Figure 3(a). The mechanism consists of a lift compartment and a counterweight, supported by a cable passing over a pulley. We wish to investigate this design with the simplifying assumptions that all frictional forces and air resistance can be ignored.

In order to model this system we will need to make some more simplifying assumptions. Figure 3(b) shows our mathematical model, which consists of two particles of masses m_1 and m_2 connected by a model string passing over a model pulley with its axle fixed at P. We assume that m_2 is greater than m_1, so when the system is released from rest, the lift (i.e. the particle of mass m_2) will begin to move down. We denote the vertical displacements of the particles of masses m_1 and m_2 from P by x_1 and x_2, respectively. Since the model string is inextensible, we have $x_1 + x_2 = \text{constant}$.

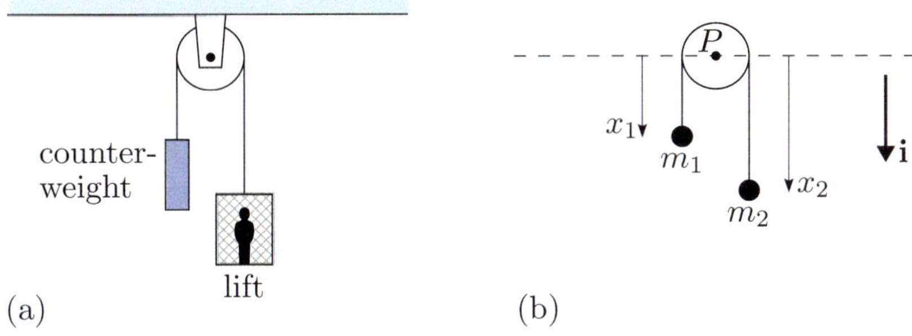

(a)

(b)

Figure 3   (a) A simple lift, and (b) a mathematical model for it

Before investigating the motion of this system, we will look at the system as a whole, in order to derive a result that will be generalised later. The force diagram in Figure 4 shows the forces acting on the two particles and on the model pulley.

Now we apply Newton's second law to each particle. The accelerations of the particles of masses m_1 and m_2 are $\ddot{x}_1\mathbf{i}$ and $\ddot{x}_2\mathbf{i}$. The forces acting on the particles are $\mathbf{W}_1 + \mathbf{T}_2$ and $\mathbf{W}_2 + \mathbf{T}_4$, where $\mathbf{W}_1 = m_1 g\mathbf{i}$ and $\mathbf{W}_2 = m_2 g\mathbf{i}$. So Newton's second law gives

$$\mathbf{W}_1 + \mathbf{T}_2 = m_1 g\mathbf{i} + \mathbf{T}_2 = m_1 \ddot{x}_1\mathbf{i}, \tag{1}$$

$$\mathbf{W}_2 + \mathbf{T}_4 = m_2 g\mathbf{i} + \mathbf{T}_4 = m_2 \ddot{x}_2\mathbf{i}. \tag{2}$$

We also observe that the axle of the pulley is stationary (i.e. in equilibrium). Introducing the reaction force $\mathbf{S}$ to balance the other forces on the pulley, we have

$$\mathbf{S} + \mathbf{T}_1 + \mathbf{T}_3 = \mathbf{0}. \tag{3}$$

On adding equations (1), (2) and (3), we obtain

$$\mathbf{S} + \mathbf{T}_1 + \mathbf{T}_2 + \mathbf{T}_3 + \mathbf{T}_4 + m_1 g\mathbf{i} + m_2 g\mathbf{i} = m_1 \ddot{x}_1\mathbf{i} + m_2 \ddot{x}_2\mathbf{i}.$$

The left-hand side of this equation simplifies because we have a model string, which implies that $\mathbf{T}_1 + \mathbf{T}_2 = \mathbf{0}$ and $\mathbf{T}_3 + \mathbf{T}_4 = \mathbf{0}$. So we obtain

$$\mathbf{S} + m_1 g\mathbf{i} + m_2 g\mathbf{i} = m_1 \ddot{x}_1\mathbf{i} + m_2 \ddot{x}_2\mathbf{i}.$$

The right-hand side of this equation is almost of the mass times acceleration form that we expect from Newton's second law. A little rearrangement puts the equation in this form:

$$\mathbf{S} + m_1 g\mathbf{i} + m_2 g\mathbf{i} = \underbrace{(m_1 + m_2)}_{\text{the total mass}} \frac{m_1 \ddot{x}_1 + m_2 \ddot{x}_2}{m_1 + m_2}\,\mathbf{i}. \tag{4}$$

The significance of equation (4) will become clear shortly, but for the moment just notice that the right-hand side is the total mass of the system multiplied by an expression that has the dimensions of acceleration.

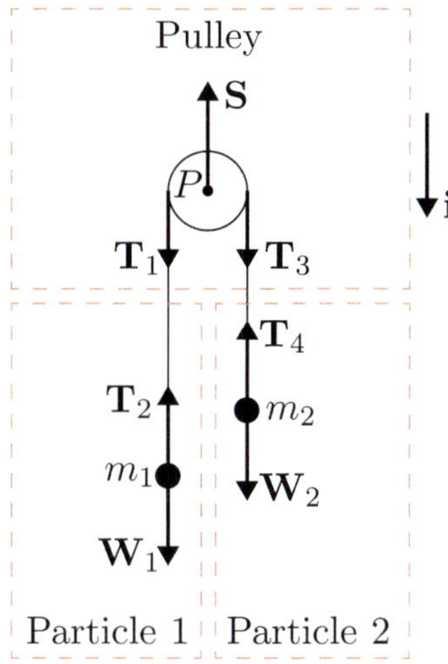

Figure 4   Force diagram for the simple lift

Exercise 1

This exercise completes the analysis of the simple design for a lift described above. Use the notation of Figure 4.

(a) Use the facts that the model string is inextensible (giving $x_1 + x_2 = \text{constant}$, $|\mathbf{T}_1| = |\mathbf{T}_2|$ and $|\mathbf{T}_3| = |\mathbf{T}_4|$) and that the pulley is a model pulley (so $|\mathbf{T}_1| = |\mathbf{T}_3|$) together with equations (1) and (2) to determine the acceleration of the lift, $\ddot{x}_2$.

(b) Use the result of part (a) and equation (2) to determine the magnitude of the tension in the cable.

(c) Use the result of part (b) and equation (3) to determine the magnitude of the reaction force on the pulley, $\mathbf{S}$.

(d) Suppose that the lift and the counterweight are each of mass $1000\,\text{kg}$, and that a person of mass $65\,\text{kg}$ steps into the lift. Calculate the numerical value of the acceleration of the lift and also the velocity of the lift after it starts from rest and travels $100\,\text{m}$. Comment on the suitability of the lift design.

A drum major's mace

In this example we attempt to model the motion of a drum major's mace. You may have seen a military band with the drum major striding out in front, twirling a long rod with a heavy weight at one end: his mace. Every few strides, he twirls the mace and then hurls it high into the air; it then appears to follow a complicated path until he finally catches it and continues with the twirling.

Recall from Unit 2 that a model rod is a rigid body with length but no breadth or depth. In this case we have a *light* model rod, so it has no mass either.

We model the mace as two particles A and B of unequal masses m_1 and m_2, connected by a light model rod, where l is the distance between the two masses, as in Figure 5(a). Our purpose is to try to discover some pattern in the behaviour of the mace (and perhaps provide some useful advice to would-be drum majors).

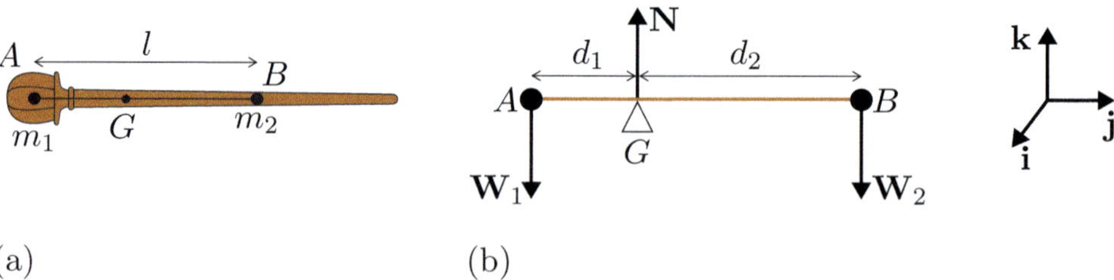

(a)   (b)

Figure 5   (a) A drum major's mace. (b) A force diagram for the mace at rest on its balance point.

As you will see later, G is the centre of mass of the system.

We assume that the drum major holds the mace at its 'balance point' G, by which we mean that if we were to balance the model rod, plus the two particles, on a model pivot at G, then the rod would be in equilibrium.

The equilibrium condition for rigid bodies tells us that the sum of the torques on the rod when balanced in this way is zero. This situation and the associated forces are shown in Figure 5(b).

You met torques and the equilibrium condition for rigid bodies in Unit 2.

Take the origin for the torques at G, so that the torque due to $\mathbf{N}$ is zero. The position vectors of A and B are $\mathbf{r}_1 = -d_1\mathbf{j}$ and $\mathbf{r}_2 = d_2\mathbf{j}$. The weights are $\mathbf{W}_1 = -m_1 g\mathbf{k}$ and $\mathbf{W}_2 = -m_2 g\mathbf{k}$. The torque equilibrium equation is

$$\mathbf{r}_1 \times \mathbf{W}_1 + \mathbf{r}_2 \times \mathbf{W}_2 = (-d_1\mathbf{j}) \times (-m_1 g\mathbf{k}) + (d_2\mathbf{j}) \times (-m_2 g\mathbf{k}) = \mathbf{0}.$$

This simplifies to

$$d_1 m_1 \mathbf{i} - d_2 m_2 \mathbf{i} = \mathbf{0},$$

and resolving in the $\mathbf{i}$-direction gives $d_1 m_1 - d_2 m_2 = 0$, or

$$d_1/d_2 = m_2/m_1. \tag{5}$$

From this we see that the ratio d_1/d_2 will be small if m_1 is much larger than m_2. In other words, if the head of the mace is very heavy, then the balance point will be near this end.

Drum majors hold their maces near the end, whereas drum majorettes hold their batons near the centre. Presumably this is because a drum majorette's baton is more centrally balanced.

We now suppose that the drum major throws the mace into the air and at the same time gives it a flick in such a way that the mace turns end-over-end in a vertical plane. If we ignore the effects of air resistance, Figure 6 shows the forces acting on the mace and its position at time t (in terms of position vectors relative to an origin O, the point at which the mace was released). The rod gives rise to a force $\mathbf{R}_1$ on particle A and a force $\mathbf{R}_2$ on particle B.

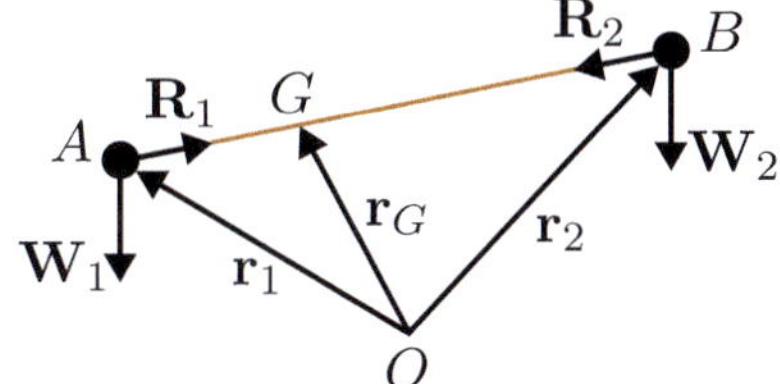

Figure 6   Force diagram for the mace in the air

It turns out that the position of G is of crucial importance. It will be useful later if we express the position vector of G in terms of the position vectors of A and B. Consider the displacement vectors $\overrightarrow{AG} = \mathbf{r}_G - \mathbf{r}_1$ and $\overrightarrow{GB} = \mathbf{r}_2 - \mathbf{r}_G$ (see Figure 6). We know that $|\overrightarrow{AG}| = d_1$ and $|\overrightarrow{GB}| = d_2$, so $|\overrightarrow{AG}|/|\overrightarrow{GB}| = d_1/d_2 = m_2/m_1$, by equation (5). Hence $m_1 \overrightarrow{AG} = m_2 \overrightarrow{GB}$, or equivalently,

$$m_1(\mathbf{r}_G - \mathbf{r}_1) = m_2(\mathbf{r}_2 - \mathbf{r}_G),$$

which can be rearranged to give the position vector of G as

$$\mathbf{r}_G = \frac{m_1\mathbf{r}_1 + m_2\mathbf{r}_2}{m_1 + m_2}. \tag{6}$$

Differentiating equation (6) twice with respect to time, we obtain the following expression for the acceleration of the point G in terms of the accelerations $\ddot{\mathbf{r}}_1$ and $\ddot{\mathbf{r}}_2$ of the points A and B:

$$\ddot{\mathbf{r}}_G = \frac{m_1\ddot{\mathbf{r}}_1 + m_2\ddot{\mathbf{r}}_2}{m_1 + m_2}. \tag{7}$$

The total force on particle A is $\mathbf{R}_1 + \mathbf{W}_1$, and the total force on particle B is $\mathbf{R}_2 + \mathbf{W}_2$. So, applying Newton's second law to particles A and B in turn, we obtain

$$\mathbf{R}_1 + \mathbf{W}_1 = \mathbf{R}_1 - m_1 g\mathbf{k} = m_1\ddot{\mathbf{r}}_1,$$
$$\mathbf{R}_2 + \mathbf{W}_2 = \mathbf{R}_2 - m_2 g\mathbf{k} = m_2\ddot{\mathbf{r}}_2.$$

Adding these equations gives

$$\mathbf{R}_1 - m_1 g\mathbf{k} + \mathbf{R}_2 - m_2 k\mathbf{k} = m_1\ddot{\mathbf{r}}_1 + m_2\ddot{\mathbf{r}}_2.$$

But as the mace is modelled by a light model rod, $\mathbf{R}_1 + \mathbf{R}_2 = \mathbf{0}$, so this equation simplifies to

$$-m_1 g\mathbf{k} - m_2 g\mathbf{k} = m_1\ddot{\mathbf{r}}_1 + m_2\ddot{\mathbf{r}}_2.$$

This can be rearranged as

$$-(m_1 + m_2)g\mathbf{k} = (m_1 + m_2)\frac{m_1\ddot{\mathbf{r}}_1 + m_2\ddot{\mathbf{r}}_2}{m_1 + m_2}.$$

Substituting from equation (7), we obtain

$$\ddot{\mathbf{r}}_G = -g\mathbf{k},$$

which is the equation of motion of the point G.

This result may appear to be simple, but it is really quite surprising. It tells us that the balance point G – the centre of mass of the mace – behaves as if it were a particle acted on solely by the force of gravity. Thus if the initial velocity of G is vertically upwards, then G will simply move up and down in a vertical line. However, if the initial velocity has a horizontal component, then G will follow the parabolic trajectory of a projectile. In both cases, the other points on the mace may move along much more complicated trajectories. Figure 7 illustrates the simple parabolic trajectory of G and the more complicated trajectory of another point on the mace when there is a horizontal component to the initial velocity.

What does all this mean for the drum major? First, if he wants to hold the mace horizontally, then he should hold it at its balance point G, where he will not need to apply any torques to keep it level. Second, if he wants to be able to catch it after throwing it into the air, then catching it at G would be best, as G is the only point whose trajectory in the air he can easily predict.

So the drum major knows where to hold and catch his mace. But how should he throw it? When he is stationary, it is clear that the initial upwards velocity of G should be vertical; but what if he is marching? Should he throw the mace upwards and slightly forwards? The answer is no, he should simply throw it upwards. This is because when the drum major is marching (at a constant velocity) holding the mace, the mace has a constant horizontal component of velocity that is the same as that of the drum major. Then the horizontal distance travelled by the mace, after being thrown upwards, will be the same as that travelled by the marching drum major. So the mace should return to hand level just as the drum major's hand gets there. If the mace were thrown slightly forwards, then the drum major would have to increase his marching pace to match, or else the mace would, embarrassingly, fall to the ground. However, when throwing the mace (vertically upwards), the drum major needs to make sure that he keeps his marching velocity constant in order to catch the mace safely. It would be a mistake, for example, to throw the mace while marching and then stop to wait for it to fall, or to change direction while the mace is in the air.

See Unit 3 for details about projectiles. We assume here that air resistance can be neglected.

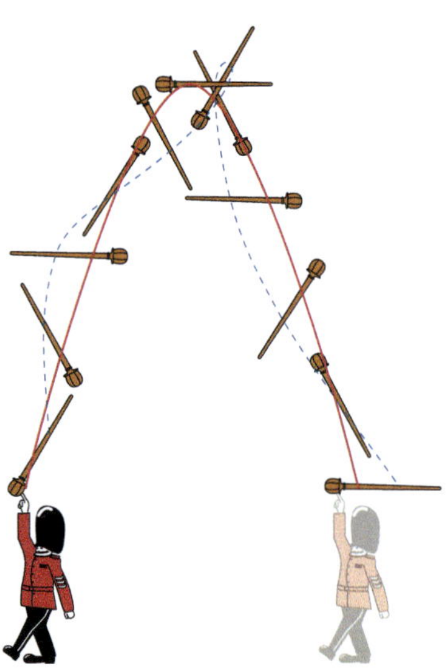

Figure 7   Trajectory of the balance point

Vibrations of a nitrogen molecule

A nitrogen molecule consists of two nitrogen atoms, each of mass m, joined by a chemical bond, as illustrated in Figure 8(a). The atoms vibrate relative to each other along the direction of this bond. The molecule can be modelled as a pair of particles joined by a model spring, which exerts a force $\mathbf{H}_1$ on the left-hand particle, labelled A, and a force $\mathbf{H}_2$ on the right-hand particle, labelled B, as illustrated in Figure 8(b). Here we are concerned with the vibrations of the molecule in isolation and hence do not consider the effects of gravity (for example, you could consider that the molecule is vibrating in deep space, far away from any gravitational influences). So the total force acting on the *system* is $\mathbf{F} = \mathbf{H}_1 + \mathbf{H}_2$.

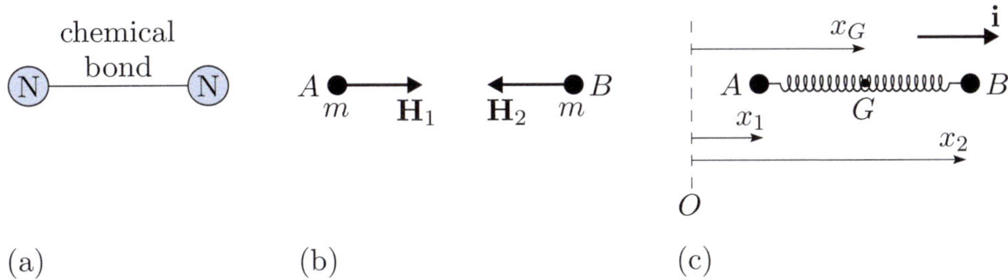

Figure 8   (a) A nitrogen molecule, (b) its force diagram, and (c) a spring–mass system model for vibrations of the molecule

You saw how to model the vibrations of such a system in terms of normal modes in Unit 11. Here we want to look at what information we can glean from the motion of the centre G of the spring, which is the geometric centre of this symmetric system. We begin as usual by defining axes: we choose a fixed origin O and a direction $\mathbf{i}$ aligned with the spring. We measure x_1, x_2 and x_G along this direction, as shown in Figure 8(c).

As you will see later, the centre of the spring is the centre of mass of the system.

To derive the equation of motion of G, we start by expressing the position vector of G relative to O. From equation (6) with $m_1 = m_2$, we have $\mathbf{r}_G = (\mathbf{r}_1 + \mathbf{r}_2)/2$.

Applying Newton's second law to particles A and B in turn gives

$$\mathbf{H}_1 = m\ddot{\mathbf{r}}_1, \quad \mathbf{H}_2 = m\ddot{\mathbf{r}}_2.$$

Proceeding as before, we add together the above equations to obtain

$$\mathbf{H}_1 + \mathbf{H}_2 = m\ddot{\mathbf{r}}_1 + m\ddot{\mathbf{r}}_2.$$

But $\mathbf{H}_1$ and $\mathbf{H}_2$ are forces exerted by the same model spring, so $\mathbf{H}_1 = -\mathbf{H}_2$. Hence the left-hand side of the above equation is zero, so

$$\mathbf{0} = m(\ddot{\mathbf{r}}_1 + \ddot{\mathbf{r}}_2).$$

You saw this before, in Unit 11.

We now write the right-hand side of this equation in the form of Newton's second law, that is, a mass multiplied by the acceleration of a single point. To do this, we differentiate $\mathbf{r}_G = (\mathbf{r}_1 + \mathbf{r}_2)/2$ twice to obtain $\ddot{\mathbf{r}}_G = (\ddot{\mathbf{r}}_1 + \ddot{\mathbf{r}}_2)/2$.

Substituting for $\ddot{\mathbf{r}}_1 + \ddot{\mathbf{r}}_2$ in the above equation then gives

$$\underbrace{\mathbf{0}}_{\text{total force}} = \underbrace{2m}_{\text{total mass}} \times \underbrace{\ddot{\mathbf{r}}_G}_{\text{acceleration}}.$$

Now $\mathbf{r}_G = x_G \mathbf{i}$, since the motion is one-dimensional. So $\ddot{x}_G = 0$, and this can be integrated to give $\dot{x}_G = \text{constant}$, that is, G moves at a constant speed along the direction of the axis of the spring. The net result is that the geometric centre of the molecule moves just like a particle with no forces acting on it, that is, it either remains at rest or travels with constant speed along a straight line (Newton's first law). It is worth emphasising that the motion of G remains the same irrespective of how much the atoms at the ends of the bond vibrate. We have not discussed the motion of the individual particles, which will depend on the forces acting on them.

Newton's first law was stated in Units 2 and 3.

1.2 Some general conclusions

Now let us try to extract some general principles from the three examples of two-particle systems discussed in the previous subsection.

The first point to note is that when we obtained an expression for the total force on the system, certain forces on the components of the system balance due to properties of model springs and rods, and Newton's third law. The forces that balance in this way are known as **internal forces** of the system, while the remaining forces are referred to as **external forces**.

The internal forces for the three examples from the previous subsection are shown as yellow arrows in Figure 9, and the external forces are shown as blue arrows. Notice that internal forces occur in equal and opposite pairs (i.e. each force in the pair has the same magnitude but opposite direction to the other). We have altered the notation of the previous subsection slightly in Figure 9 to reflect this.

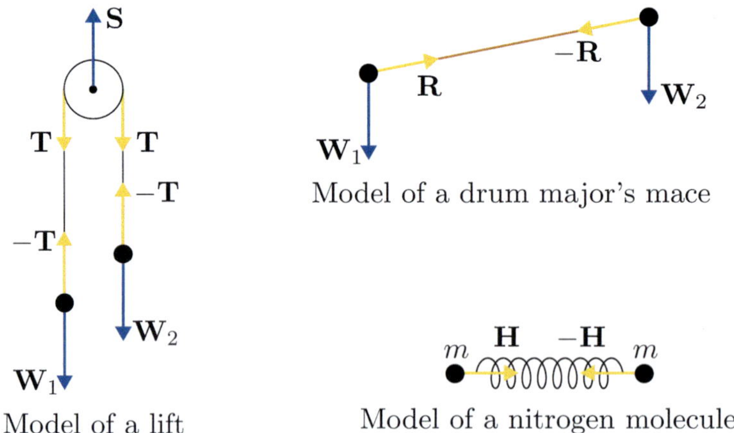

Figure 9   Internal and external forces

Exercise 2

Consider the system of weights, pulleys and strings shown in Figure 10(a). Assume that the pulleys are model pulleys (i.e. light and frictionless) and that the strings are model strings (i.e. light and inextensible). Modelling the weights as particles, we can therefore model the system and the forces acting on it as shown in Figure 10(b).

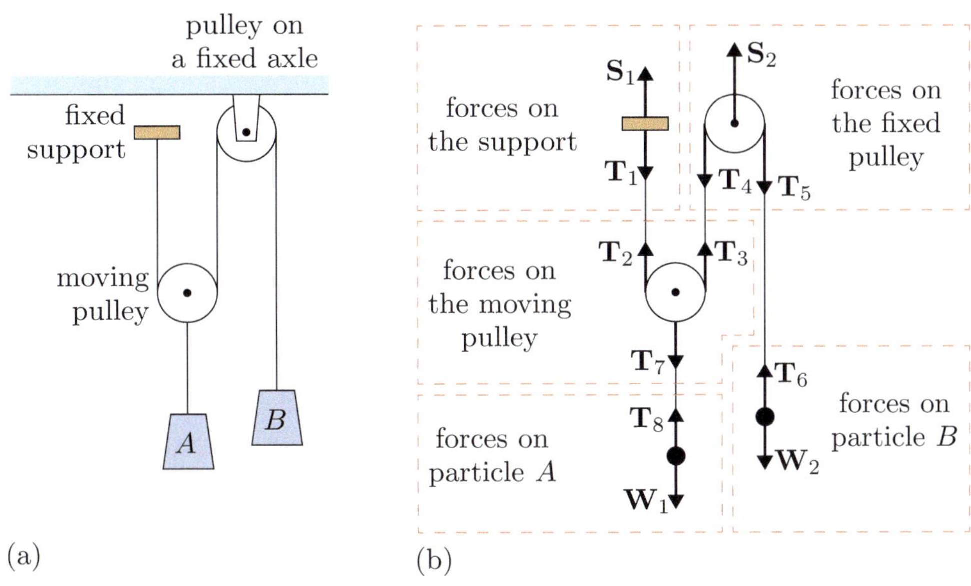

Figure 10   (a) A system of weights and pulleys, and (b) its force diagram

Decide which of the forces are internal and which are external. Hence write down the total force acting on the whole system.

In all three examples of two-particle systems in the previous subsection, we not only obtained an expression for the total force $\mathbf{F}$ on the system, but also used Newton's second law to obtain an equation involving that force. In the lift example we obtained equation (4):

$$\mathbf{F} = (m_1 + m_2)\,\frac{m_1 \ddot{x}_1 + m_2 \ddot{x}_2}{m_1 + m_2}\,\mathbf{i}.$$

If we write $\mathbf{r}_1 = x_1 \mathbf{i}$ and $\mathbf{r}_2 = x_2 \mathbf{i}$, so that $\mathbf{r}_1$ and $\mathbf{r}_2$ are the position vectors of the lift and counterweight particles relative to the point P in Figure 3, then we obtain

$$\mathbf{F} = (m_1 + m_2)\,\frac{m_1 \ddot{\mathbf{r}}_1 + m_2 \ddot{\mathbf{r}}_2}{m_1 + m_2}.$$

This is exactly the equation that we obtained for the drum major's mace, where again $\mathbf{r}_1$ and $\mathbf{r}_2$ are the position vectors of the two particles.

For the nitrogen molecule we obtained

$$\mathbf{0} = 2m\,\ddot{\mathbf{r}}_G,$$

where $\mathbf{r}_G$ is the position vector of the geometric centre of the position of the two particles, $\mathbf{r}_G = (\mathbf{r}_1 + \mathbf{r}_2)/2$. At first sight this looks different, but it is of the same form because there are no external forces acting (so $\mathbf{F} = \mathbf{0}$), both particles have the same mass (so $m_1 + m_2 = 2m$), and also

$$\frac{m_1\ddot{\mathbf{r}}_1 + m_2\ddot{\mathbf{r}}_2}{m_1 + m_2} = \frac{m\ddot{\mathbf{r}}_1 + m\ddot{\mathbf{r}}_2}{m + m} = \frac{\ddot{\mathbf{r}}_1 + \ddot{\mathbf{r}}_2}{2} = \ddot{\mathbf{r}}_G.$$

In summary, for each two-particle system, the right-hand side of the Newton's second law equation is a product of the total mass of the system and a term of the form

$$\frac{m_1\ddot{\mathbf{r}}_1 + m_2\ddot{\mathbf{r}}_2}{m_1 + m_2},$$

which is the acceleration of the point G, with position vector

$$\mathbf{r}_G = \frac{m_1\mathbf{r}_1 + m_2\mathbf{r}_2}{m_1 + m_2},$$

within that system. We call this point the **centre of mass**.

In the case of symmetric systems (where the two particles have equal mass), such as the nitrogen molecule, the centre of mass is at the geometric centre. For systems where the particles are joined by a light model rod, such as the drum major's mace, the centre of mass is at the balance point.

Exercise 3

The mass of the Earth is approximately 5.97×10^{24} kg, and its diameter is about 1.27×10^4 km, while the mass of the Moon is approximately 7.34×10^{22} kg, and its diameter is about 3.5×10^3 km. The mean distance from the centre of the Earth to the centre of the Moon is about 3.84×10^5 km.

How far above the Earth's surface would you estimate the centre of mass of the Earth and the Moon to lie?

The equations obtained by applying Newton's second law to our examples, which led to the definition of centre of mass, are all of the form $\mathbf{F} = M\ddot{\mathbf{r}}_G$, where $\mathbf{F}$ is the total force on the system and M is its total mass. We should now verify that this equation holds for the centre of mass of *any* two-particle system.

Consider a general system consisting of two particles, labelled 1 and 2, of masses m_1 and m_2, and with position vectors $\mathbf{r}_1$ and $\mathbf{r}_2$ relative to an origin O. In general, each particle is subject to internal and external forces. Let us denote the resultant of the external forces on each particle by $\mathbf{E}_1$ and $\mathbf{E}_2$, respectively. The internal force on each particle is exerted by the other particle by means of a model string, a model rod or a model spring, or in some other way. Let us denote the internal force exerted on particle 1 by particle 2 by $\mathbf{I}_{12}$, and similarly denote the internal force exerted on particle 2 by particle 1 by $\mathbf{I}_{21}$. The situation is illustrated in Figure 11.

Applying Newton's second law to each particle, we obtain

$$\mathbf{E}_1 + \mathbf{I}_{12} = m_1\ddot{\mathbf{r}}_1, \quad \mathbf{E}_2 + \mathbf{I}_{21} = m_2\ddot{\mathbf{r}}_2.$$

We can add these equations to obtain

$$\mathbf{E}_1 + \mathbf{I}_{12} + \mathbf{E}_2 + \mathbf{I}_{21} = m_1\ddot{\mathbf{r}}_1 + m_2\ddot{\mathbf{r}}_2.$$

Since $\mathbf{I}_{12}$ and $\mathbf{I}_{21}$ are internal forces, we can apply Newton's third law to obtain $\mathbf{I}_{12} + \mathbf{I}_{21} = \mathbf{0}$. So the above equation simplifies to

$$\mathbf{F}^{\text{ext}} = m_1\ddot{\mathbf{r}}_1 + m_2\ddot{\mathbf{r}}_2, \tag{8}$$

where $\mathbf{F}^{\text{ext}} = \mathbf{E}_1 + \mathbf{E}_2$ is the sum of the external forces.

The centre of mass of this system is defined by the position vector

$$\mathbf{r}_G = \frac{m_1\mathbf{r}_1 + m_2\mathbf{r}_2}{m_1 + m_2} = \frac{m_1\mathbf{r}_1 + m_2\mathbf{r}_2}{M},$$

where $M = m_1 + m_2$ is the total mass of the system.

Differentiating twice gives

$$\ddot{\mathbf{r}}_G = \frac{m_1\ddot{\mathbf{r}}_1 + m_2\ddot{\mathbf{r}}_2}{M}. \tag{9}$$

From equations (8) and (9), we have

$$\mathbf{F}^{\text{ext}} = M\ddot{\mathbf{r}}_G,$$

which is an important result that is worth re-stating.

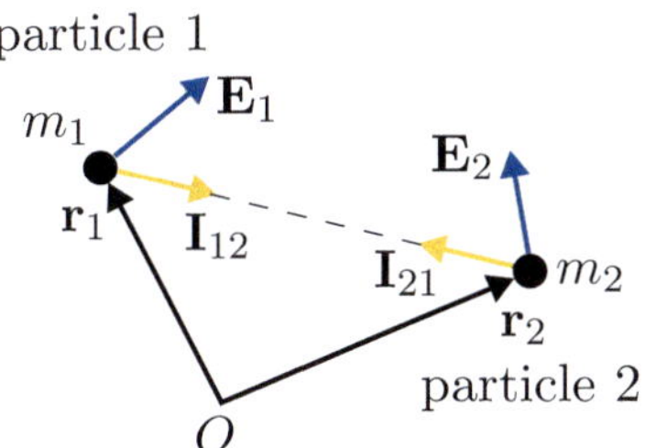

Figure 11   Internal and external forces

> ### Motion of the centre of mass of a two-particle system
>
> The motion of the centre of mass G with position vector $\mathbf{r}_G$ of a two-particle system of total mass M subject to external forces with sum $\mathbf{F}^{\text{ext}}$ is given by
>
> $$\mathbf{F}^{\text{ext}} = M\ddot{\mathbf{r}}_G. \tag{10}$$

Example 1

Two particles of masses m_1 and m_2 are attached to a model spring of natural length l_0 and stiffness k (as shown in Figure 12). The particle–spring system moves vertically.

Determine the equation of motion of the centre of mass of the system. What would be the effect on this equation of motion if the stiffness of the spring were doubled?

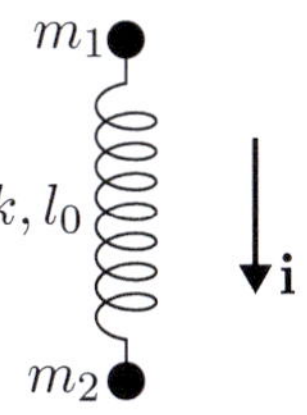

Figure 12   Two particles joined by a model spring

Solution

The spring forces are internal forces. The only external forces are the weights of the two particles, given by $m_1 g\mathbf{i}$ and $m_2 g\mathbf{i}$. Thus equation (10) gives

$$(m_1 + m_2)g\mathbf{i} = (m_1 + m_2)\ddot{\mathbf{r}}_G, \tag{11}$$

where $\mathbf{r}_G$ is the position vector of the centre of mass of the system. This simplifies to $\ddot{\mathbf{r}}_G = g\mathbf{i}$, so the centre of mass moves as if it is a particle falling under gravity.

Since the spring forces are internal forces, they have no effect on the equation of motion (11), so changing the stiffness of the spring will have no effect on the equation of motion of the centre of mass of the system.

Exercise 4

Two objects of masses m_1 and m_2, connected by a light model rod, are sliding down an inclined plane in the direction of the axis of the rod, as shown in Figure 13. The coefficient of sliding friction between each object and the plane is μ'. The objects may be modelled as particles and the connecting rod as a model rod.

(a) Draw a force diagram for the system, and identify any internal forces.

(b) Determine the equation of motion for the centre of mass of this system.

(c) How would your solution to part (b) change in the following situations?

- The rod is replaced by a light spring.

- The rod is removed, leaving just two disconnected objects.

(d) Compare your equation of motion in part (b) with that for a single particle of mass $m_1 + m_2$ sliding down the same inclined plane.

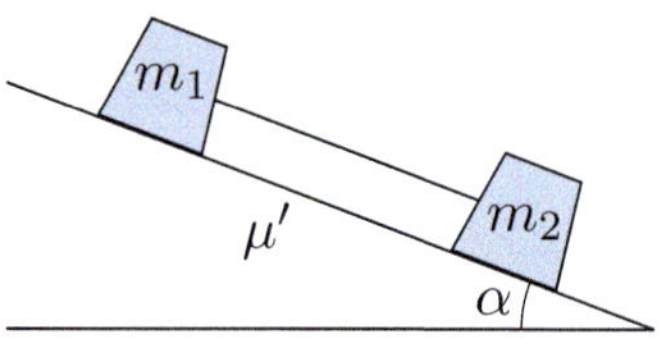

Figure 13   Two objects connected by a light model rod, on an inclined plane

The two-particle system in Exercise 4 is actually a rigid body and does not rotate, so its motion is completely described by the equation of motion of its centre of mass. The drum major's mace is also a rigid body, but it is subject to torques to make it rotate as the drum major twirls it. Nevertheless, the motion of the centre of mass still provides useful information for the drum major. In other cases, knowing about the motion of the centre of mass may not be directly useful. However, as the following example illustrates, it can form the basis for an analysis of the complete motion of such a system.

Example 2

Figure 14(a) shows two particles A and B, each of mass m, lying on a smooth horizontal surface and connected by a model spring of natural length $2l_0$ and stiffness k.

Initially, the particles are at rest a distance $2l_0$ apart, so the spring is unstretched, and with centre of mass at the origin. Then a constant horizontal force $\mathbf{P}$ is applied to B in the direction AB, as shown in Figure 14(b). At time t, the particles are a distance $2d$ apart.

It is important to measure coordinates from a *fixed* point O. It is tempting to measure distances from the position of G or A or B, but as the whole system is moving, none of these is appropriate.

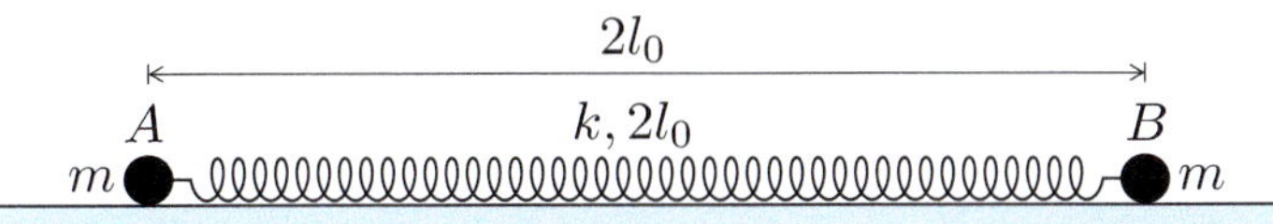

Figure 14   A system of two masses connected by a spring, subjected to a constant force $\mathbf{P}$ at one end: (a) unstretched spring, (b) stretched spring

(a) Find the position of the centre of mass at time t.

(b) Find a differential equation for d.

(c) Find the x-coordinates of A and B at time t.

(d) Interpret your solution.

Solution

(a) The force diagrams are shown in Figure 14(b). The two spring forces are internal forces, so equation (10) gives

$$\mathbf{P} + \mathbf{N}_1 + \mathbf{W}_1 + \mathbf{N}_2 + \mathbf{W}_2 = 2m\ddot{x}_G\mathbf{i}.$$

But both particles are in vertical equilibrium, so $\mathbf{N}_1 + \mathbf{W}_1 = \mathbf{0}$ and $\mathbf{N}_2 + \mathbf{W}_2 = \mathbf{0}$. Thus resolving in the $\mathbf{i}$-direction, we find that the centre of mass moves to the right with constant acceleration

$$\ddot{x}_G = \frac{|\mathbf{P}|}{2m}. \tag{12}$$

Since the system starts from rest with the centre of mass at the origin, integrating twice gives

$$x_G = \frac{|\mathbf{P}|}{4m}\, t^2.$$

Note that $\mathbf{N}_1$ and $\mathbf{W}_1$ are not internal forces, so we cannot invoke Newton's third law to obtain $\mathbf{N}_1 + \mathbf{W}_1 = \mathbf{0}$; this equation is true because the particles are not moving vertically.

Alternatively, we can use the constant acceleration equation $x = x_0 + v_0 t + \frac{1}{2}a_0 t^2$ from Unit 3, with $x_0 = 0$, $v_0 = 0$ and $a_0 = \ddot{x}_G$.

If we started by applying Newton's second law to particle A, we would obtain the same differential equation for d.

(b) Applying Newton's second law to particle B gives

$$\mathbf{H}_2 + \mathbf{P} = m\ddot{x}_2\mathbf{i}.$$

Now we need to model the forces. We are given that the force $\mathbf{P}$ acts in the direction AB, so $\mathbf{P} = |\mathbf{P}|\mathbf{i}$. Also, by Hooke's law, for a spring of length l, natural length $2l_0$ and stiffness k, and a unit vector $\widehat{\mathbf{s}}$ in the direction from the particle to the centre of the spring, we have

$$\mathbf{H}_2 = k(l - 2l_0)\,\widehat{\mathbf{s}} = k(2d - 2l_0)(-\mathbf{i}).$$

Therefore we obtain

$$-2k(d - l_0)\mathbf{i} + |\mathbf{P}|\mathbf{i} = m\ddot{x}_2\mathbf{i}.$$

Resolving in the $\mathbf{i}$-direction and rearranging gives

$$m\ddot{x}_2 + 2kd = 2kl_0 + |\mathbf{P}|. \tag{13}$$

To proceed further, we need to express the acceleration $\ddot{x}_2$ of particle B in terms of d and known parameters. We start by noting that throughout the motion, the centre of mass is at the point

$$x_G\mathbf{i} = \frac{mx_1\mathbf{i} + mx_2\mathbf{i}}{2m} = \tfrac{1}{2}(x_1 + x_2)\mathbf{i},$$

that is, the centre of mass is always halfway between A and B.

This gives $x_2 = x_G + d$, which we can differentiate twice to obtain $\ddot{x}_2 = \ddot{x}_G + \ddot{d}$. Now we can use equation (12) to obtain $\ddot{x}_2 = |\mathbf{P}|/(2m) + \ddot{d}$, which is the desired expression for $\ddot{x}_2$ in terms of d and known parameters. Substituting this into equation (13) and rearranging gives

$$m\ddot{d} + 2kd = 2kl_0 + |\mathbf{P}|/2, \tag{14}$$

which is the required differential equation for d.

See Unit 9.

(c) You should recognise equation (14) as the differential equation for simple harmonic motion, which has solution

$$d = C_1\cos(\omega t) + C_2\sin(\omega t) + l_0 + |\mathbf{P}|/(4k),$$

where $\omega = \sqrt{2k/m}$ and the values of the constants C_1 and C_2 are determined by initial conditions. When $t = 0$, before the force $\mathbf{P}$ is applied, we have $d = l_0$ and $\dot{d} = 0$. This gives $C_1 = -|\mathbf{P}|/(4k)$ and $C_2 = 0$, so

$$d = -\frac{|\mathbf{P}|}{4k}\cos(\omega t) + l_0 + \frac{|\mathbf{P}|}{4k}.$$

We use $x_2 = x_G + d$ together with $x_G = |\mathbf{P}|t^2/(4m)$ and the above expression for d, to obtain

$$x_2 = \frac{|\mathbf{P}|}{4m}t^2 + \left(l_0 + \frac{|\mathbf{P}|}{4k} - \frac{|\mathbf{P}|}{4k}\cos(\omega t)\right).$$

Similarly, $x_1 = x_G - d$, so we have

$$x_1 = \frac{|\mathbf{P}|}{4m}t^2 - \left(l_0 + \frac{|\mathbf{P}|}{4k} - \frac{|\mathbf{P}|}{4k}\cos(\omega t)\right).$$

(d) The equations for x_1 and x_2 show that particle A oscillates according to the function $(|\mathbf{P}|/(4k))\cos(\omega t)$ about a point a fixed distance $l_0 + |\mathbf{P}|/(4k)$ to the left of the moving position $x_G = |\mathbf{P}|t^2/(4m)$ of the centre of mass, and that particle B oscillates according to the function $-(|\mathbf{P}|/(4k))\cos(\omega t) = (|\mathbf{P}|/(4k))\cos(\omega t + \pi)$ about a point the same fixed distance to the right of the moving position of the centre of mass.

The oscillations are simple harmonic with the same angular frequency $\omega = \sqrt{2k/m}$ and amplitude $|\mathbf{P}|/(4k)$, but are phase-opposed (in the terminology of Unit 11) because the terms (in large brackets above) describing the oscillations have different signs; this means that, relative to the centre of mass, the particles are always moving in opposite directions.

The situation is illustrated in Figure 15.

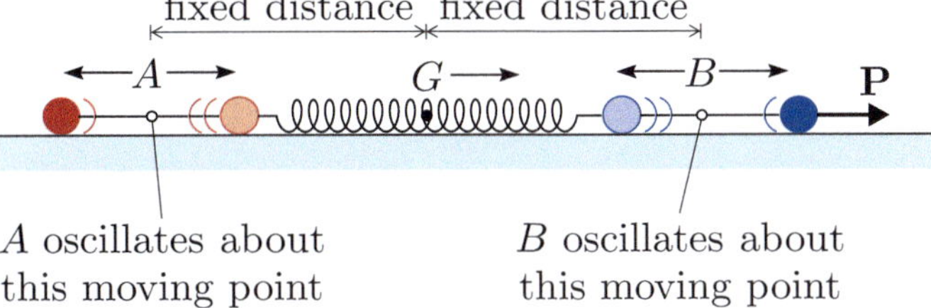

Figure 15   The effect of applying a force $\mathbf{P}$ to one end of a spring–mass system

Exercise 5

Suppose that the model spring in Example 2 is replaced by a light model rod of length $2l_0$.

What can you say about the motion of the centre of mass of this new system, and about the motion of A and B? What forces are exerted by the rod on particles A and B?

Exercise 6

Two particles A and B, each of mass m, move along the x-axis, and although each particle exerts a force on the other, there are no external forces. Initially, A is at rest at the origin, while B is at $x = 3$ and moving away from the origin with speed $6\,\mathrm{m\,s^{-1}}$.

If particle B returns to its initial position at time $t = 0.5$, where is A at this time?

Exercise 7

Long ago and in a galaxy far, far away, an alien was alone in his spaceship, floating aimlessly through the void. He cursed his luck for running out of fuel, but then came up with a brilliant idea. Fortunately, he weighed half as much as the entire ship and he was quick on his (many) feet, so, starting from the back of the ship, he would run as fast as he could and hurl himself at the forward bulkhead. If he did this a few thousand times, he thought, the ship would gradually pick up speed and he would eventually arrive home. Comment on his chances of success.

Exercise 8

Two particles A and B, each of mass $50\,\mathrm{kg}$, are at rest on a smooth horizontal surface and connected by a model string, which is stretched tightly between them. The breaking strain of the string is $10\,\mathrm{N}$ (i.e. it cannot sustain a tension greater than this value). A horizontal force $\mathbf{P}$ is applied to B in the direction AB, as shown in Figure 16.

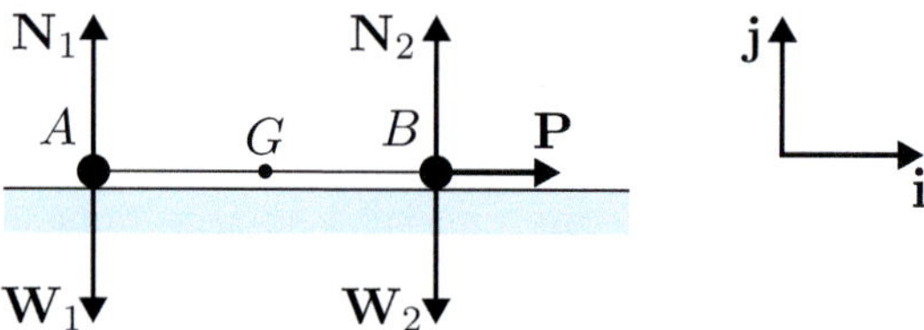

Figure 16   Two particles connected by a model string

What is the greatest magnitude of the force $\mathbf{P}$ that can be applied to B before the string breaks?

Exercise 9

Two particles A and B slide along a straight rough horizontal wire, and the coefficient of sliding friction between each of the particles and the wire is $\mu' = 0.4$. Particle A has mass $3\,\mathrm{kg}$ and particle B has mass $0.5\,\mathrm{kg}$, and each particle exerts a force on the other.

Find the acceleration of the centre of mass when both particles move along the wire in the same direction.

Exercise 10

Two particles move along a smooth straight horizontal track, and each exerts a force on the other. The first particle has mass $4\,\mathrm{kg}$ and experiences no external forces. The second particle has mass $1\,\mathrm{kg}$ and is pulled along the track by an external force of magnitude $20\,\mathrm{N}$. It is found that the two particles accelerate along the track a fixed distance apart.

Find the common acceleration of the particles and the magnitude of the internal forces.

2 Many-particle systems

In this section we generalise the ideas of Section 1 to systems of more than two particles. We also look at how to find the centre of mass of a variety of different systems and solid objects.

2.1 Motion of the centre of mass

We generalise here the results of Subsection 1.2 to systems with more than two particles. In fact, we consider a system of n particles, where n may be very large.

> An n-**particle system** is a system modelled so that the total mass of the system is divided among n particles.

We assume that particle i has mass m_i and that its position vector is $\mathbf{r}_i$ (relative to some fixed origin O). We begin our discussion by extending the notion of the centre of mass to a system of n particles.

> The **centre of mass** of an n-particle system, whose particles have masses $m_1, m_2, \ldots, m_n$, and position vectors $\mathbf{r}_1, \mathbf{r}_2, \ldots, \mathbf{r}_n$ relative to an origin O, has position vector
>
> $$\mathbf{r}_G = \frac{m_1\mathbf{r}_1 + m_2\mathbf{r}_2 + \cdots + m_n\mathbf{r}_n}{m_1 + m_2 + \cdots + m_n} = \frac{\sum_{i=1}^{n} m_i\mathbf{r}_i}{M} \tag{15}$$
>
> relative to O, where $M = \sum_{i=1}^{n} m_i$ is the total mass of the system.

Exercise 11

Determine the centre of mass of each of the systems of particles in Figure 17, where each particle has the same mass m.

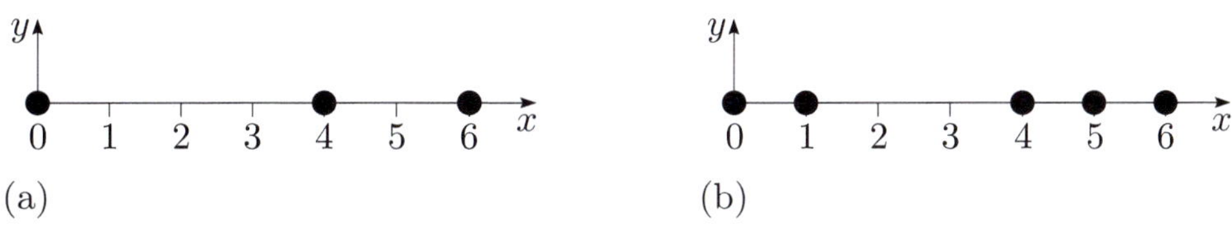

(a)   (b)

Figure 17   Particles on the x-axis

Each particle may be subject to a number of external forces, but these can always be added together to give a single resultant force, so we may assume that particle i is acted on by a single external force $\mathbf{E}_i$. The total external force on the whole system will then be

$$\mathbf{F}^{\text{ext}} = \mathbf{E}_1 + \mathbf{E}_2 + \cdots + \mathbf{E}_n = \sum_{i=1}^{n} \mathbf{E}_i. \tag{16}$$

In addition to the external forces, we assume that each particle exerts a force on every other (and that this force acts along the line joining them). Each particle is therefore acted on by $n-1$ internal forces. We denote the force acting on particle i due to particle j by $\mathbf{I}_{ij}$. Then (from Newton's third law) we have

$$\mathbf{I}_{ij} = -\mathbf{I}_{ji}. \tag{17}$$

It will make the summations that follow easier to write if we also introduce a term $\mathbf{I}_{ii}$ (the force exerted by the ith particle on itself), which we assume to be zero.

Now we apply Newton's second law to the ith particle and obtain

$$\underbrace{\mathbf{E}_i}_{\substack{\text{external force} \\ \text{on the } i\text{th particle}}} + \underbrace{\sum_{j=1}^{n} \mathbf{I}_{ij}}_{\substack{\text{sum of the internal} \\ \text{forces on the} \\ i\text{th particle}}} = \underbrace{m_i}_{\substack{\text{mass of the} \\ i\text{th particle}}} \times \underbrace{\ddot{\mathbf{r}}_i.}_{\substack{\text{acceleration of} \\ \text{the } i\text{th particle}}}$$

Adding together all of the equations like this, for all choices of i, we obtain

$$\underbrace{\sum_{i=1}^{n} \mathbf{E}_i}_{\substack{\text{sum of all the} \\ \text{external forces} \\ \text{on the system}}} + \underbrace{\sum_{i=1}^{n}\sum_{j=1}^{n} \mathbf{I}_{ij}}_{\substack{\text{sum of all the} \\ \text{internal forces} \\ \text{in the system}}} = \sum_{i=1}^{n} m_i \ddot{\mathbf{r}}_i. \tag{18}$$

The first term on the left-hand side is simply $\mathbf{F}^{\text{ext}}$, by equation (16). The second term, involving a double summation, looks rather complicated. But it simply represents the sum of all the internal forces in the system. From equation (17) we know that these forces balance in pairs, or are zero when $i = j$. The net result is that this term is zero. We now use equation (15) to simplify the right-hand side of equation (18), and we are left with the following important result.

> ### Motion of the centre of mass of an n-particle system
>
> The motion of the centre of mass $\mathbf{r}_G$ of an n-particle system of total mass M subject to external forces with sum $\mathbf{F}^{\text{ext}}$ is given by
>
> $$\mathbf{F}^{\text{ext}} = M\ddot{\mathbf{r}}_G. \tag{19}$$

Notice that this formula is identical to the one for the two-particle case given by equation (10).

Thus the centre of mass of an n-particle system moves as if it is a particle with mass the same as the total mass of the system and with all the external forces on the system applied to this particle.

Equation (19) is critical to the modelling of real objects as single particles. It tells us that an object composed of a very large number of components (modelled as particles) can be modelled by a single particle. However, if we do this, then the single-particle model will predict only the motion of the centre of mass. For example, consider the diver mentioned in the Introduction, moving under gravity alone. As you know from Unit 3, the single-particle model predicts that the particle (i.e. the diver's centre of mass) moves along a parabolic path. This model does not give any information about the motion of any other part of the diver, which may move along a complicated path (see Figure 1).

2.2   Locating the centre of mass

In this subsection you will see how to find the centre of mass of a variety of systems and objects. One method for finding the centre of mass is to use the definition directly (equation (15)). If the system is symmetric, then we can locate the centre of mass by finding the geometric centre (this was the method used in Unit 2).

Try the following exercise to find the centre of mass of two systems.

Exercise 12

Determine the centre of mass of each of the systems of particles in Figure 18, where each particle has the same mass m.

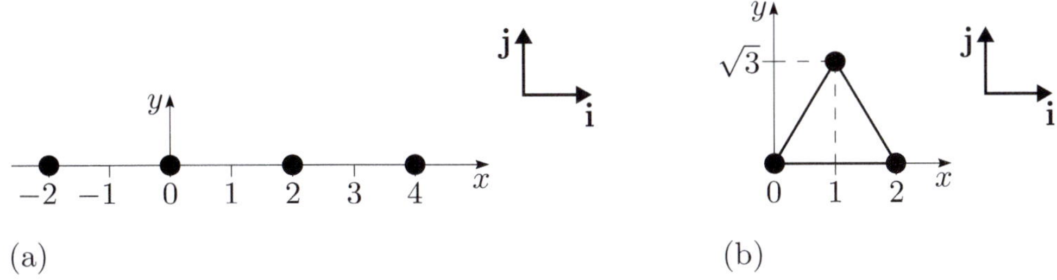

Figure 18   Two systems of particles

So far, we have looked at finding the centres of mass only of systems of particles. But you saw in Unit 2 that sometimes it is more appropriate to model an object as a rigid body. To find the centre of mass of a rigid body, we simply suppose that the object is made up of a very large number of very small chunks of material, each of which can be modelled as a particle – that is, we model the object as a system of particles. Then we can apply equation (15) to find its centre of mass. However, this poses questions about how many particles we should use, what mass to give them, and where to locate them.

Recall that *density* is mass per unit volume.

In general, this is a difficult problem. But there are many objects whose physical properties can be used to help in this process. In particular, many objects are of *uniform density*, in that the mass of any small chunk of the object is proportional to its volume. Such an object is said to be *homogeneous*.

The use of 'homogeneous' here is different from its use in the context of differential equations.

A **homogeneous rigid body** is a rigid body of uniform density.

For a symmetric homogeneous rigid body, we can see at once that the centre of mass is at the geometric centre. We exploited this fact in Unit 2 in locating the centres of mass of objects with some simple geometric shapes. Figure 19 shows six examples of symmetric homogeneous rigid bodies with their centres of mass G marked. Notice that the centre of mass in each case is located on an axis of symmetry – this is true in general of any symmetric homogeneous rigid body.

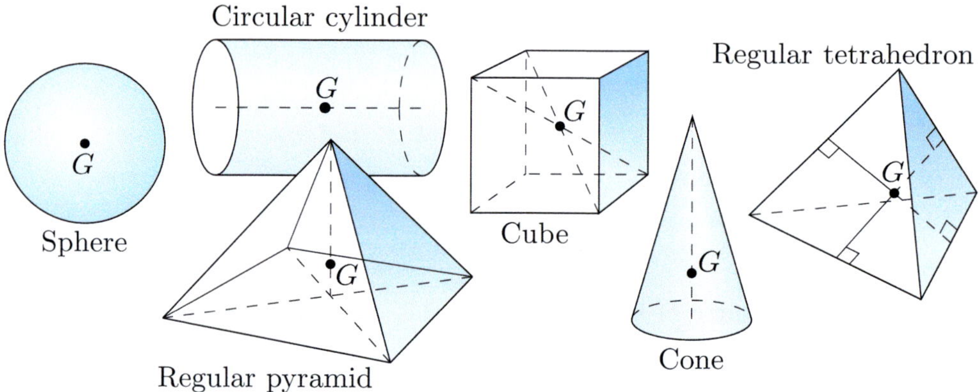

Figure 19   Six common solids, where G is the centre of mass in each case

A homogeneous rigid body that is not symmetric can often be broken down into symmetric parts. The centre of mass of each symmetric part is at its geometric centre. If we model each part as a particle of appropriate mass located at its geometric centre, then we can use equation (15) to find the centre of mass of the whole rigid body.

Example 3

Determine the centre of mass of the L-shaped piece of uniform cardboard shown in Figure 20.

Solution

Since the cardboard is uniform, its centre of mass must lie halfway through its thickness, so we can ignore the thickness and treat this as a two-dimensional problem.

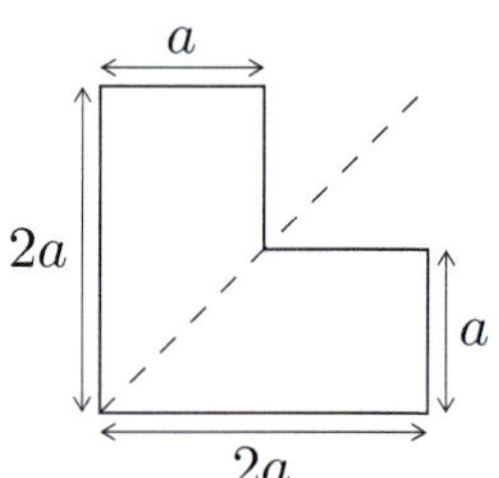

Figure 20   An L-shaped region

Thus we model the cardboard as a homogeneous two-dimensional rigid body. We know that its centre of mass must lie on the axis of symmetry shown in Figure 20. To find where on this axis it lies, we divide the L-shape into two parts, a square and a rectangle, as in Figure 21. Since we have homogeneity, their individual centres of mass must be at their geometric centres G_1 and G_2, which have position vectors

$$\mathbf{r}_{G_1} = \tfrac{1}{2}a\mathbf{i} + \tfrac{3}{2}a\mathbf{j},$$
$$\mathbf{r}_{G_2} = a\mathbf{i} + \tfrac{1}{2}a\mathbf{j},$$

with respect to O, where $\mathbf{i}$ and $\mathbf{j}$ are Cartesian unit vectors in the positive x- and y-directions. We now model each part as a particle: the square as a particle of mass m (say) at $\mathbf{r}_{G_1}$, and the rectangle as a particle of mass $2m$ (since the rectangle is twice the size of the square) at $\mathbf{r}_{G_2}$. Equation (15) then gives the centre of mass as

$$\mathbf{r}_G = \frac{m\left(\tfrac{1}{2}a\mathbf{i} + \tfrac{3}{2}a\mathbf{j}\right) + 2m\left(a\mathbf{i} + \tfrac{1}{2}a\mathbf{j}\right)}{3m}$$
$$= \tfrac{5}{6}a\mathbf{i} + \tfrac{5}{6}a\mathbf{j},$$

which lies on the axis of symmetry (the line $y = x$), as expected.

Exercise 13

The shape in Figure 20 can be divided into three squares as shown in Figure 22. Use this division of the shape to find its centre of mass.

It is often the case that a three-dimensional problem can be reduced to one in two dimensions by taking account of certain uniformities or symmetries in the problem, as in Example 3. But even then there is no guarantee that we can divide the resulting plane figure into parts whose geometric centres we can find easily. Squares and rectangles are no problem, as Example 3 illustrates. But what about triangles?

Figure 23(a) shows an isosceles triangle, and clearly its geometric centre lies on its axis of symmetry. Figure 23(b) shows the shape divided into a large number of very thin horizontal strips, and Figure 23(c) shows these strips pushed to the right to form another triangle. The centre of each strip lies on the broken line in Figure 23(c), so the geometric centre of this triangular shape lies on this line. Such a line passes through a vertex of the triangle and bisects the opposite side, and is known as a *median* of the triangle. This argument works for thin strips parallel to any of the three sides of the triangle. It follows that the geometric centre must lie on all three medians, that is, the geometric centre must be the point of intersection of the medians, as shown in Figure 23(d). Any triangle may be constructed in this way, so this result holds for all triangles.

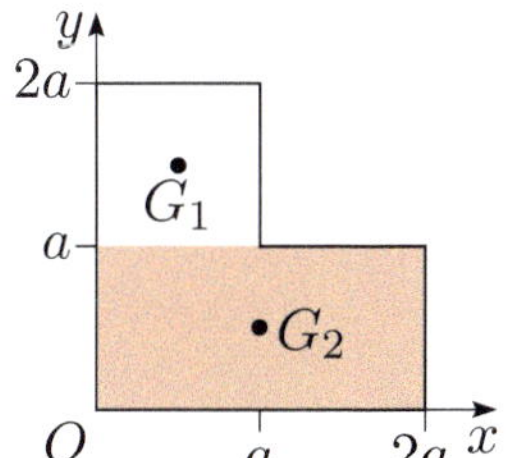

Figure 21   The L-shaped region as two rectangles

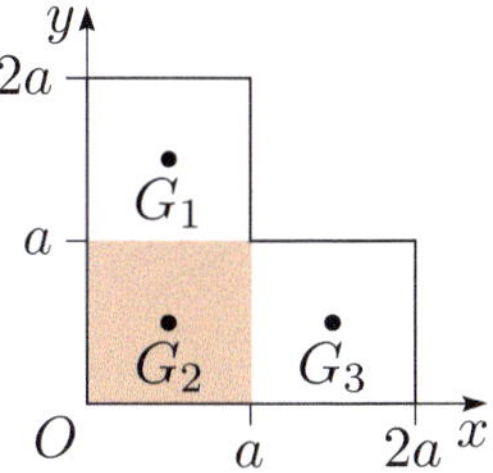

Figure 22   The L-shaped region as three squares

You saw other examples of this in Unit 2.

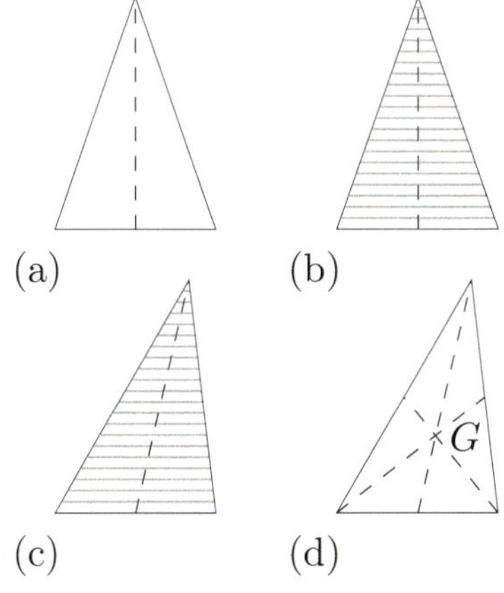

Figure 23   For a triangle, the centre of mass lies on the median

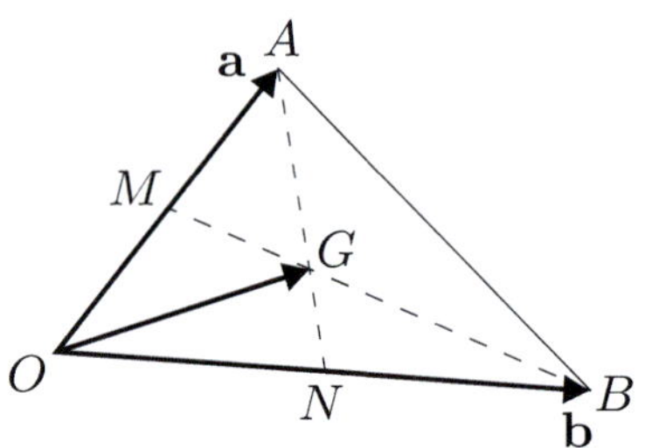

Figure 24   Finding the centre of mass of a triangle

We can find the precise location of this point of intersection by using vectors. Figure 24 shows an arbitrary triangle OAB, and we know that the geometric centre G lies at the intersection of the medians AN and BM. It follows from the triangle rule for adding vectors (defined in Unit 2) that for some numbers λ and μ, we have

$$\overrightarrow{OG} = \mathbf{a} + \lambda\,\overrightarrow{AN} = \mathbf{b} + \mu\,\overrightarrow{BM}. \tag{20}$$

But $\overrightarrow{AN} = -\mathbf{a} + \tfrac{1}{2}\mathbf{b}$ and $\overrightarrow{BM} = -\mathbf{b} + \tfrac{1}{2}\mathbf{a}$, so equation (20) gives

$$\mathbf{a} + \lambda\left(-\mathbf{a} + \tfrac{1}{2}\mathbf{b}\right) = \mathbf{b} + \mu\left(-\mathbf{b} + \tfrac{1}{2}\mathbf{a}\right).$$

Equating the coefficients of $\mathbf{a}$ and $\mathbf{b}$ gives

$$1 - \lambda = \tfrac{1}{2}\mu,$$
$$1 - \mu = \tfrac{1}{2}\lambda,$$

so $\lambda = \mu = \tfrac{2}{3}$. Thus the geometric centre lies on a median at the point two-thirds of the distance from the vertex to the centre of the opposite side, or equivalently, one-third of the way up a median from its base.

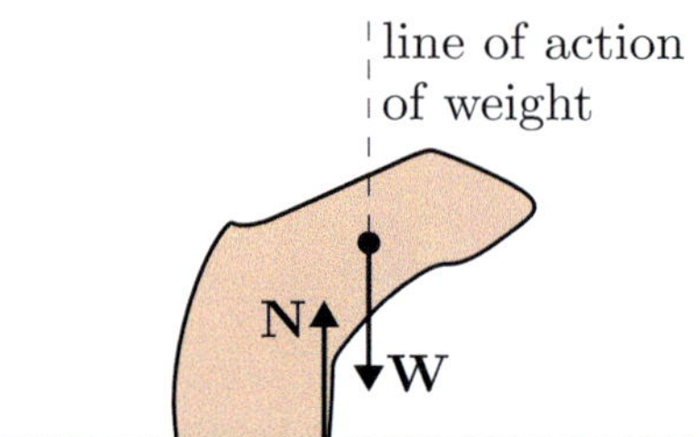

Figure 25   Toppling condition

Now we mention one simple application of the centre of mass. There is a simple test to determine whether an object with a square base resting on a flat horizontal surface will topple over. Consider the situation shown in Figure 25. The forces acting on the object here are its weight $\mathbf{W}$ and the normal reaction force $\mathbf{N}$ due to the horizontal surface, as shown.

Recall from Unit 2 that the weight acts through the centre of mass of the object. The point of action of the normal reaction force can be any point of contact between the object and the surface. If the object does not topple, then the equilibrium conditions of Unit 2 apply, that is, both the resultant force and the resultant torque must be zero. The condition that the resultant force must be zero fixes the magnitude of the normal reaction and can always be satisfied. For the resultant torque to be zero requires that the two forces have the same line of action. So the line of action of the weight of the object must pass through the base of the object, that is, the centre of mass must be vertically above the base. This gives the following useful result.

> **Toppling condition**
>
> Consider an object with a square base resting on a flat horizontal surface. The object will topple over if and only if the centre of mass of the object is not vertically above the base.

Try to apply this result in the following exercise.

Exercise 14

Figure 26(a) shows the design for a sculpture. It is to be made of concrete of uniform density, to be 1 m thick, 3 m high, and to have constant cross-sectional shape as shown in Figure 26(b), where $a = \sqrt{3}/2 \simeq 0.866$.

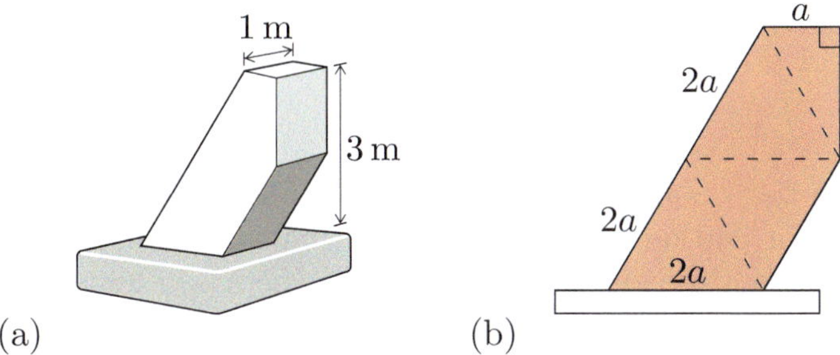

(a)   (b)

Figure 26   Design for a sculpture

The sculpture is intended to rest on a horizontal surface, as shown in the figure. Will it topple over?

You have seen how to exploit the geometry of certain homogeneous rigid bodies to determine their centres of mass. In some cases, however, we may not be able to subdivide the object into parts whose geometric centres are known or easily found. In such cases we need to resort to integration to help us to find the centre of mass, using the methods of Unit 17. We explore this in the next subsection.

2.3 Centres of mass for laminas

We now turn to the calculation of centres of mass using area integrals. How can we find the centre of mass of a lamina such as a semicircular plate that is bounded by curved lines? We start with the definition of centre of mass. The position vector $\mathbf{r}_G = (x_G, y_G)$ of the centre of mass of a system of n particles each with mass m_i and position vector $\mathbf{r}_i$ is

Area integrals are defined in Unit 17.

$$\mathbf{r}_G = \frac{\sum_{i=1}^{n} m_i \mathbf{r}_i}{M} = x_G \mathbf{i} + y_G \mathbf{j}, \tag{21}$$

where M is the total mass. If the particles lie in the (x, y)-plane, then we can project equation (21) onto the unit vectors $\mathbf{i}$ and $\mathbf{j}$ to obtain the coordinates of the centre of mass:

$$\mathbf{r}_G \cdot \mathbf{i} = x_G = \frac{\sum_{i=1}^{n} m_i x_i}{M} \quad \text{and} \quad \mathbf{r}_G \cdot \mathbf{j} = y_G = \frac{\sum_{i=1}^{n} m_i y_i}{M}. \tag{22}$$

Now, expressions (22) are for the centre of mass of a system of n discrete particles lying in the (x, y)-plane. Suppose that the number of particles n increases indefinitely. Then in the limit, the particles will coalesce to form a lamina, and the summations in (22) will become integrals and give the centre of mass of the lamina.

Let us look at the expression for x_G in detail. First, consider the denominator. This is the total mass of the lamina and can be considered as either the limit of the sum of n particles each of mass m_i or as the limit of the sum of area elements $\delta x_j\,\delta y_k$ each of surface density $f(x_j, y_k)$. So

$$M = \lim_{n\to\infty} \sum_{i=1}^{n} m_i = \lim_{p\to\infty} \left(\sum_{j=1}^{p} \left(\lim_{q\to\infty} \sum_{k=1}^{q} f(x_j, y_k)\,\delta y_k \right) \delta x_j \right)$$

$$= \int_S f(x, y)\,dA,$$

where S is the region of the (x, y)-plane occupied by the lamina. In a similar way, the numerator can be considered as the limit as $n \to \infty$ for the particle model and as both $p \to \infty$ and $q \to \infty$ for the area element model. So we have

$$\lim_{n\to\infty} \sum_{i=1}^{n} m_i x_i = \lim_{p\to\infty} \left(\sum_{j=1}^{p} \left(\lim_{q\to\infty} \sum_{k=1}^{q} f(x_j, y_k)\,\delta y_k \right) x_j\,\delta x_j \right)$$

$$= \int_S x\, f(x, y)\,dA.$$

> The coordinates (x_G, y_G) of the **centre of mass of a lamina** S are
>
> $$x_G = \frac{\int_S xf\,dA}{M} = \frac{\int_S xf\,dA}{\int_S f\,dA} \quad \text{and} \quad y_G = \frac{\int_S yf\,dA}{M} = \frac{\int_S yf\,dA}{\int_S f\,dA}.$$
>
> For a uniform lamina, the surface density f is a constant and can be taken outside the integral.

Example 4

Determine the position of the centre of mass of the uniform semicircular plate in Figure 27.

Solution

The y-axis is a symmetry axis of the semicircle, so we must have $x_G = 0$. To find the y-coordinate, we must evaluate $\int_S yf\,dA$. We can use polar coordinates with $y = r\sin\theta$ and surface density $f = $ constant. Then the area integral is

$$\int_S yf\,dA = f \int_{\theta=0}^{\theta=\pi} \left(\int_{r=0}^{r=a} r\sin\theta\, r\,dr \right) d\theta$$

$$= f \int_{\theta=0}^{\theta=\pi} \left[\tfrac{1}{3}r^3 \right]_{r=0}^{r=a} \sin\theta\, d\theta$$

$$= \tfrac{1}{3}fa^3 \left[-\cos\theta \right]_{\theta=0}^{\theta=\pi} = \tfrac{2}{3}fa^3.$$

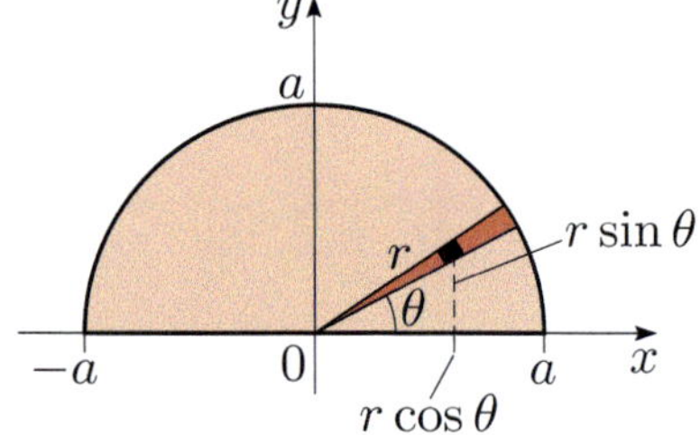

Figure 27

Recall that $\delta A = r\,\delta r\,\delta\theta$.

The mass is the area $\tfrac{1}{2}\pi a^2$ times the (constant) surface density f.

The mass M of the semicircle is $\tfrac{1}{2}\pi a^2 f$, so we have $y_G = 4a/3\pi$. Hence the centre of mass of the semicircular plate is $\mathbf{r} = 4a\mathbf{j}/3\pi$.

Exercise 15

Confirm the answer for the coordinate y_G in Example 4 by working in Cartesian coordinates (see Figure 28).

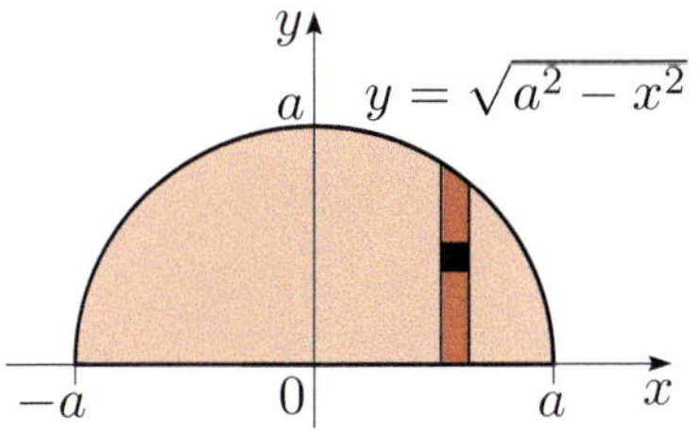

Exercise 16

Confirm the result of the previous subsection, that the centre of mass of an isosceles triangle is one-third of the way up a median from its base, by evaluating an area integral.

Figure 28

2.4 Potential energy

In this short subsection we look at another important application of the centre of mass – calculating potential energy.

Consider the homogeneous solid block shown in two positions in Figure 29. Clearly the block has a lower potential energy in the position shown on the right than in that shown on the left, but how are we to calculate it?

Figure 29   Two positions for a block

Let us return to the general system of particles introduced in Subsection 2.1, where the ith particle has mass m_i and position vector $\mathbf{r}_i$ (relative to some fixed origin O), but this time imagine the particles to lie in the Earth's gravitational field. Choose a Cartesian coordinate system using x, y and z, with origin at O and z-axis pointing vertically upwards, and take corresponding Cartesian unit vectors $\mathbf{i}$, $\mathbf{j}$ and $\mathbf{k}$. The height of the ith particle above the (x, y)-plane is the $\mathbf{k}$-component of $\mathbf{r}_i$, given by $\mathbf{r}_i \cdot \mathbf{k}$, so the potential energy of the ith particle (relative to the datum O) is $m_i g(\mathbf{r}_i \cdot \mathbf{k})$. From equation (15) we have

$$\sum_{i=1}^{n} m_i \mathbf{r}_i = M\mathbf{r}_G = M(x_G \mathbf{i} + y_G \mathbf{j} + h\mathbf{k}).$$

Taking the dot product of each side of this equation with $\mathbf{k}$, and then multiplying by g, we obtain

$$\sum_{i=1}^{n} m_i g(\mathbf{r}_i \cdot \mathbf{k}) = Mg(\mathbf{r}_G \cdot \mathbf{k}) = Mgh, \tag{23}$$

where h is the height of the centre of mass of the system above O. The left-hand side of this equation is the total potential energy of the system of particles, and the right-hand side is the potential energy of a single particle of mass M (equal to the total mass of the system) placed at the centre of mass. Equation (23) thus tells us that to find the potential energy of a system of particles or of a homogeneous rigid body of total mass M, all we need to do is to find the height h of the centre of mass above the datum and then use the formula Mgh.

Exercise 17

If the homogeneous solid block in Figure 29 has mass M, height $4h$, width h and depth h, calculate the change in its potential energy in the two positions shown in Figure 29.

Exercise 18

A uniform ball of mass M is dropped from rest and falls under gravity. What is the speed of its centre of mass when the ball has fallen a distance h? What is the change in the potential energy when the ball has fallen a distance h?

Exercise 19

Find the centre of mass of the homogeneous rigid body shown in Figure 30. Will the object topple over when it is placed on a horizontal surface as shown on the right in Figure 30?

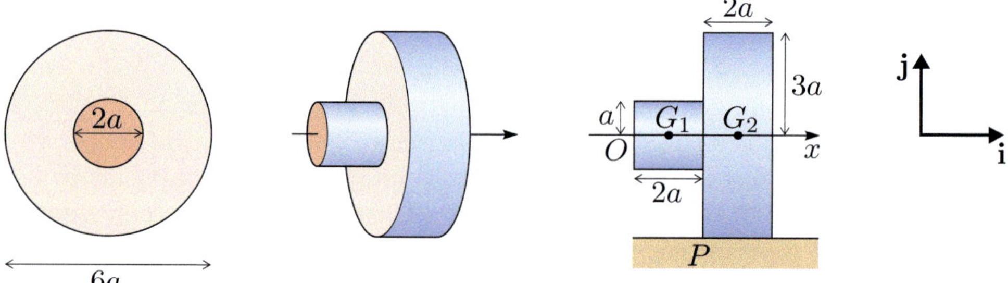

Figure 30   Three images of a rigid body

Exercise 20

Figure 31 shows the Great Pyramid of Cheops, with its centre of mass G a quarter of its height h above the square base. Originally the pyramid was 147 m high and built on a square base with sides of length 230 m. The stones of the pyramid are made of limestone.

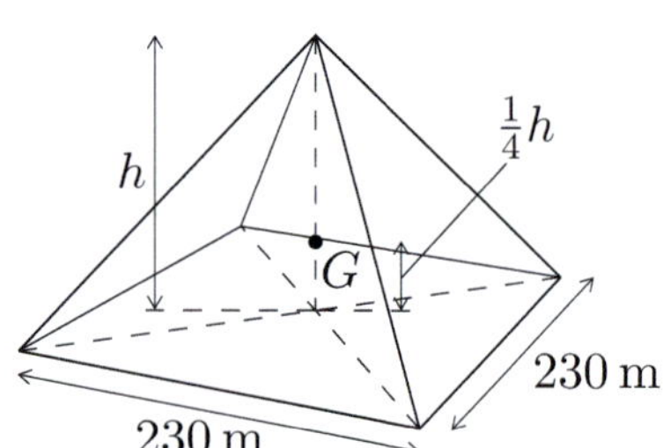

Figure 31   Great Pyramid of Cheops

(a) Assuming that the density of limestone is approximately $2500\,\text{kg m}^{-3}$, what is the approximate mass of the pyramid? (*Hint*: Recall that the volume of a pyramid is $\frac{1}{3} \times$ base area $\times$ vertical height.)

(b) Estimate the total energy required to lift all the stones of the pyramid into place.

(c) Given that a man can lift approximately 1000 kg of stones through a height of 1 m in a day, estimate how long a gang of 1000 men would have taken to lift all the stones of the pyramid into place.

3 Collisions

In this section we are concerned with the behaviour of objects when they collide. We begin in Subsection 3.1 by defining the concept of linear momentum, and we obtain a result that tells us when linear momentum is conserved. In Subsection 3.2 we go on to define and explore elastic and inelastic collisions.

3.1 Conservation of linear momentum

Suppose that two balls A and B, with position vectors $\mathbf{r}_1$ and $\mathbf{r}_2$ relative to a fixed origin, are moving far out in space, away from any outside influence, so there are no external forces. Let m_1 be the mass of ball A, and let m_2 be the mass of ball B. The balls collide, and just after the collision A has velocity $\dot{\mathbf{R}}_1$ and B has velocity $\dot{\mathbf{R}}_2$, as shown in Figure 32. (We will denote correspondingly the position vectors of A and B, relative to the origin, after the collision by $\mathbf{R}_1$ and $\mathbf{R}_2$.)

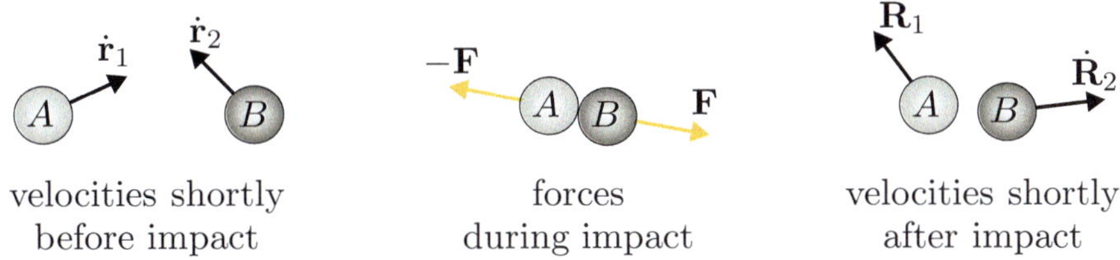

velocities shortly before impact

forces during impact

velocities shortly after impact

Figure 32   Two balls collide

The system under consideration consists of the two balls A and B, which we model as two particles. The impact happens over a very small interval of time. During that time interval, B experiences a force $\mathbf{F}$ (due to A) and, by Newton's third law, A experiences an equal and opposite force $-\mathbf{F}$ (due to B). These forces may vary over the very small time interval when the balls are in contact, but the important point to appreciate is that they are internal forces. Thus, because there are no external forces, the discussion in Section 2 tells us that the centre of mass of the system moves with constant velocity before, during and after the collision.

If we denote the velocity of the centre of mass before and after the collision by $\dot{\mathbf{r}}_G$ and $\dot{\mathbf{R}}_G$, respectively, then we have $\dot{\mathbf{r}}_G = \dot{\mathbf{R}}_G$, where

$$\dot{\mathbf{r}}_G = \frac{m_1\dot{\mathbf{r}}_1 + m_2\dot{\mathbf{r}}_2}{m_1 + m_2} \quad \text{and} \quad \dot{\mathbf{R}}_G = \frac{m_1\dot{\mathbf{R}}_1 + m_2\dot{\mathbf{R}}_2}{m_1 + m_2}.$$

It follows that

$$m_1\dot{\mathbf{r}}_1 + m_2\dot{\mathbf{r}}_2 = m_1\dot{\mathbf{R}}_1 + m_2\dot{\mathbf{R}}_2. \tag{24}$$

The quantity

$$(\text{mass of } A) \times (\text{velocity of } A) + (\text{mass of } B) \times (\text{velocity of } B)$$

has been unchanged by the collision, and this leads us to make the following general definitions.

The SI units for the magnitude of linear momentum are $\mathrm{kg\,m\,s^{-1}}$.

Here *linear* is being used in a translational sense as an object moves through space with velocity $\dot{\mathbf{r}}$. 'Linear momentum' is often abbreviated to 'momentum' when the meaning is clear.

The **linear momentum p** of a particle with position vector $\mathbf{r}$ is a vector quantity given as the product of its mass m and its velocity $\dot{\mathbf{r}}$, so

$$\mathbf{p} = m\dot{\mathbf{r}}.$$

The **linear momentum P** of an n-particle system is the vector sum of the linear momenta of the individual particles, so

$$\mathbf{P} = \sum_{i=1}^{n} m_i \dot{\mathbf{r}}_i = M\dot{\mathbf{r}}_G, \tag{25}$$

where m_i and $\mathbf{r}_i$ represent the mass and position of particle i, M is the total mass of the system, and $\mathbf{r}_G$ is its centre of mass.

Thus equation (24) tells us that the linear momentum of our two-ball system with no external forces is the same before and after the collision, that is, it is *conserved*.

We can link linear momentum to the motion of the centre of mass of a system, given by equation (19) as $\mathbf{F}^{\text{ext}} = M\ddot{\mathbf{r}}_G$. Differentiating equation (25) with respect to time gives $\dot{\mathbf{P}} = M\ddot{\mathbf{r}}_G$, so equation (19) becomes

$$\mathbf{F}^{\text{ext}} = \dot{\mathbf{P}}. \tag{26}$$

This tells us that the sum of the external forces on an n-particle system is equal to the rate of change of linear momentum of the system. The case when there are no external forces is particularly simple, for then we have $\dot{\mathbf{P}} = \mathbf{0}$, so the linear momentum is constant throughout the motion.

As we observed above, the forces acting during a collision are *internal* forces, thus do not affect equations (19) and (26). Therefore equation (19) tells us that the motion of the centre of mass of an n-particle system is unaffected by any collisions between the particles. Equation (26) tells us that the rate of change of linear momentum of the system is also unaffected by any collisions.

In the case where there is no net external force, equation (26) also tells us that the linear momentum remains constant despite any collisions, as you saw in the two-ball example. Let us now consider what equation (26) tells us about the linear momentum (rather than its rate of change) when our two-ball system is subject to external forces. With this aim, suppose now that the two balls are in the Earth's gravitational field, as shown in Figure 33, where there are no other external forces.

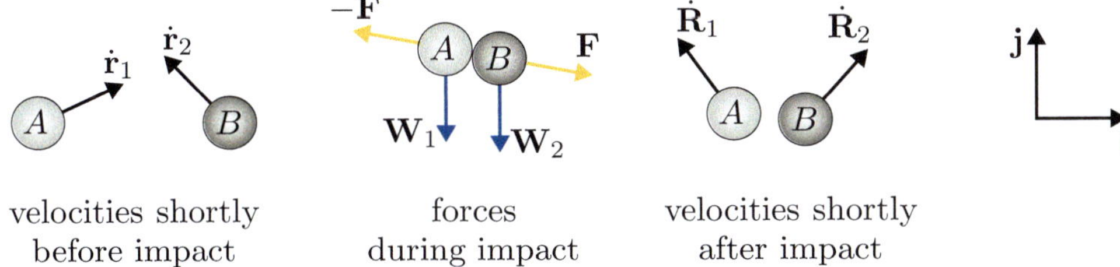

Figure 33   Two balls collide, subject to gravity

In this case there is a net external force, the combined weight of the balls, so $\mathbf{F}^{\mathrm{ext}} = \mathbf{W}_1 + \mathbf{W}_2 = (m_1 + m_2)g(-\mathbf{j})$, which is constant, and equation (26) becomes $-(m_1 + m_2)g\mathbf{j} = \dot{\mathbf{P}}$. The collision occurs over a very small time interval of length T. If we integrate this equation with respect to time over this time interval, we obtain

$$\int_0^T -(m_1 + m_2)g\mathbf{j}\,dt = \int_0^T \dot{\mathbf{P}}\,dt = \big[\mathbf{P}\big]_0^T,$$

and it follows that $-(m_1 + m_2)Tg\mathbf{j} = \mathbf{P}_{\mathrm{after}} - \mathbf{P}_{\mathrm{before}}$. If we suppose that the collision is instantaneous, so $T = 0$ (i.e. if we take the limit as $T \to 0$), then $\mathbf{P}_{\mathrm{after}} = \mathbf{P}_{\mathrm{before}}$; in other words, the linear momentum is conserved.

This result generalises to any n-particle system. Suppose that we have such a system on which the resultant of the external forces is $\mathbf{F}^{\mathrm{ext}}$. Over a very small time interval of length T, we can take $\mathbf{F}^{\mathrm{ext}}$ to be constant, so integrating equation (26) with respect to time gives

$$\int_0^T \mathbf{F}^{\mathrm{ext}}\,dt = \int_0^T \dot{\mathbf{P}}\,dt = \big[\mathbf{P}\big]_0^T,$$

thus $\mathbf{F}^{\mathrm{ext}}T = \mathbf{P}_{\mathrm{after}} - \mathbf{P}_{\mathrm{before}}$. Hence, for an instantaneous collision, with $T = 0$ (i.e. if we take the limit as $T \to 0$), we have $\mathbf{P}_{\mathrm{after}} = \mathbf{P}_{\mathrm{before}}$.

This means that in the presence of external forces, the total linear momentum of an n-particle system may change over time, but the instantaneous collisions of the particles within the system have no effect on the total linear momentum.

Principle of conservation of linear momentum

The total linear momentum of an n-particle system is not affected by (instantaneous) collisions among the particles. Furthermore, in the absence of external forces, the total linear momentum of the system remains constant.

By modelling a rigid body as a system of particles (as we did in Section 2), this principle can be extended to systems involving rigid bodies and/or particles. In particular, it means that if two rigid bodies collide, then the collision causes no change in the total linear momentum of the bodies.

Example 5

A railway truck A of mass $50\,000\,\text{kg}$ rolls down a slight incline and collides with a stationary truck B of mass $30\,000\,\text{kg}$. After the collision, the trucks are coupled together (so A and B have the same velocity).

If the velocity of A immediately before the collision is $2\mathbf{i}$ (in m s^{-1}), what is the combined speed of the trucks immediately after the collision (assuming that the collision is instantaneous)?

Solution

We ignore the motion of the wheels and model the trucks as rigid bodies. Let $v\mathbf{i}$ be the velocity of the trucks after impact. Then, using the notation $\mathbf{P}_{\text{before}}$ and $\mathbf{P}_{\text{after}}$ for the total linear momentum before and after impact, we have

$$\mathbf{P}_{\text{before}} = 50\,000 \times 2\mathbf{i},$$
$$\mathbf{P}_{\text{after}} = (50\,000 + 30\,000) \times v\mathbf{i}.$$

It follows from the principle of conservation of linear momentum that $80\,000v = 100\,000$, so $v = 1.25$. Hence the combined speed after impact is $1.25\,\text{m s}^{-1}$.

Exercise 21

A railway engine of mass M, moving on a straight horizontal track, collides with a stationary truck of mass m. The truck becomes attached to the engine, and both move off together.

Express the speed v of the engine and truck immediately after the collision in terms of M, m and the speed u of the engine immediately prior to the collision.

3.2   Elastic and inelastic collisions

Sometimes in a collision between objects, the total kinetic energy of the system before and after the collision remains the same, that is, energy is conserved. However, in other cases, some of the kinetic energy may be transformed into other forms of energy by the collision. We use different terms for the two types of collision.

Notice that here we use the terms 'elastic' and 'inelastic' in a technical sense that is somewhat different from their everyday usage.

If the kinetic energy of a system is the same before and after a collision within the system, then the collision is said to be **elastic**; otherwise, it is said to be **inelastic**.

Example 6

Suppose that a snooker ball A, of mass m, moving with speed $5\,\mathrm{m\,s^{-1}}$ in a straight line across a smooth snooker table, collides elastically with a similar ball B, of the same mass, at rest. Suppose that A hits B head on, that is, they both move along the same straight line after impact as A was moving along before impact.

Modelling each ball as a particle, predict the speeds of the balls after the collision.

Solution

Define an x-axis along the line of motion of the balls, with positive direction in the direction of motion of A before the collision. Then the velocities of A and B before the collision can be written as $\dot{\mathbf{r}}_1 = \dot{x}_1\mathbf{i}$ and $\dot{\mathbf{r}}_2 = \dot{x}_2\mathbf{i}$, and their velocities after the collision as $\dot{\mathbf{R}}_1 = \dot{X}_1\mathbf{i}$ and $\dot{\mathbf{R}}_2 = \dot{X}_2\mathbf{i}$, where $\mathbf{i}$ is a unit vector in the positive x-direction.

By the principle of conservation of linear momentum, we have

$$m\dot{x}_1\mathbf{i} + m\dot{x}_2\mathbf{i} = m\dot{X}_1\mathbf{i} + m\dot{X}_2\mathbf{i}.$$

Dividing by m and resolving in the $\mathbf{i}$-direction gives

$$\dot{x}_1 + \dot{x}_2 = \dot{X}_1 + \dot{X}_2,$$

which, since $\dot{x}_1 = 5$ and $\dot{x}_2 = 0$, gives

$$\dot{X}_1 + \dot{X}_2 = 5. \tag{27}$$

The kinetic energies of the balls A and B before impact are $\frac{1}{2}m\dot{x}_1^2 = \frac{25}{2}m$ and $\frac{1}{2}m\dot{x}_2^2 = 0$, and the kinetic energies after impact are $\frac{1}{2}m\dot{X}_1^2$ and $\frac{1}{2}m\dot{X}_2^2$. Since the collision is elastic, kinetic energy is conserved, so we have

$$\frac{25}{2}m = \frac{1}{2}m\dot{X}_1^2 + \frac{1}{2}m\dot{X}_2^2,$$

hence

$$\dot{X}_1^2 + \dot{X}_2^2 = 25. \tag{28}$$

From equation (27), $\dot{X}_2 = 5 - \dot{X}_1$. Substituting for $\dot{X}_2$ in equation (28), we obtain

$$\dot{X}_1^2 + (5 - \dot{X}_1)^2 = 25,$$

which simplifies to

$$\dot{X}_1(\dot{X}_1 - 5) = 0.$$

So there are two possibilities: either $\dot{X}_1 = 5$ or $\dot{X}_1 = 0$. Substituting $\dot{X}_1 = 5$ into equation (27) gives $\dot{X}_2 = 0$, which is physically impossible (since B is in front of A). So $\dot{X}_1 = 0$ and $\dot{X}_2 = 5$ (again from equation (27)). In other words, after the collision, ball A comes to rest, while ball B moves across the table with a speed equal to the original speed of ball A.

Exercise 22

A railway truck of mass $15\,000$ kg is moving at $4\,\mathrm{m\,s^{-1}}$ along a straight track and collides with a stationary truck of mass $10\,000$ kg. After impact the trucks move together along the track.

(a) What is the combined speed of the trucks after impact?

(b) Is the collision elastic?

In Example 6 and Exercise 22, the motion before and after impact was along the same straight line. Also, only one object was moving before impact. In the next example there is still only one object moving before impact, but the directions of motion of the two objects after impact are along different straight lines. This sort of collision, referred to as an *oblique* collision, occurs, for example, when one ball strikes a glancing blow on another, as illustrated in Figure 34.

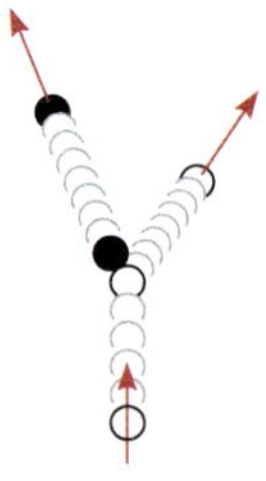

Figure 34 Oblique collision

Example 7

A white snooker ball A, moving parallel to the sides of the table with speed u, collides obliquely with the stationary black ball B (which is equidistant from two of the end pockets), as shown in Figure 35. As a result, the black ball moves off at an angle of $\frac{\pi}{4}$ to the original direction of motion of the white ball, and ends up in the pocket at the top of Figure 35.

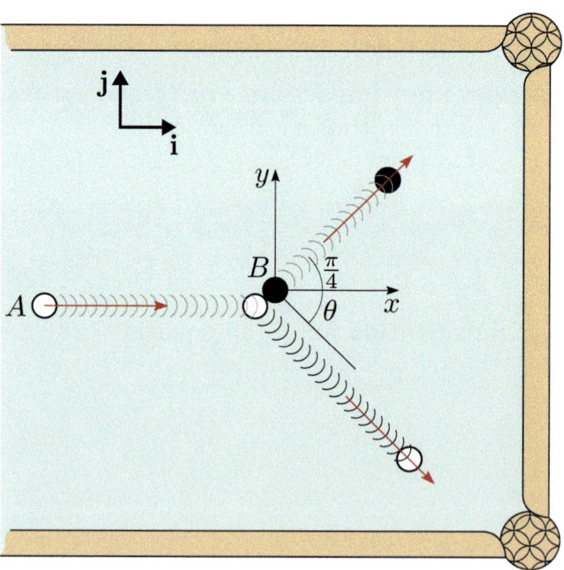

Figure 35   Balls on a snooker table

If the collision is elastic and each ball has mass m, find the speeds of the balls after impact, and decide if the white ball is likely to enter a pocket.

Solution

We choose axes as shown in Figure 35, with origin at the original position of the black ball. Modelling the balls as particles and using $\dot{\mathbf{r}}_1, \dot{\mathbf{r}}_2$ and

$\dot{\mathbf{R}}_1, \dot{\mathbf{R}}_2$ for the velocities of A, B just before and just after impact, respectively, we have

$$\dot{\mathbf{r}}_1 = u\mathbf{i} \quad \text{and} \quad \dot{\mathbf{r}}_2 = \mathbf{0},$$

where $\mathbf{i}$ and $\mathbf{j}$ are Cartesian unit vectors in the positive x- and y-directions, respectively.

We are told that the direction of motion of the black ball after impact is at an angle of $\frac{\pi}{4}$ to the x-axis, so

$$\dot{\mathbf{R}}_2 = \left(|\dot{\mathbf{R}}_2| \cos \tfrac{\pi}{4}\right)\mathbf{i} + \left(|\dot{\mathbf{R}}_2| \sin \tfrac{\pi}{4}\right)\mathbf{j} = \tfrac{1}{\sqrt{2}}|\dot{\mathbf{R}}_2|\mathbf{i} + \tfrac{1}{\sqrt{2}}|\dot{\mathbf{R}}_2|\mathbf{j} = V\mathbf{i} + V\mathbf{j},$$

where $V = |\dot{\mathbf{R}}_2|/\sqrt{2}$. Also, if θ is the angle made with the x-axis by the white ball's direction of motion after impact, then

$$\dot{\mathbf{R}}_1 = \left(|\dot{\mathbf{R}}_1| \cos \theta\right)\mathbf{i} - \left(|\dot{\mathbf{R}}_1| \sin \theta\right)\mathbf{j} = V_x\mathbf{i} - V_y\mathbf{j},$$

We write $-V_y\mathbf{j}$ because we expect the $\mathbf{j}$-component of the white ball to be negative.

where $V_x = |\dot{\mathbf{R}}_1| \cos \theta$ and $V_y = |\dot{\mathbf{R}}_1| \sin \theta$ denote the components of the white ball's velocity after impact.

The principle of conservation of linear momentum gives

$$mu\mathbf{i} = m(V_x\mathbf{i} - V_y\mathbf{j}) + m(V\mathbf{i} + V\mathbf{j}).$$

Resolving in the $\mathbf{i}$- and $\mathbf{j}$-directions in turn, and dividing by m, leads to

$$V_x = u - V \quad \text{and} \quad V_y = V. \tag{29}$$

As the collision is elastic, kinetic energy is conserved, so we have

$$\begin{aligned}
\tfrac{1}{2}mu^2 &= \tfrac{1}{2}m|\dot{\mathbf{R}}_1|^2 + \tfrac{1}{2}m(\sqrt{2}V)^2 \\
&= \tfrac{1}{2}m\left((V_x\mathbf{i} - V_y\mathbf{j}) \cdot (V_x\mathbf{i} - V_y\mathbf{j})\right) + mV^2 \\
&= \tfrac{1}{2}m(V_x^2 + V_y^2) + mV^2.
\end{aligned}$$

If we substitute the values for V_x and V_y obtained in equations (29) and divide by $\tfrac{1}{2}m$, then this equation becomes

$$u^2 = (u - V)^2 + V^2 + 2V^2,$$

which simplifies to $uV = 2V^2$.

We know that $V \neq 0$, therefore $V = \tfrac{1}{2}u$, and equations (29) give $V_x = V_y = \tfrac{1}{2}u$. Hence the speeds of the white ball and the black ball after impact are, respectively,

$$|\dot{\mathbf{R}}_1| = \sqrt{V_x^2 + V_y^2} = \tfrac{1}{\sqrt{2}}u \quad \text{and} \quad |\dot{\mathbf{R}}_2| = \sqrt{2}V = \tfrac{1}{\sqrt{2}}u.$$

Also,

$$\tan \theta = \left(|\dot{\mathbf{R}}_1| \sin \theta\right)/\left(|\dot{\mathbf{R}}_1| \cos \theta\right) = V_y/V_x = 1,$$

hence $\theta = \frac{\pi}{4}$, so the white ball moves off at an angle of $\frac{\pi}{4}$ to its original direction of motion, towards the bottom pocket shown in Figure 35.

However, although the white ball is travelling in the general direction of the bottom pocket, notice from Figure 35 that its diagonal motion does not start at exactly the same point as that of the black ball, but rather starts slightly below and to the left of it. So although the white ball goes close to the bottom pocket, we need to know more about the size of snooker balls, the dimensions of snooker tables and the geometry of snooker table pockets before we can make any firm conclusion about whether it enters the bottom pocket. (Even with all this information, we still could not be sure because our two-particle model takes no account of aspects of the physical situation such as the rolling motion of the balls.)

Exercise 23

You could try an experiment at home with two identical coins and a piece of paper on a flat surface. Draw a circle on the paper round one coin. Place the other coin on the paper a little way from the first coin, and flick it towards the stationary coin in the circle. Measure the angle between the paths of the coins after the collision.

Show that if two identical balls of mass m collide in an elastic collision, where one is initially stationary and the other is travelling with velocity $\mathbf{u}$, then immediately after the collision, either one ball becomes stationary or the balls go off at right angles to each other.

(*Hint*: Write the initial kinetic energy of the ball as $\frac{1}{2}m|\mathbf{u}|^2 = \frac{1}{2}m\mathbf{u} \cdot \mathbf{u}$.)

Exercise 24

A white snooker ball, travelling with speed u, collides with a stationary green ball. As a result, the green ball moves off at an angle of $\frac{\pi}{3}$ to the original direction of motion of the white ball.

Model the balls as particles of equal mass m, and assume that the collision is elastic. Determine the direction of motion of the white ball after the collision. (*Hint*: Use the result of the previous exercise.) Also, find the velocity of each ball after the collision.

Exercise 25

A particle of mass m_1 travels in the positive x-direction with speed u. Another particle of mass m_2 travels in the positive y-direction with the same speed u. The particles collide and then move off together with the same velocity.

Find the velocity of the particles after the collision. Is the collision elastic?

Exercise 26

A particle of mass m_1 collides with a particle of mass m_2. The velocities of the particles are $\mathbf{v}_1$ and $\mathbf{v}_2$, respectively, before impact, and $\mathbf{V}_1$ and $\mathbf{V}_2$, respectively, after impact.

(a) If $m_1 = m_2 = 3$, $\mathbf{v}_1 = 2\mathbf{i}$, $\mathbf{v}_2 = \mathbf{0}$, $\mathbf{V}_1 = \mathbf{i} + \mathbf{j}$ and $\mathbf{V}_2 = \mathbf{i} - \mathbf{j}$, show that the collision is elastic.

(b) If $m_1 = 1$, $m_2 = 3$, $\mathbf{v}_1 = 2\mathbf{i} + \mathbf{j}$, $\mathbf{v}_2 = \mathbf{j}$ and $\mathbf{V}_1 = \mathbf{V}_2 = \frac{1}{2}\mathbf{i} + \mathbf{j}$, show that the collision is inelastic, and find the decrease in kinetic energy due to the collision, assuming that all quantities are measured in SI units.

4 Newton's law of restitution

Here we look at the relationship between the relative velocities of objects before and after collisions.

Although some collisions are elastic, usually we expect there to be some transfer of energy when objects collide, to heat energy or sound energy, for example. So most collisions are inelastic. To obtain information about velocities after an inelastic collision, where kinetic energy is not conserved, we can use *Newton's law of restitution*, a law that is well supported by empirical evidence.

The starting point for this law is the experimental observation that if you drop a ball onto a flat fixed horizontal solid surface, then the height to which the ball rebounds appears to be a fixed fraction of the original height. If the ball is dropped from a height H and rebounds to a height h, then it appears that

$$h = cH,$$

where c is a constant (with $0 \leq c \leq 1$) depending on the material of both the ball and the surface. If $c = 1$, then the ball rebounds to its original height; if $c = 0$, then the ball does not bounce. A steel ball dropped onto a steel plate would provide an example of a collision where c is close to 1; on the other hand, a ball of a material like putty would hardly bounce, so c would be almost 0.

If we model the ball as a particle of mass m, then (relative to a datum at the surface) the potential energy of the ball as it is dropped is mgH, and its potential energy at the top of the first bounce is $mgh = cmgH$. So with each successive bounce, a fixed fraction of the ball's energy is lost (we will see later where it has gone). Let $\mathbf{i}$ be a unit vector in the upwards direction. Suppose that the velocities are $\dot{x}\mathbf{i}$ (with $\dot{x} < 0$) and $\dot{X}\mathbf{i}$ (with $\dot{X} > 0$) just before and after the first impact, as shown in Figure 36. We can apply the law of conservation of mechanical energy (see Unit 9) to the one-dimensional motion of the ball before impact and after impact (but not during impact, as then there is an extra force due to the impact). The law applied to the motion before impact gives

$$\tfrac{1}{2}m\dot{x}^2 = mgH,$$

and after impact it gives

$$\tfrac{1}{2}m\dot{X}^2 = mgh = cmgH.$$

Therefore $\dot{X}^2 = c\dot{x}^2$, so $\dot{X}\mathbf{i} = -\sqrt{c}\,\dot{x}\mathbf{i}$ (since $\dot{X}$ and $\dot{x}$ have opposite signs). We normally replace the constant $\sqrt{c}$ by e, which is known as the *coefficient of restitution* between the ball and the surface, so

$$\dot{X}\mathbf{i} = -e\dot{x}\mathbf{i}.$$

This means that the velocity of the ball after impact is $-e$ times its velocity before impact.

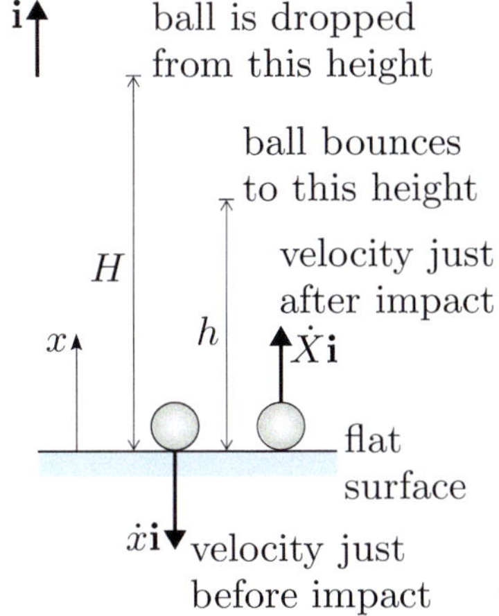

Figure 36   A ball bouncing off a flat surface

This result holds not only for balls bouncing on surfaces, but also for any instantaneous collision between any two objects moving along the same line, in which case it takes the form

$$\text{relative velocity after impact}$$
$$= -e \times \text{relative velocity before impact}. \tag{30}$$

Notice that if $e = 1$, then the speed is the same before and after impact, hence so is the kinetic energy. Therefore $e = 1$ corresponds to an elastic collision, while $e < 1$ corresponds to an inelastic one.

Exercise 27

Suppose that there is a collision between a ball A of mass m_1, which is moving with velocity $\dot{r}_1$, and a ball B of mass m_2, which is moving with velocity $\dot{r}_2$. Suppose also that both balls are moving along the same straight line, and that the coefficient of restitution between the balls is e.

(a) Find the velocities of the balls after impact in terms of the velocities before impact.

(b) Use your results to provide an alternative solution to Example 6.

(c) Show that if the balls coalesce, then $e = 0$.

Exercise 28

A ball of mass m_1 moves with speed u across a smooth table and collides head on with a stationary ball of mass m_2 (so the motion before and after impact takes place along the same straight line).

Given that the coefficient of restitution between the balls is e, and assuming that the balls can be modelled as particles, show that the kinetic energy lost from the system due to the collision is

$$\frac{m_1 m_2 (1 - e^2) u^2}{2(m_1 + m_2)}.$$

Consider a ball bouncing on a smooth fixed horizontal solid surface. When the ball is dropped onto the surface, the only forces acting on the ball at the moment of impact are the weight of the ball and the normal reaction of the surface on the ball. These forces are vertical and act along the line of motion, so their effect changes the velocity of the ball but not its line of motion, which remains vertical.

There is no friction as the surface is smooth.

Suppose now that the ball is not dropped but is thrown so that it strikes the surface at an angle, as in Figure 37. At the moment of impact, the forces acting on the ball are still just the weight and the normal reaction, and these are both vertical. So the impact affects just the vertical component of the ball's velocity – there is no force with a horizontal component to affect the horizontal component of the ball's velocity.

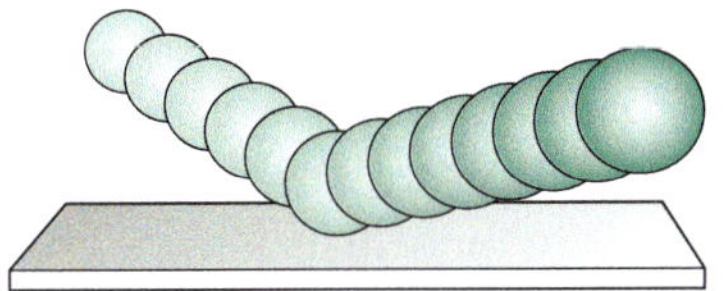

Figure 37   A ball striking a surface at an angle

If the velocities before and after impact are $\dot{\mathbf{r}} = \dot{x}\mathbf{i} + \dot{y}\mathbf{j}$ and $\dot{\mathbf{R}} = \dot{X}\mathbf{i} + \dot{Y}\mathbf{j}$, where $\mathbf{i}$ and $\mathbf{j}$ are horizontal and vertical Cartesian unit vectors, then we have $\dot{X} = \dot{x}$. Also, applying equation (30) to the vertical components of the velocity, we have $\dot{Y} = -e\dot{y}$, where e is the coefficient of restitution between the ball and the surface. The kinetic energy just before impact is $\frac{1}{2}m\left(\dot{x}^2 + \dot{y}^2\right)$, and the kinetic energy just after impact is
$$\tfrac{1}{2}m\left(\dot{X}^2 + \dot{Y}^2\right) = \tfrac{1}{2}m\left(\dot{x}^2 + e^2\dot{y}^2\right).$$

As before, these ideas can be extended to collisions between any two objects, one or both of which may be moving, provided that the areas of contact between the objects during impact are both smooth (i.e. frictionless) and lie on the same tangent plane, as illustrated in Figure 38. In our mathematical models of colliding objects, the objects are often spheres or planes, in which case the proviso about the areas of contact lying on the same tangent plane is automatically satisfied, but the ideas can be applied to collisions between objects of other shapes too. We will also need to assume that neither object is rotating, or at least that any rotation can be ignored – this assumption is reasonable in many cases. (The assumption that an object can be modelled as a particle implies that any rotation of the object can be ignored – see Unit 3.)

As for the ball bouncing obliquely on a surface, only the component of the velocity perpendicular to the tangent plane of contact is affected by the collision. At a point of contact, the direction perpendicular to the common tangent plane is referred to as the **common normal**; this is illustrated in Figure 38. Thus we have the following result, which as we noted earlier is well supported by empirical evidence.

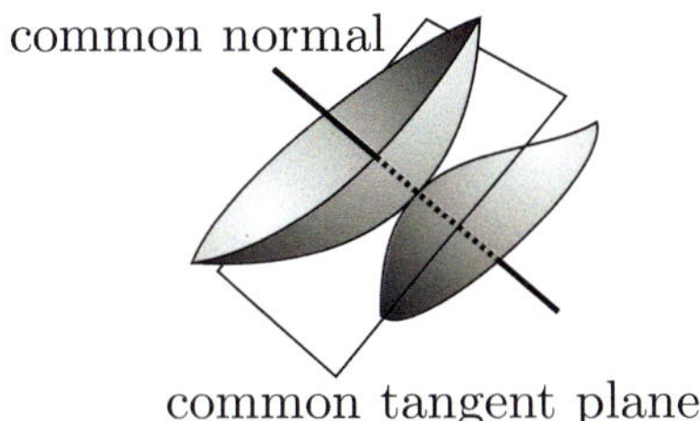

Figure 38   Collision between two smooth objects

If there is significant rotation, then other forces come into play and affect the results.

Newton's law of restitution

In an (instantaneous) collision between two smooth non-rotating objects, where the area of contact at the moment of impact lies on a common tangent plane, the velocities parallel to the tangent plane remain unchanged before and after impact. Also, we have

$$\begin{pmatrix} \text{relative velocity component} \\ \text{parallel to the common} \\ \text{normal after impact} \end{pmatrix} = -e \times \begin{pmatrix} \text{relative velocity component} \\ \text{parallel to the common} \\ \text{normal before impact} \end{pmatrix},$$

where e is the **coefficient of restitution** for a collision between the two objects.

To see how we can make use of this law, suppose that two smooth balls A and B of the same radius, of masses m_1 and m_2, are moving along a smooth horizontal surface before colliding. Assume that any rotation can be ignored. The situation is illustrated in Figure 39, which also shows the common tangent plane at the point of impact and Cartesian unit vectors $\mathbf{i}$ and $\mathbf{j}$ in the direction of the tangent plane and perpendicular to it (in the direction of the common normal). Figure 39 also shows the velocities of the balls before and after impact.

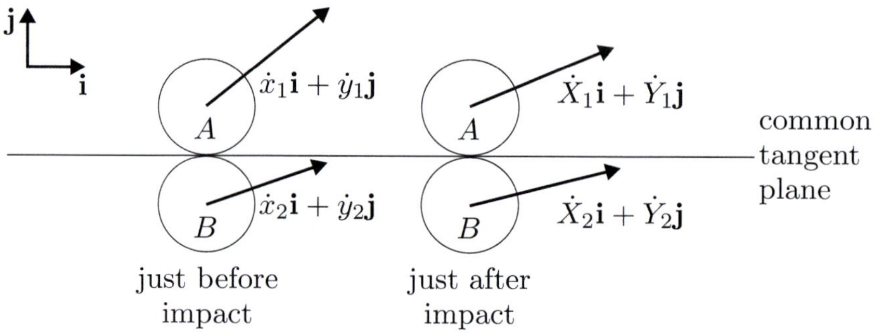

Figure 39   Oblique impact between two balls

Since there are four components of velocity after the collision, we need four equations to determine them. Two equations come from the tangential conditions, one from conservation of linear momentum in the $\mathbf{j}$-direction, and one from Newton's law of restitution.

From the tangential conditions and Newton's law of restitution, we have

$$\dot{x}_1\mathbf{i} = \dot{X}_1\mathbf{i}, \quad \dot{x}_2\mathbf{i} = \dot{X}_2\mathbf{i}, \quad (\dot{Y}_1 - \dot{Y}_2)\mathbf{j} = -e(\dot{y}_1 - \dot{y}_2)\mathbf{j}, \tag{31}$$

where e is the coefficient of restitution for the balls. Also, the principle of conservation of linear momentum gives

$$m_1(\dot{x}_1\mathbf{i} + \dot{y}_1\mathbf{j}) + m_2(\dot{x}_2\mathbf{i} + \dot{y}_2\mathbf{j}) = m_1(\dot{X}_1\mathbf{i} + \dot{Y}_1\mathbf{j}) + m_2(\dot{X}_2\mathbf{i} + \dot{Y}_2\mathbf{j}). \tag{32}$$

Resolving in the $\mathbf{j}$-direction in equations (31) and (32) gives

$$\dot{Y}_1 - \dot{Y}_2 = -e(\dot{y}_1 - \dot{y}_2),$$
$$m_1\dot{y}_1 + m_2\dot{y}_2 = m_1\dot{Y}_1 + m_2\dot{Y}_2.$$

You solved the equivalent equations using xs rather than ys in Exercise 27.

These equations can be solved for $\dot{Y}_1$ and $\dot{Y}_2$ to give

$$\dot{Y}_1 = \frac{(m_1 - em_2)\dot{y}_1 + m_2(1 + e)\dot{y}_2}{m_1 + m_2},$$
$$\dot{Y}_2 = \frac{m_1(1 + e)\dot{y}_1 + (m_2 - em_1)\dot{y}_2}{m_1 + m_2}.$$

So, given e and the velocities before impact, we can find the velocities after impact.

Example 8

Figure 40 shows the head of a golf club A just before and just after it strikes a golf ball B. Just before impact, the ball is stationary and the head of the club is moving horizontally with speed u.

The mass of A is $9m$ and the mass of B is m, the coefficient of restitution between the club and the ball is 0.8, and the face of the club is inclined at an angle α, as shown in the figure.

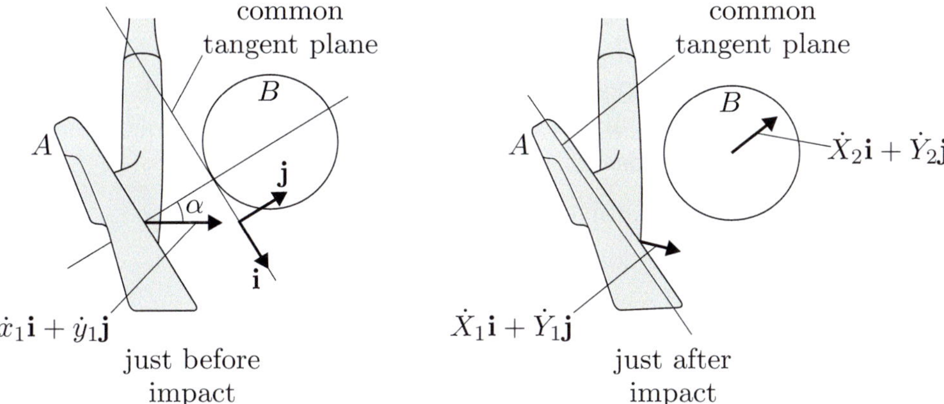

Figure 40 Golf club striking a ball

Model the club head and the ball as smooth objects, and ignore any rotational effects. Estimate the loss of kinetic energy caused by the collision.

The 'spin' imparted to a golf ball by a golf club can be very important, as any golfer will tell you, but we ignore it here.

Solution

Choosing the Cartesian unit vectors as shown in Figure 40, we can write the velocity of A just before impact as

$$\dot{x}_1\mathbf{i} + \dot{y}_1\mathbf{j} = (u\sin\alpha)\mathbf{i} + (u\cos\alpha)\mathbf{j},$$

while the velocity of B just before impact is $\mathbf{0}$.

The common tangent plane is in the $\mathbf{i}$-direction. From the tangential conditions and Newton's law of restitution, we have

$$\dot{X}_1\mathbf{i} = \dot{x}_1\mathbf{i}, \quad \dot{X}_2\mathbf{i} = \dot{x}_2\mathbf{i}, \quad (\dot{Y}_1 - \dot{Y}_2)\mathbf{j} = -e(\dot{y}_1 - \dot{y}_2)\mathbf{j}.$$

Resolving in the $\mathbf{i}$- and $\mathbf{j}$-directions, and substituting in known values, we obtain

$$\dot{X}_1 = \dot{x}_1 = u\sin\alpha,$$
$$\dot{X}_2 = \dot{x}_2 = 0,$$
$$\dot{Y}_1 - \dot{Y}_2 = -0.8(u\cos\alpha - 0) = -0.8u\cos\alpha. \tag{33}$$

Also, the principle of conservation of linear momentum gives

$$9m(\dot{x}_1\mathbf{i} + \dot{y}_1\mathbf{j}) + \mathbf{0} = 9m(\dot{X}_1\mathbf{i} + \dot{Y}_1\mathbf{j}) + m(\dot{X}_2\mathbf{i} + \dot{Y}_2\mathbf{j}),$$

which, resolving in the $\mathbf{j}$-direction, using known values and dividing through by m, gives

$$9u\cos\alpha = 9\dot{Y}_1 + \dot{Y}_2. \tag{34}$$

Solving equations (33) and (34) for $\dot{Y}_1$ and $\dot{Y}_2$, we obtain

$$\dot{Y}_1 = 0.82u\cos\alpha, \quad \dot{Y}_2 = 1.62u\cos\alpha.$$

We now have enough information to be able to compare the kinetic energy just after impact with the kinetic energy just before impact. The kinetic energy of the system just before impact is

$$\tfrac{1}{2}(9m)u^2 = 4.5mu^2.$$

The velocity of the club head just after impact is
$\dot{X}_1\mathbf{i} + \dot{Y}_1\mathbf{j} = (u\sin\alpha)\mathbf{i} + (0.82u\cos\alpha)\mathbf{j}$, and the velocity of the ball just after impact is $\dot{X}_2\mathbf{i} + \dot{Y}_2\mathbf{j} = 0\mathbf{i} + (1.62u\cos\alpha)\mathbf{j}$, so their combined kinetic energy just after impact is

$$\tfrac{1}{2}(9m)(u^2\sin^2\alpha + (0.82)^2u^2\cos^2\alpha) + \tfrac{1}{2}m(1.62)^2u^2\cos^2\alpha$$
$$= (4.5\sin^2\alpha + 4.338\cos^2\alpha)mu^2$$
$$= (4.5 - 0.162\cos^2\alpha)mu^2.$$

So the model estimates that the kinetic energy of the system has decreased by $0.162mu^2\cos^2\alpha$.

In Unit 9 we showed that the total mechanical energy of certain one-particle systems is conserved, but in this section we have shown that the mechanical energy decreases in an inelastic collision. What has happened to this energy?

It is a basic assumption of physics that the total energy of a system is conserved. So either our model is inadequate, or some of the mechanical energy has been converted into another form of energy. Actually, it is a bit of both. During a collision, a little of the mechanical energy is converted into sound energy, and some is converted into heat energy; but much of the 'missing' energy still exists in the form of mechanical energy – it is just that our particle model is too crude to detect it.

For the golf ball, some of the 'missing' energy is in the spinning motion of the ball.

Exercise 29

Such a collision is sometimes called a *plastic* collision.

Two smooth non-rotating balls A and B of equal mass m slide on a frictionless horizontal table and undergo an inelastic collision with coefficient of restitution $e = 0$. Before the collision, A has speed u while B is stationary. After the collision, B moves off at an angle $\frac{\pi}{4}$ measured clockwise from the direction of approach of A. Let $\mathbf{i}$ be a unit vector in the direction of motion of B. Let $\mathbf{j}$ be the unit vector obtained by rotating $\mathbf{i}$ through $\frac{\pi}{2}$ anticlockwise in the plane of the horizontal table.

Find the velocities of both balls after the collision. (*Hint*: The vector $\mathbf{j}$ lies in the common tangent plane for the collision.)

Exercise 30

A rubber ball is dropped from rest onto a horizontal floor, and after bouncing twice it rebounds to half its original height.

Calculate the coefficient of restitution between the ball and the floor.

Learning outcomes

After studying this unit, you should be able to:

- find the centre of mass of a system of particles, a lamina, and some homogeneous rigid bodies
- determine the motion of the centre of mass of a system of particles
- use the centre of mass to calculate the potential energy of a homogeneous rigid body
- apply the principle of conservation of linear momentum to collisions between objects
- use the change in kinetic energy during a collision to determine whether or not the collision is elastic
- apply Newton's law of restitution to collisions between objects.

Solutions to exercises

Solution to Exercise 1

(a) Resolving the forces into components gives $\mathbf{T}_2 = -|\mathbf{T}_2|\mathbf{i}$ and $\mathbf{T}_4 = -|\mathbf{T}_4|\mathbf{i}$. Resolving equations (1) and (2) in the $\mathbf{i}$-direction then gives

$$m_1 g - |\mathbf{T}_2| = m_1 \ddot{x}_1,$$
$$m_2 g - |\mathbf{T}_4| = m_2 \ddot{x}_2.$$

To progress further, we need to relate $\ddot{x}_1$ and $\ddot{x}_2$, and to relate $|\mathbf{T}_2|$ and $|\mathbf{T}_4|$.

Using the assumption that the cable is inextensible gives $x_1 + x_2 = \text{constant}$, as quoted in the question. Differentiating this twice with respect to t gives the desired relation between the accelerations of the lift and the counterweight: $\ddot{x}_1 = -\ddot{x}_2$.

To find relationships between the magnitudes of the tension forces, we use the assumptions that the cable is a model string and the pulley is a model pulley to obtain $|\mathbf{T}_2| = |\mathbf{T}_1|$ (model string), $|\mathbf{T}_1| = |\mathbf{T}_3|$ (model pulley) and $|\mathbf{T}_3| = |\mathbf{T}_4|$ (model string again). Putting these together gives $|\mathbf{T}_2| = |\mathbf{T}_4|$, which we can substitute into the above equations to obtain

$$m_1 g - |\mathbf{T}_2| = -m_1 \ddot{x}_2,$$
$$m_2 g - |\mathbf{T}_2| = m_2 \ddot{x}_2.$$

Eliminating $|\mathbf{T}_2|$ by subtracting the first equation from the second gives

$$m_2 g - m_1 g = m_2 \ddot{x}_2 + m_1 \ddot{x}_2.$$

Rearrangement gives the acceleration of the lift:

$$\ddot{x}_2 = \frac{(m_2 - m_1)g}{m_1 + m_2}.$$

(b) Substituting for $\ddot{x}_2$ in equation (2) gives

$$m_2 g\, \mathbf{i} + \mathbf{T}_4 = m_2 \left(\frac{(m_2 - m_1)g}{m_1 + m_2} \right) \mathbf{i}.$$

Rearranging gives

$$\mathbf{T}_4 = m_2 g \left(\frac{m_2 - m_1}{m_1 + m_2} - 1 \right) \mathbf{i}$$
$$= m_2 g \left(\frac{m_2 - m_1 - m_1 - m_2}{m_1 + m_2} \right) \mathbf{i}$$
$$= -\frac{2 m_1 m_2 g}{m_1 + m_2} \mathbf{i}.$$

From part (a) we have $|\mathbf{T}_1| = |\mathbf{T}_2| = |\mathbf{T}_3| = |\mathbf{T}_4|$, so the magnitude of the tension in the cable is $2 m_1 m_2 g/(m_1 + m_2)$.

(c) Resolving equation (3) in the **i**-direction, where $\mathbf{S} = |\mathbf{S}|(-\mathbf{i})$, gives

$$-|\mathbf{S}| + |\mathbf{T}_1| + |\mathbf{T}_3| = 0.$$

From part (a) we have $|\mathbf{T}_1| = |\mathbf{T}_2| = |\mathbf{T}_3| = |\mathbf{T}_4|$, so

$$|\mathbf{S}| = 2|\mathbf{T}_2| = \frac{4m_1 m_2 g}{m_1 + m_2}.$$

(d) Substituting the given values $m_1 = 1000$ and $m_2 = 1065$ into the equation for the acceleration of the lift in part (a) gives

$$\ddot{x}_2 = 65g/2065 \simeq 0.31.$$

The speed of the lift after it has travelled $100\,\mathrm{m}$ can be calculated using the constant acceleration formula $v^2 = v_0^2 + 2a_0(x - x_0)$, where v_0 is the initial speed of the lift (zero in this case). Substituting in the values gives

$$v^2 \simeq 0^2 + 2 \times 0.31 \times 100 = 62.$$

So the speed of the lift after travelling $100\,\mathrm{m}$ is $\sqrt{62} \simeq 7.9\,\mathrm{m\,s}^{-1}$.

This speed is more than fast enough to give the occupant a nasty bump as the lift hits the floor, despite the acceleration being quite small! This explains why such a design is impractical: a lift that travels with constant acceleration (either up or down) could be quite dangerous.

This constant acceleration formula was derived in Unit 3, Subsection 1.2.

Solution to Exercise 2

The internal forces are $\mathbf{T}_1$, $\mathbf{T}_2$, $\mathbf{T}_3$, $\mathbf{T}_4$, $\mathbf{T}_5$, $\mathbf{T}_6$, $\mathbf{T}_7$ and $\mathbf{T}_8$. The external forces are $\mathbf{S}_1$, $\mathbf{S}_2$, $\mathbf{W}_1$ and $\mathbf{W}_2$.

The total force is $\mathbf{F} = \mathbf{S}_1 + \mathbf{S}_2 + \mathbf{W}_1 + \mathbf{W}_2$.

Solution to Exercise 3

Since the radii of the Earth and the Moon are small compared with their distance apart, we can model each as a particle so that we have a two-particle system. The system is pictured in the diagram below (not to scale!).

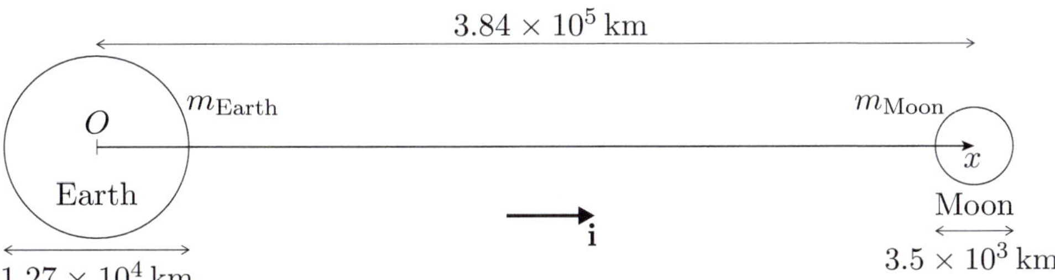

If we choose an x-axis joining the centre of the Earth to the centre of the Moon, as shown in the figure, with the centre of the Earth at the origin, then using the definition of centre of mass we have, working in kilometres,

$$\mathbf{r}_G = \frac{m_{\text{Earth}}\mathbf{r}_{\text{Earth}} + m_{\text{Moon}}\mathbf{r}_{\text{Moon}}}{m_{\text{Earth}} + m_{\text{Moon}}}$$

$$= \frac{(5.97 \times 10^{24}) \times \mathbf{0} + (7.34 \times 10^{22}) \times (3.84 \times 10^5)\mathbf{i}}{5.97 \times 10^{24} + 7.34 \times 10^{22}}$$

$$\simeq 4660\mathbf{i}.$$

So the distance from the centre of the Earth to the centre of mass of the Earth/Moon system is about $4660\,\text{km}$. But the radius of the Earth is about $6350\,\text{km}$, so the centre of mass of this system is about $1690\,\text{km}$ *below* the Earth's surface.

Solution to Exercise 4

(a) The force diagram is shown below.

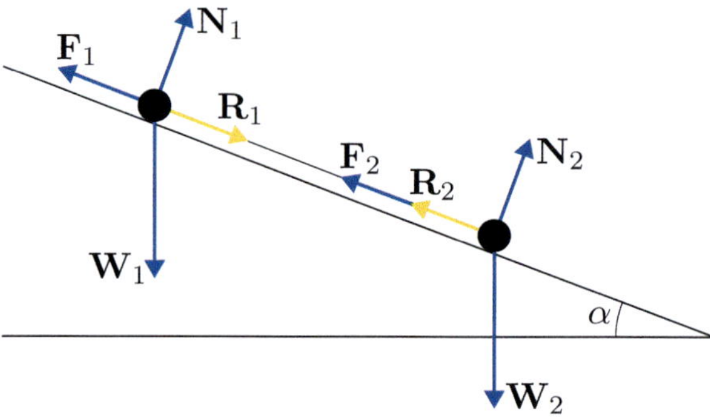

The only internal forces are those due to the rod, $\mathbf{R}_1$ and $\mathbf{R}_2$. The remaining forces – the weights $\mathbf{W}_1$ and $\mathbf{W}_2$, the normal reaction forces $\mathbf{N}_1$ and $\mathbf{N}_2$, and the friction forces $\mathbf{F}_1$ and $\mathbf{F}_2$ – are all external forces.

(b) We define axes as shown below, with origin at the initial position of the upper particle.

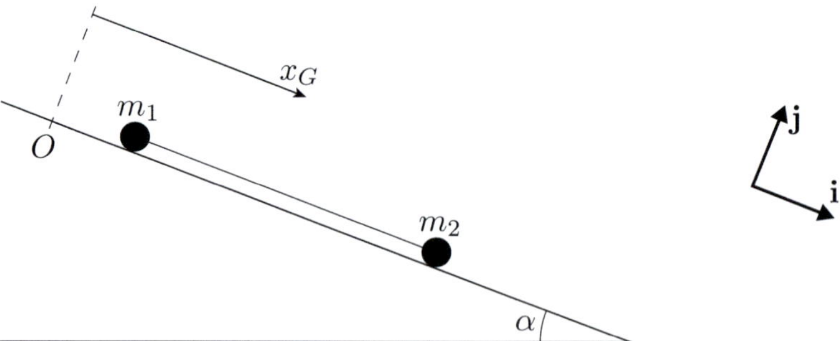

Equation (10), $\mathbf{F}^{\text{ext}} = M\ddot{\mathbf{r}}_G$, gives the equation of motion

$$\mathbf{W}_1 + \mathbf{W}_2 + \mathbf{N}_1 + \mathbf{N}_2 + \mathbf{F}_1 + \mathbf{F}_2 = (m_1 + m_2)\ddot{\mathbf{r}}_G,$$

where $\mathbf{r}_G = x_G\mathbf{i}$ is the position vector of the centre of mass relative to O. We have

$$\begin{aligned}
\mathbf{W}_1 &= m_1 g(\sin\alpha\,\mathbf{i} - \cos\alpha\,\mathbf{j}),\\
\mathbf{W}_2 &= m_2 g(\sin\alpha\,\mathbf{i} - \cos\alpha\,\mathbf{j}),\\
\mathbf{N}_1 &= |\mathbf{N}_1|\mathbf{j}, \quad \mathbf{N}_2 = |\mathbf{N}_2|\mathbf{j},\\
\mathbf{F}_1 &= -|\mathbf{F}_1|\mathbf{i}, \quad \mathbf{F}_2 = -|\mathbf{F}_2|\mathbf{i}.
\end{aligned}$$

Also, $|\mathbf{F}_1| = \mu'|\mathbf{N}_1|$ and $|\mathbf{F}_2| = \mu'|\mathbf{N}_2|$.

Thus the equation of motion becomes

$$\begin{aligned}
m_1 g(\sin\alpha\,\mathbf{i} - \cos\alpha\,\mathbf{j}) &+ m_2 g(\sin\alpha\,\mathbf{i} - \cos\alpha\,\mathbf{j})\\
&+ |\mathbf{N}_1|\mathbf{j} + |\mathbf{N}_2|\mathbf{j} - \mu'|\mathbf{N}_1|\mathbf{i} - \mu'|\mathbf{N}_2|\mathbf{i} = (m_1 + m_2)\ddot{x}_G\mathbf{i}.
\end{aligned}$$

Resolving in the $\mathbf{i}$- and $\mathbf{j}$-directions in turn gives the two equations

$$\begin{aligned}
(m_1 + m_2)g\sin\alpha - \mu'(|\mathbf{N}_1| + |\mathbf{N}_2|) &= (m_1 + m_2)\ddot{x}_G,\\
-(m_1 + m_2)g\cos\alpha + |\mathbf{N}_1| + |\mathbf{N}_2| &= 0.
\end{aligned}$$

Hence, from the second equation, we have

$$|\mathbf{N}_1| + |\mathbf{N}_2| = (m_1 + m_2)g\cos\alpha.$$

Substituting this into the first equation gives

$$(m_1 + m_2)g(\sin\alpha - \mu'\cos\alpha) = (m_1 + m_2)\ddot{x}_G,$$

which simplifies to

$$\ddot{x}_G = g(\sin\alpha - \mu'\cos\alpha).$$

This is the required equation of motion for the centre of mass.

(c) Changing the rod to a light spring would change only the internal forces; the external forces would remain the same. Similarly, removing the rod would not change the external forces. So in both cases the equation of motion of the centre of mass will be as in part (b).

(d) You saw in Unit 3 (e.g. Example 6 and Exercise 19) that the equation of motion of a particle sliding down an inclined plane is

$$a = g(\sin\alpha - \mu'\cos\alpha),$$

where a is the acceleration of the particle and α is the angle of incline. This is identical to the equation of motion for the centre of mass found in part (b) with $\ddot{x}_G$ replaced by a. Thus the motion of the centre of mass of a two-particle system sliding down an inclined plane is the same as the motion of a single particle sliding down the plane.

Solution to Exercise 5

The equation $\mathbf{P} = 2m\ddot{x}_G\mathbf{i}$ still holds, so $\ddot{x}_G = |\mathbf{P}|/(2m)$. But now d is constant, so $x_1 = x_G - d$ and $x_2 = x_G + d$ give $\ddot{x}_1 = \ddot{x}_G = \ddot{x}_2$. Hence A, B and G all move with the same fixed acceleration $|\mathbf{P}|/(2m)$. The system moves as a rigid body and there is no oscillatory motion.

We denote the forces on particles A and B due to the model rod by $\mathbf{R}_1$ and $\mathbf{R}_2$. Applying Newton's second law to particle A gives

$$\mathbf{R}_1 = m\ddot{x}_1\mathbf{i} = m\ddot{x}_G\mathbf{i} = \tfrac{1}{2}|\mathbf{P}|\mathbf{i}.$$

Also, since $\mathbf{R}_2 = -\mathbf{R}_1$, we have $\mathbf{R}_2 = -\tfrac{1}{2}|\mathbf{P}|\mathbf{i}$. So the forces exerted by the rod on A and B have half the magnitude of $\mathbf{P}$ and are directed from each particle towards the centre of the rod.

Solution to Exercise 6

There are no external forces, so from equation (10) we have $\ddot{\mathbf{r}}_G = \mathbf{0}$, and it follows that $\dot{x}_G = c$ (a constant). Now

$$x_G = \frac{mx_1 + mx_2}{m + m} = \frac{x_1 + x_2}{2},$$

where $\mathbf{r}_1 = x_1\mathbf{i}$ and $\mathbf{r}_2 = x_2\mathbf{i}$ are the position vectors of A and B. So

$$\dot{x}_G = \frac{\dot{x}_1 + \dot{x}_2}{2} = c.$$

Initially, we have $\dot{x}_1 = 0$ and $\dot{x}_2 = 6$, so $c = 3$. Integrating the equation $\dot{x}_G = 3$ with respect to time gives $x_G = 3t + 1.5$ (because the centre of mass is initially at $x = 1.5$). Substituting $t = 0.5$ into this gives $x_G = (3 \times 0.5) + 1.5 = 3$ and $x_2 = 3$, therefore we must also have $x_1 = 3$, that is, the particles collide at $t = 0.5$.

Solution to Exercise 7

Not a chance! If we consider the ship and the alien as a two-particle system, then from Newton's third law, any force that the alien exerts on the ship is met by an equal and opposite force from the ship on the alien. In other words, nothing that the alien does can affect the external forces acting on the system, so he cannot change the motion of the centre of mass of the two-particle system.

Solution to Exercise 8

We can use notation similar to that of Example 2, using $x_G\mathbf{i}$ as the position vector of the centre of mass G (which is midway between A and B), and $x_1\mathbf{i}, x_2\mathbf{i}$ as the position vectors of A and B, all with respect to a horizontal x-axis with origin at the position of A before the force is applied.

As in Example 2, the motion is in a horizontal plane, so the vertical forces must sum to zero, that is, $\mathbf{N}_1 = -\mathbf{W}_1$ and $\mathbf{N}_2 = -\mathbf{W}_2$. So from equation (10) we have

$$\mathbf{P} = \mathbf{F}^{\text{ext}} = (m_1 + m_2)\ddot{\mathbf{r}}_G = 100\ddot{x}_G\mathbf{i}.$$

The string is taut and inextensible, so A, B and G have the same acceleration, and in particular $\ddot{x}_2\mathbf{i} = \ddot{x}_G\mathbf{i}$. Let $\mathbf{T}$ be the force exerted by the string on B. Then applying Newton's second law to B gives

$$\mathbf{P} + \mathbf{T} = m_2\ddot{x}_2\mathbf{i} = 50\ddot{x}_G\mathbf{i} = \tfrac{1}{2}\mathbf{P}.$$

It follows that $\mathbf{T} = -\tfrac{1}{2}\mathbf{P}$, so if the string is not to break, the magnitude of $\mathbf{P}$ must not exceed $20\,\text{N}$.

Solution to Exercise 9

Choose the wire to lie along the x-axis, and suppose that the particles are moving in the positive direction (left to right). As shown in the figure below, the forces acting on the particles are the weights $\mathbf{W}_1$ and $\mathbf{W}_2$, the normal reactions of the wire $\mathbf{N}_1$ and $\mathbf{N}_2$, and friction forces $\mathbf{F}_1$ and $\mathbf{F}_2$ (which act at right angles to the other forces).

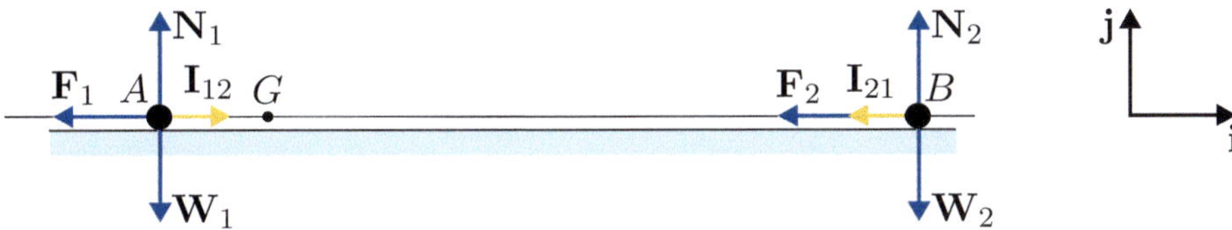

We have

$$\mathbf{N}_1 = -\mathbf{W}_1, \quad \mathbf{N}_2 = -\mathbf{W}_2,$$
$$|\mathbf{F}_1| = \mu'|\mathbf{N}_1|, \quad |\mathbf{F}_2| = \mu'|\mathbf{N}_2|.$$

If particle A has mass m_1 and particle B has mass m_2, then the friction force on A is $\mathbf{F}_1 = -\mu'm_1g\mathbf{i}$ and the friction force on B is $\mathbf{F}_2 = -\mu'm_2g\mathbf{i}$. Since the normal reactions and weights balance each other, the total external force is

$$\mathbf{F}^{\text{ext}} = \mathbf{F}_1 + \mathbf{F}_2 = -\mu'(m_1 + m_2)g\mathbf{i}.$$

Hence, by equation (10), we have

$$-\mu'(m_1 + m_2)g\mathbf{i} = (m_1 + m_2)\ddot{\mathbf{r}}_G,$$

giving $\ddot{\mathbf{r}}_G = -\mu'g\mathbf{i} = -0.4g\mathbf{i}$. So the centre of mass accelerates at $0.4g\,\text{m s}^{-2}$ in the direction opposite to that of the motion of the particles. (Notice that the masses of the particles do not affect this result.)

Solution to Exercise 10

We choose the x-axis to lie along the track in the direction of the external force – see the figure below.

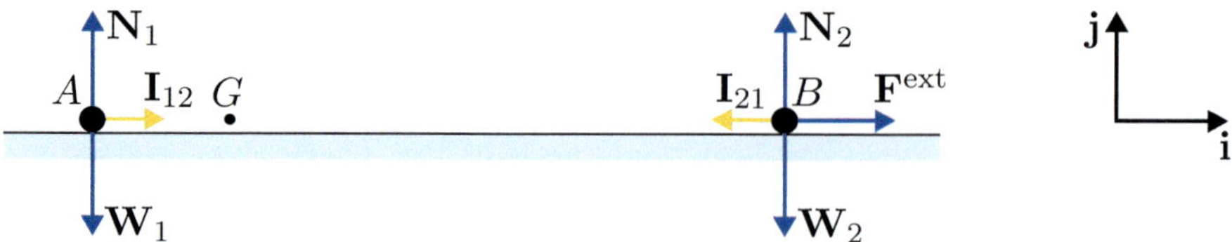

Using equation (10) with $\mathbf{F}^{\text{ext}} = 20\mathbf{i}$ and total mass $M = 5$, we have $20\mathbf{i} = 5\ddot{\mathbf{r}}_G$, so $\ddot{\mathbf{r}}_G = 4\mathbf{i}$. Therefore the common acceleration of the particles is $4\,\mathrm{m\,s}^{-2}$.

Applying Newton's second law to the first particle, we obtain

$$\text{internal force} = 4 \times 4\mathbf{i} = 16\mathbf{i},$$

so the internal forces have magnitude $16\,\mathrm{N}$.

Solution to Exercise 11

(a) If we let $\mathbf{i}$ denote a unit vector in the positive x-direction, then we have particles of mass m at the points $\mathbf{r}_1 = \mathbf{0}$, $\mathbf{r}_2 = 4\mathbf{i}$ and $\mathbf{r}_3 = 6\mathbf{i}$, relative to the origin. From equation (15) we have

$$\mathbf{r}_G = \frac{m\mathbf{r}_1 + m\mathbf{r}_2 + m\mathbf{r}_3}{m + m + m} = \frac{\mathbf{0} + 4\mathbf{i} + 6\mathbf{i}}{3} = \tfrac{10}{3}\mathbf{i}.$$

(b) Similarly,

$$\mathbf{r}_G = \frac{\mathbf{0} + \mathbf{i} + 4\mathbf{i} + 5\mathbf{i} + 6\mathbf{i}}{5} = \tfrac{16}{5}\mathbf{i}.$$

Solution to Exercise 12

(a) Let $\mathbf{i}$ be a unit vector in the positive x-direction. The most obvious way to calculate the centre of mass is to use equation (15), giving

$$\mathbf{r}_G = \frac{m \times (-2\mathbf{i}) + m \times \mathbf{0} + m \times 2\mathbf{i} + m \times 4\mathbf{i}}{m + m + m + m} = \frac{4m\mathbf{i}}{4m},$$

so $\mathbf{r}_G = \mathbf{i}$.

Alternatively, in this case the centre of mass can also be found by inspection, since the system is symmetric about the point $(1, 0)$. So this is the centre of mass, that is, $\mathbf{r}_G = \mathbf{i}$.

(b) To find the position vector of the centre of mass, relative to the origin, consider Cartesian unit vectors $\mathbf{i}$ and $\mathbf{j}$ in the positive x- and y-directions. Now use equation (15) to obtain

$$\mathbf{r}_G = \frac{m\mathbf{r}_1 + m\mathbf{r}_2 + m\mathbf{r}_3}{m + m + m} = \frac{\mathbf{0} + 2\mathbf{i} + (\mathbf{i} + \sqrt{3}\mathbf{j})}{3} = \mathbf{i} + \tfrac{1}{\sqrt{3}}\mathbf{j}.$$

Alternatively, you could have used symmetry. In this case notice that the particles are at the corners of an equilateral triangle (Pythagoras' theorem tells us that the sloping sides are each of length 2). So the centre of mass is at the geometric centre of the triangle.

If you know that the geometric centre of an equilateral triangle is one-third of the way up the perpendicular from a base to an apex, then you can write down the equation immediately as

$$\mathbf{r}_G = \mathbf{i} + \tfrac{1}{3}\left(\sqrt{3}\mathbf{j}\right) = \mathbf{i} + \tfrac{1}{\sqrt{3}}\mathbf{j}.$$

Solution to Exercise 13

The squares are of equal size and hence of equal mass, m say, with centres of mass at

$$\mathbf{r}_{G_1} = \tfrac{1}{2}a\mathbf{i} + \tfrac{3}{2}a\mathbf{j}, \quad \mathbf{r}_{G_2} = \tfrac{1}{2}a\mathbf{i} + \tfrac{1}{2}a\mathbf{j}, \quad \mathbf{r}_{G_3} = \tfrac{3}{2}a\mathbf{i} + \tfrac{1}{2}a\mathbf{j}.$$

Hence, using equation (15), the centre of mass of the whole shape is at position vector

$$\mathbf{r}_G = \frac{m\left(\tfrac{1}{2}a\mathbf{i} + \tfrac{3}{2}a\mathbf{j}\right) + m\left(\tfrac{1}{2}a\mathbf{i} + \tfrac{1}{2}a\mathbf{j}\right) + m\left(\tfrac{3}{2}a\mathbf{i} + \tfrac{1}{2}a\mathbf{j}\right)}{3m} = \tfrac{5}{6}a\mathbf{i} + \tfrac{5}{6}a\mathbf{j},$$

which, as expected, is the same as the result obtained in Example 3.

Solution to Exercise 14

The uniform density and constant cross-sectional area mean that the problem reduces to a two-dimensional one. One way to find the centre of mass of the two-dimensional cross-section is to divide it into triangles as in Figure 26(b), and to find the centre of mass of each triangle (see the figure below).

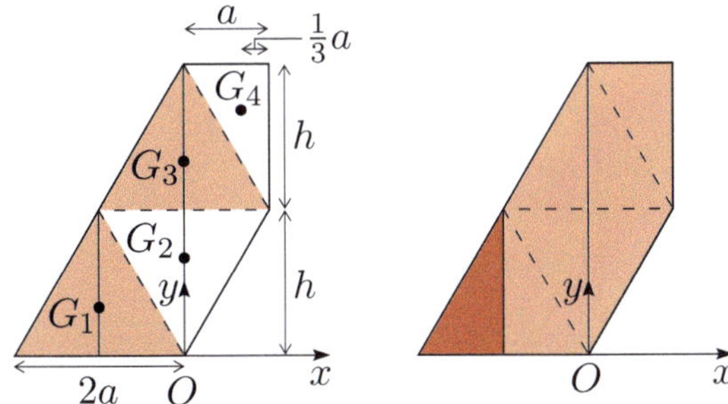

Relative to the origin at O shown above, the position vectors of the centres of mass G_1, G_2, G_3 and G_4 are

$$\mathbf{r}_{G_1} = -a\mathbf{i} + \tfrac{1}{3}h\mathbf{j}, \quad \mathbf{r}_{G_2} = \tfrac{2}{3}h\mathbf{j}, \quad \mathbf{r}_{G_3} = \tfrac{4}{3}h\mathbf{j}, \quad \mathbf{r}_{G_4} = \tfrac{2}{3}a\mathbf{i} + \tfrac{5}{3}h\mathbf{j},$$

where h is half the height of the sculpture (as shown), and $\mathbf{i}$ and $\mathbf{j}$ are Cartesian unit vectors in the positive x- and y-directions.

If we model the equilateral triangles as particles of mass m (say) at G_1, G_2 and G_3, then the fourth triangle can be modelled as a particle of mass $\frac{1}{2}m$ at G_4. Then equation (15) gives the centre of mass of the cross-section as

$$\mathbf{r}_G = \frac{m(-a\mathbf{i} + \frac{1}{3}h\mathbf{j}) + m(\frac{2}{3}h\mathbf{j}) + m(\frac{4}{3}h\mathbf{j}) + \frac{1}{2}m(\frac{2}{3}a\mathbf{i} + \frac{5}{3}h\mathbf{j})}{m + m + m + \frac{1}{2}m}$$

$$= \frac{-\frac{2}{3}a\mathbf{i} + \frac{19}{6}h\mathbf{j}}{\frac{7}{2}}$$

$$= -\frac{4}{21}a\mathbf{i} + \frac{19}{21}h\mathbf{j}.$$

Since $\mathbf{r}_G$ lies to the left of O, over the base, we find that the sculpture will not topple over.

Alternatively, you could notice that the lightly shaded part of the cross-section on the right above is symmetric about the y-axis, so its centre of mass will lie on the y-axis. Hence the whole cross-section, with the darker shaded triangle added, must have its centre of mass to the left of O, so the sculpture will not topple over.

Solution to Exercise 15

The area integral is

$$\int_S yf \, dA = f \int_{x=-a}^{x=a} \left(\int_{y=0}^{y=\sqrt{a^2-x^2}} y \, dy \right) dx$$

$$= f \int_{x=-a}^{x=a} \left[\tfrac{1}{2}y^2 \right]_{y=0}^{y=\sqrt{a^2-x^2}} dx$$

$$= \tfrac{1}{2} f \int_{-a}^{a} (a^2 - x^2) \, dx$$

$$= \tfrac{1}{2} f \left[a^2 x - \tfrac{1}{3}x^3 \right]_{-a}^{a}$$

$$= \tfrac{2}{3} f a^3,$$

and since $M = \frac{1}{2}\pi a^2 f$, $y_G = 4a/3\pi$ as in Example 4.

Solution to Exercise 16

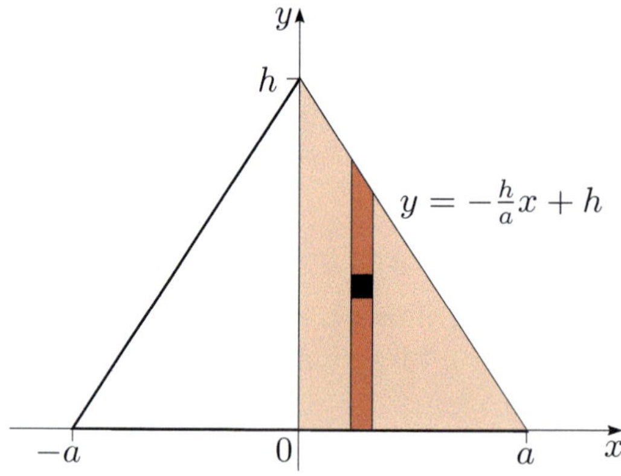

We have $x_G = 0$ by symmetry. Also by symmetry, the y-coordinates of the centres of mass of the two triangles in the positive and negative x-regions are the same (and equal to the y-coordinate of the centre of mass of the whole triangle), so we only need to find y_G for the triangle in the positive x-region. This is given by

$$y_G = \frac{1}{M} f \int_{x=0}^{x=a} \left(\int_{y=0}^{y=-\frac{h}{a}x+h} y\, dy \right) dx,$$

where the mass M of the half-triangle is $\frac{1}{2}ahf$. Thus

$$y_G = \frac{2}{ah} \int_{x=0}^{x=a} \frac{1}{2} \left(-\frac{h}{a}x + h \right)^2 dx = \tfrac{1}{3}h.$$

Note that
$$\int (\alpha x + \beta)^2 \, dx = \frac{(\alpha x + \beta)^3}{3\alpha} + C.$$

Solution to Exercise 17

Taking a datum at ground level, the height of the centre of mass above the datum is $2h$ when upright and $\frac{1}{2}h$ when lying down. So the change in potential energy is $Mg \times 2h - Mg \times \frac{1}{2}h = \frac{3}{2}Mgh$.

Solution to Exercise 18

Let the x-axis point downwards, and let $\mathbf{i}$ be a unit vector in this direction. The total external force on the ball is its weight $Mg\mathbf{i}$, so from equation (19), we have

$$Mg\mathbf{i} = M\ddot{\mathbf{r}}_G,$$

which gives $\ddot{\mathbf{r}}_G = g\mathbf{i}$, so the centre of mass of the ball accelerates downwards with the acceleration due to gravity. We have $\mathbf{r}_G = x_G\mathbf{i}$ and

$$\ddot{x}_G = g.$$

So the acceleration is constant and we may use the constant acceleration formula $v^2 = v_0^2 + 2a_0(x - x_0)$ from Unit 3. In this case we have

$$\dot{x}_G^2 = \dot{x}_G(0)^2 + 2gh = 2gh,$$

since the ball starts from rest. So the speed of the ball after falling a distance h is $\sqrt{2gh}$.

(Notice that these equations are exactly what we would have obtained if we had modelled the ball as a particle, using the energy methods of Unit 9.)

The change in the potential energy is $-Mgh$.

Solution to Exercise 19

The solid shape is composed of two cylinders with radii a and $3a$, each with thickness $2a$. The centre of mass lies on the axis of symmetry (the x-axis in Figure 30). The positions of the centres of mass of the two cylinders, G_1 and G_2, relative to O are given by

$$\mathbf{r}_{G_1} = a\mathbf{i} \quad \text{and} \quad \mathbf{r}_{G_2} = 3a\mathbf{i}.$$

The volume of the smaller cylinder is given by

$$V_1 = \pi a^2 \times 2a = 2\pi a^3,$$

while the volume of the larger cylinder is given by

$$V_2 = \pi(3a)^2 \times 2a = 18\pi a^3.$$

If the density of the material is f, then the corresponding masses are

$$m_1 = 2\pi f a^3 \quad \text{and} \quad m_2 = 18\pi f a^3.$$

From equation (15) we have

$$\mathbf{r}_G = \frac{2\pi f a^3(a\mathbf{i}) + 18\pi f a^3(3a\mathbf{i})}{20\pi f a^3} = \tfrac{14}{5}a\mathbf{i}.$$

The centre of mass is a little to the left of G_2, as we would expect. The object will not topple over because $14a/5 > 2a$, so the centre of mass is to the right of the point P in Figure 30 (the leftmost point of contact with the surface).

Solution to Exercise 20

(a) The volume V of a pyramid is given by

$$V = \tfrac{1}{3} \times \text{base area} \times \text{vertical height},$$

so in this case we have (in m^3)

$$V = \tfrac{1}{3} \times (230)^2 \times 147 = 2\,592\,100.$$

Hence the total mass (in kg) is approximately

$$2\,592\,100 \times 2500 \simeq 6.48 \times 10^9.$$

(b) The centre of mass is a quarter of the height above the base, which in this case is $h/4$. The total energy required to lift all the stones from ground level is (in J)

$$Mg(h/4) = (6.48 \times 10^9) \times 9.81 \times 147/4$$
$$\simeq 2.34 \times 10^{12}.$$

(c) According to the question, the energy that a man can expend in a day is $1000g$ J. So the total number of days required by 1000 men is approximately

$$\frac{2.34 \times 10^{12}}{1000 \times 9.81 \times 1000} \simeq 2.4 \times 10^5.$$

A gang of 1000 men would therefore take about 652 years. (This figure is certainly a low estimate, since it takes no account of friction in the system used to lift the stones, or the time taken to construct the lifting mechanism. However, we can be reasonably sure that it would have taken 1000 men more than 600 years to lift the stones to the appropriate heights, which gives some idea of the scale of the enterprise.)

Solution to Exercise 21

If we take $\mathbf{i}$ to be a unit vector in the direction of motion, then the total linear momentum of the system just before the collision is $\mathbf{P}_{\text{before}} = Mu\mathbf{i} + m\mathbf{0} = Mu\mathbf{i}$, and the total linear momentum just after the collision is $\mathbf{P}_{\text{after}} = (M + m)v\mathbf{i}$. If the collision is instantaneous, then the principle of conservation of linear momentum tells us that

$$Mu\mathbf{i} = (M + m)v\mathbf{i}.$$

Resolving in the $\mathbf{i}$-direction and rearranging, we obtain

$$v = Mu/(M + m).$$

Solution to Exercise 22

(a) We define an x-axis along the track in the direction of motion of the first truck before impact. Let the trucks have masses m_1 and m_2, and velocities before impact $\dot{\mathbf{r}}_1 = \dot{x}_1\mathbf{i}$ and $\dot{\mathbf{r}}_2 = \dot{x}_2\mathbf{i}$. Then the total linear momentum before impact is

$$m_1\dot{x}_1\mathbf{i} + m_2\dot{x}_2\mathbf{i} = 15\,000 \times 4\mathbf{i} + 10\,000 \times \mathbf{0} = 60\,000\mathbf{i}.$$

After impact, the combined mass of $25\,000\,\text{kg}$ moves with speed V, say, so the linear momentum is $25\,000V\mathbf{i}$. By the principle of conservation of linear momentum,

$$60\,000\mathbf{i} = 25\,000V\mathbf{i},$$

so $V = 2.4$. Thus the combined speed after impact is $2.4\,\text{m}\,\text{s}^{-1}$.

(b) The kinetic energy before impact is

$$\tfrac{1}{2}m_1\dot{x}_1^2 + \tfrac{1}{2}m_2\dot{x}_2^2 = \tfrac{1}{2} \times 15\,000 \times 4^2 = 120\,000,$$

while after impact the kinetic energy is

$$\tfrac{1}{2} \times 25\,000 \times 2.4^2 = 72\,000,$$

so the collision is not elastic.

Solution to Exercise 23

Let the velocities of the balls after collision be $\mathbf{U}$ and $\mathbf{V}$. By conservation of linear momentum,

$$m\mathbf{u} = m\mathbf{U} + m\mathbf{V},$$

so

$$\mathbf{u} = \mathbf{U} + \mathbf{V}.$$

Since the collision is elastic, there is no loss of kinetic energy during the collision, and we have

$$\tfrac{1}{2}m\mathbf{u} \cdot \mathbf{u} = \tfrac{1}{2}m\mathbf{U} \cdot \mathbf{U} + \tfrac{1}{2}m\mathbf{V} \cdot \mathbf{V},$$

so

$$\mathbf{u} \cdot \mathbf{u} = \mathbf{U} \cdot \mathbf{U} + \mathbf{V} \cdot \mathbf{V}.$$

Since $\mathbf{u} = \mathbf{U} + \mathbf{V}$, we also have

$$\mathbf{u} \cdot \mathbf{u} = (\mathbf{U} + \mathbf{V}) \cdot (\mathbf{U} + \mathbf{V}) = \mathbf{U} \cdot \mathbf{U} + 2\mathbf{U} \cdot \mathbf{V} + \mathbf{V} \cdot \mathbf{V}.$$

Hence we must have $\mathbf{U} \cdot \mathbf{V} = 0$. This can be achieved only if $\mathbf{U} = \mathbf{0}$, $\mathbf{V} = \mathbf{0}$, or the angle between $\mathbf{U}$ and $\mathbf{V}$ is $\pi/2$, as required.

Solution to Exercise 24

Choose the positive x-direction to be along the direction of motion of the white ball before the collision, and the positive y-direction to correspond to a positive $\mathbf{j}$-component for the velocity of the green ball after the collision, where $\mathbf{i}$ and $\mathbf{j}$ are Cartesian unit vectors in the positive x- and y-directions. The situation is illustrated in the figure below.

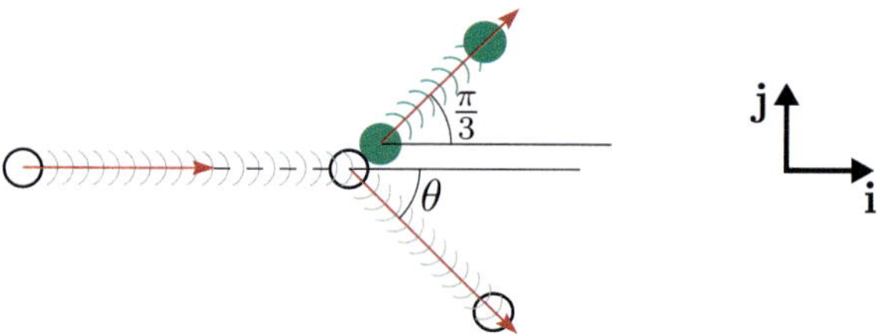

From Exercise 23, since the collision is elastic and the balls are identical apart from their colour, the white ball moves at right angles to the green ball, so we can deduce that $\theta = \pi/6$.

If $\dot{\mathbf{R}}_1$ and $\dot{\mathbf{R}}_2$ are the velocities of the white and green balls, respectively, after the collision, then we have

$$\begin{aligned}
\dot{\mathbf{R}}_1 &= (|\dot{\mathbf{R}}_1| \cos \theta)\mathbf{i} + (|\dot{\mathbf{R}}_1| \sin \theta)(-\mathbf{j}) \\
&= U \cos \tfrac{\pi}{6} \, \mathbf{i} - U \sin \tfrac{\pi}{6} \, \mathbf{j} \\
&= \tfrac{\sqrt{3}}{2} U \mathbf{i} - \tfrac{1}{2} U \mathbf{j},
\end{aligned}$$

where $U = |\dot{\mathbf{R}}_1|$ is the speed of the white ball, and

$$\begin{aligned}
\dot{\mathbf{R}}_2 &= (|\dot{\mathbf{R}}_2| \cos \tfrac{\pi}{3})\mathbf{i} + (|\dot{\mathbf{R}}_2| \sin \tfrac{\pi}{3})\mathbf{j} \\
&= \tfrac{1}{2} V \mathbf{i} + \tfrac{\sqrt{3}}{2} V \mathbf{j},
\end{aligned}$$

where $V = |\dot{\mathbf{R}}_2|$ is the speed of the green ball. Also, if $\dot{\mathbf{r}}_1$ and $\dot{\mathbf{r}}_2$ are the velocities of the white and green balls before the collision, then we have

$$\dot{\mathbf{r}}_1 = u\mathbf{i} \quad \text{and} \quad \dot{\mathbf{r}}_2 = \mathbf{0}.$$

Therefore, by the principle of conservation of linear momentum,

$$mu\mathbf{i} = m \left(\tfrac{\sqrt{3}}{2} U \mathbf{i} - \tfrac{1}{2} U \mathbf{j} \right) + m \left(\tfrac{1}{2} V \mathbf{i} + \tfrac{\sqrt{3}}{2} V \mathbf{j} \right).$$

Resolving in the $\mathbf{i}$- and $\mathbf{j}$-directions, and dividing by m, gives

$$u = \tfrac{\sqrt{3}}{2} U + \tfrac{1}{2} V, \quad 0 = -\tfrac{1}{2} U + \tfrac{\sqrt{3}}{2} V.$$

Thus $U = \sqrt{3} V$, hence we have $V = \tfrac{1}{2} u$ and $U = \tfrac{\sqrt{3}}{2} u$, so

$$\dot{\mathbf{R}}_1 = u \left(\tfrac{3}{4}\mathbf{i} - \tfrac{\sqrt{3}}{4}\mathbf{j} \right) \quad \text{and} \quad \dot{\mathbf{R}}_2 = u \left(\tfrac{1}{4}\mathbf{i} + \tfrac{\sqrt{3}}{4}\mathbf{j} \right).$$

Solution to Exercise 25

Suppose that the velocity of the particles immediately after the collision is $\mathbf{V}$. Then, from the principle of conservation of linear momentum, we have

$$m_1 u\mathbf{i} + m_2 u\mathbf{j} = (m_1 + m_2)\mathbf{V}.$$

It follows that

$$\mathbf{V} = \frac{u(m_1\mathbf{i} + m_2\mathbf{j})}{m_1 + m_2}$$
$$= \frac{m_1 u}{m_1 + m_2}\mathbf{i} + \frac{m_2 u}{m_1 + m_2}\mathbf{j}.$$

The kinetic energy before impact is $\frac{1}{2}(m_1 + m_2)u^2$, while the kinetic energy after impact is

$$\tfrac{1}{2}(m_1 + m_2)\left[\left(\frac{m_1 u}{m_1 + m_2}\right)^2 + \left(\frac{m_2 u}{m_1 + m_2}\right)^2\right] = \tfrac{1}{2}\frac{m_1^2 + m_2^2}{m_1 + m_2}u^2.$$

The change in kinetic energy is

$$\tfrac{1}{2}(m_1 + m_2)u^2 - \tfrac{1}{2}\left(\frac{m_1^2 + m_2^2}{m_1 + m_2}\right)u^2 = \tfrac{1}{2}\left(\frac{2m_1 m_2}{m_1 + m_2}\right)u^2$$
$$= \frac{m_1 m_2 u^2}{m_1 + m_2}.$$

The collision is not elastic since the loss of kinetic energy is greater than zero.

Solution to Exercise 26

(a) If $m_1 = m_2 = 3$, $\mathbf{v}_1 = 2\mathbf{i}$, $\mathbf{v}_2 = \mathbf{0}$ and $\mathbf{V}_1 = \mathbf{i} + \mathbf{j}$, $\mathbf{V}_2 = \mathbf{i} - \mathbf{j}$, then the kinetic energy just before impact is

$$\tfrac{1}{2}m_1|\mathbf{v}_1|^2 + \tfrac{1}{2}m_2|\mathbf{v}_2|^2 = \tfrac{1}{2}(3 \times 4 + 3 \times 0) = 6,$$

and the kinetic energy just after impact is

$$\tfrac{1}{2}m_1|\mathbf{V}_1|^2 + \tfrac{1}{2}m_2|\mathbf{V}_2|^2 = \tfrac{1}{2}(3 \times 2 + 3 \times 2) = 6.$$

Therefore the collision is elastic since no kinetic energy is lost during impact.

(b) If $m_1 = 1$, $m_2 = 3$, $\mathbf{v}_1 = 2\mathbf{i} + \mathbf{j}$, $\mathbf{v}_2 = \mathbf{j}$ and $\mathbf{V}_1 = \mathbf{V}_2 = \tfrac{1}{2}\mathbf{i} + \mathbf{j}$, then the kinetic energy just before impact is

$$\tfrac{1}{2}m_1|\mathbf{v}_1|^2 + \tfrac{1}{2}m_2|\mathbf{v}_2|^2 = \tfrac{1}{2}(1 \times 5 + 3 \times 1) = 4,$$

and the kinetic energy just after impact is

$$\tfrac{1}{2}m_1|\mathbf{V}_1|^2 + \tfrac{1}{2}m_2|\mathbf{V}_2|^2 = \tfrac{1}{2}(1 \times 1.25 + 3 \times 1.25) = 2.5.$$

As there is a loss of kinetic energy, the collision is inelastic.

The decrease in kinetic energy is $1.5\,\text{J}$.

Solution to Exercise 27

(a) Take the positive x-axis to be in the direction of motion of ball A before impact, and let $\mathbf{i}$ be a unit vector in this direction. Then we can write the velocities of the balls before impact as $\dot{\mathbf{r}}_1 = \dot{x}_1\mathbf{i}$ and $\dot{\mathbf{r}}_2 = \dot{x}_2\mathbf{i}$, and after impact as $\dot{\mathbf{R}}_1 = \dot{X}_1\mathbf{i}$ and $\dot{\mathbf{R}}_2 = \dot{X}_2\mathbf{i}$. Modelling the balls as particles, we can use the principle of conservation of linear momentum to obtain

$$m_1\dot{x}_1\mathbf{i} + m_2\dot{x}_2\mathbf{i} = m_1\dot{X}_1\mathbf{i} + m_2\dot{X}_2\mathbf{i}.$$

Also, equation (30) gives the relative velocity after impact as

$$\dot{X}_1\mathbf{i} - \dot{X}_2\mathbf{i} = -e(\dot{x}_1\mathbf{i} - \dot{x}_2\mathbf{i}).$$

Resolving both equations in the $\mathbf{i}$-direction gives

$$m_1\dot{x}_1 + m_2\dot{x}_2 = m_1\dot{X}_1 + m_2\dot{X}_2,$$
$$\dot{X}_1 - \dot{X}_2 = -e(\dot{x}_1 - \dot{x}_2).$$

Eliminating $\dot{X}_1$ between these equations, we obtain

$$m_1\dot{x}_1 + m_2\dot{x}_2 = m_1(\dot{X}_2 - e(\dot{x}_1 - \dot{x}_2)) + m_2\dot{X}_2$$
$$= (m_1 + m_2)\dot{X}_2 - em_1(\dot{x}_1 - \dot{x}_2),$$

so

$$\dot{X}_2 = \frac{m_1(1 + e)\dot{x}_1 + (m_2 - em_1)\dot{x}_2}{m_1 + m_2}.$$

Therefore

$$\dot{X}_1 = \dot{X}_2 - e(\dot{x}_1 - \dot{x}_2)$$
$$= \frac{(m_1 - em_2)\dot{x}_1 + m_2(1 + e)\dot{x}_2}{m_1 + m_2}.$$

So the velocities after impact are

$$\dot{\mathbf{R}}_1 = \frac{(m_1 - em_2)\dot{x}_1 + m_2(1 + e)\dot{x}_2}{m_1 + m_2}\,\mathbf{i},$$
$$\dot{\mathbf{R}}_2 = \frac{m_1(1 + e)\dot{x}_1 + (m_2 - em_1)\dot{x}_2}{m_1 + m_2}\,\mathbf{i}.$$

(b) The situation here is the same as in Example 6 with $m_1 = m_2 = m$, $\dot{x}_1 = 5$, $\dot{x}_2 = 0$ and $e = 1$ (since the collision is elastic), so we have

$$\dot{X}_1 = \frac{(m - m)5 + m(1 + 1)0}{m + m} = 0,$$
$$\dot{X}_2 = \frac{m(1 + 1)5 + (m - m)0}{m + m} = 5,$$

as in Example 6.

(c) If the balls coalesce, then $\dot{X}_1 = \dot{X}_2$, so $e(\dot{x}_1 - \dot{x}_2) = 0$. So either $e = 0$ or $\dot{x}_1 = \dot{x}_2$. However, the balls will not collide if $\dot{x}_1 = \dot{x}_2$, so $e = 0$.

Solution to Exercise 28

Let U and V be the speeds after the collision of the balls of masses m_1 and m_2, respectively. If we let $\mathbf{i}$ be a unit vector in the direction of motion, then the principle of conservation of linear momentum gives

$$m_1 u\mathbf{i} = m_1 U\mathbf{i} + m_2 V\mathbf{i},$$

so resolving in the $\mathbf{i}$-direction, we obtain

$$m_1 u = m_1 U + m_2 V.$$

Equation (30) gives the relative velocity after impact as

$$V\mathbf{i} - U\mathbf{i} = -e(\mathbf{0} - u\mathbf{i}),$$

so resolving in the $\mathbf{i}$-direction, we obtain

$$V - U = -e(0 - u) = eu.$$

Solving these equations for U and V, we find that

$$U = \frac{m_1 - em_2}{m_1 + m_2}\, u$$

and

$$V = \frac{m_1(1 + e)}{m_1 + m_2}\, u.$$

The change in kinetic energy is

$$\tfrac{1}{2}m_1 u^2 - \left(\tfrac{1}{2}m_1 U^2 + \tfrac{1}{2}m_2 V^2\right)$$

$$= \tfrac{1}{2}m_1 u^2 - \tfrac{1}{2}m_1 \left(\frac{m_1 - em_2}{m_1 + m_2}\right)^2 u^2 - \tfrac{1}{2}m_2 \left(\frac{m_1(1 + e)}{m_1 + m_2}\right)^2 u^2$$

$$= \tfrac{1}{2}m_1 u^2 - \tfrac{1}{2}\,\frac{m_1^3 - 2em_1^2 m_2 + e^2 m_1 m_2^2 + m_1^2 m_2(1 + 2e + e^2)}{(m_1 + m_2)^2}\, u^2$$

$$= \tfrac{1}{2}m_1 u^2 - \tfrac{1}{2}\,\frac{m_1^2(m_1 + m_2) + e^2 m_1 m_2(m_1 + m_2)}{(m_1 + m_2)^2}\, u^2$$

$$= \tfrac{1}{2}\,\frac{m_1(m_1 + m_2) - m_1^2 - e^2 m_1 m_2}{m_1 + m_2}\, u^2$$

$$= \frac{m_1 m_2(1 - e^2)u^2}{2(m_1 + m_2)}.$$

(Notice that if $e = 1$, then no kinetic energy is lost from the system and the collision is elastic. If $e = 0$, then the kinetic energy lost is a maximum.)

Solution to Exercise 29

We choose Cartesian unit vectors as shown in the figure below, which also shows the common tangent plane at the moment of impact, and the velocities of A and B just before and just after impact.

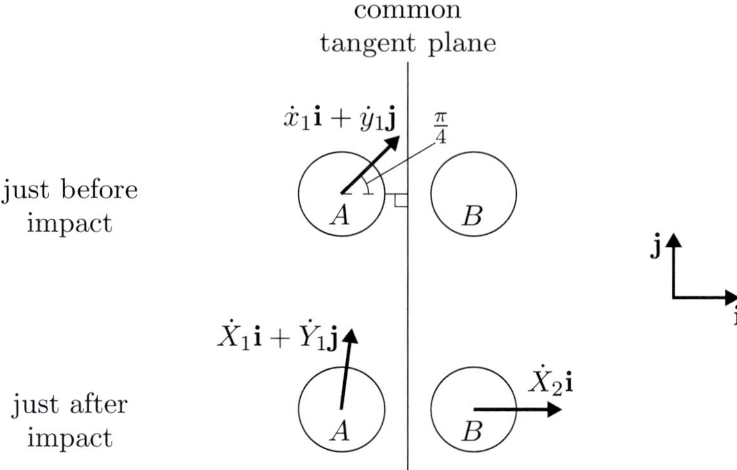

Using the fact that A has speed u just before impact while B is stationary, from the figure we see that

$$\dot{x}_1\mathbf{i} + \dot{y}_1\mathbf{j} = \left(u\cos\tfrac{\pi}{4}\right)\mathbf{i} + \left(u\sin\tfrac{\pi}{4}\right)\mathbf{j} = \tfrac{1}{\sqrt{2}}u\mathbf{i} + \tfrac{1}{\sqrt{2}}u\mathbf{j},$$

$$\dot{x}_2\mathbf{i} + \dot{y}_2\mathbf{j} = \mathbf{0}.$$

So, by Newton's law of restitution and noticing that the common normal is parallel to $\mathbf{i}$, we have

$$\dot{X}_1 - \dot{X}_2 = -e(\dot{x}_1 - \dot{x}_2) = -e\tfrac{1}{\sqrt{2}}u,$$

$$\dot{Y}_1 = \dot{y}_1 = \tfrac{1}{\sqrt{2}}u, \quad \dot{Y}_2 = \dot{y}_2 = 0.$$

Since $e = 0$, we have $\dot{X}_1 = \dot{X}_2$. Now, using the principle of conservation of linear momentum, we obtain

$$m(\dot{x}_1\mathbf{i} + \dot{y}_1\mathbf{j}) + m(\dot{x}_2\mathbf{i} + \dot{y}_2\mathbf{j}) = m(\dot{X}_1\mathbf{i} + \dot{Y}_1\mathbf{j}) + m(\dot{X}_2\mathbf{i} + \dot{Y}_2\mathbf{j}),$$

so, resolving in the $\mathbf{i}$-direction and dividing by m, we have

$$\dot{x}_1 + \dot{x}_2 = \dot{X}_1 + \dot{X}_2,$$

hence

$$\dot{X}_1 + \dot{X}_2 = \tfrac{1}{\sqrt{2}}u.$$

Therefore, since $\dot{X}_1 = \dot{X}_2$, we have

$$\dot{X}_1 = \dot{X}_2 = \tfrac{1}{2\sqrt{2}}u,$$

and the velocities just after impact are

$$\dot{X}_1\mathbf{i} + \dot{Y}_1\mathbf{j} = \tfrac{1}{2\sqrt{2}}u\mathbf{i} + \tfrac{1}{\sqrt{2}}u\mathbf{j}, \quad \dot{X}_2\mathbf{i} + \dot{Y}_2\mathbf{j} = \tfrac{1}{2\sqrt{2}}u\mathbf{i}.$$

(In the question you were told that the common tangent plane at the moment of impact is perpendicular to the subsequent direction of motion of B. This piece of information is, in fact, redundant. Ball B was originally *stationary*, so after impact it must move in the direction perpendicular to the common tangent plane, i.e. in the direction of the common normal.)

Solution to Exercise 30

Let the ball have mass m and start from height h_1. By using conservation of energy we have $mgh_1 = \frac{1}{2}mv_1^2$, where v_1 is the speed of the ball just before the first bounce. The speed of the ball just after the first bounce is given by $v_2 = ev_1$, using Newton's law of restitution, where e is the coefficient of restitution. By conservation of energy, we have $mgh_2 = \frac{1}{2}mv_2^2$, where h_2 is the maximum height after the first bounce. Putting these pieces of information together, we have

$$mgh_2 = \tfrac{1}{2}mv_2^2 = \tfrac{1}{2}m(ev_1)^2 = e^2 \times mgh_1,$$

that is, $h_2 = e^2 h_1$.

Similarly, we find that the height h_3 after the second bounce is related to the height before the second bounce by $h_3 = e^2 h_2$, so $h_3 = e^4 h_1$. But we are told that $h_3 = \frac{1}{2}h_1$, so $e^4 = \frac{1}{2}$, hence

$$e = \frac{1}{\sqrt[4]{2}} \simeq 0.84 \quad \text{(correct to 2 d.p.).}$$

Circular motion

Introduction

The theme of this unit is rotational motion. We will concentrate mainly on analysing the circular motion of a particle. This can be used to model a wide range of situations, such as a child on a swing, the pendulum of a clock and a chair-o-plane roundabout (see Figure 1) at the fairground.

Figure 1   Chair-o-plane roundabout

Many fairground rides depend for their thrills on the forces involved in circular motion. We will be able to use our model of the chair-o-plane, for example, to answer questions such as the following.

- Will a child swing out at a greater angle than a heavier adult?
- Will the people on the outside swing out at a greater angle than those on the inside?
- How does the angle at which people swing out depend on the speed of rotation of the roundabout?

These questions will be answered in Subsection 2.5.

Before we are able to answer questions such as these, we need to investigate the motion of a particle moving in a circle. In Section 1 we derive expressions for the velocity and acceleration of a particle moving in a circle. In Section 2 we apply Newton's second law to uniform circular motion, that is, motion in a circle with constant speed. In Section 3 this is extended to non-uniform circular motion. Section 4 starts to look *beyond* circular motion to the concepts that are needed in order to analyse more general rotational motion.

1 Describing circular motion

Consider an object following a circular path at constant speed. This might be a particle attached to an inextensible string swung round in a circle, or a vehicle following a circular path on a roundabout or part of a circle when cornering. In this section we analyse this motion. We do so first in Cartesian coordinates and subsequently in polar coordinates, after an intervening subsection revising the differentiation of vector functions.

1.1 Circular motion in Cartesian coordinates

In this subsection we take a first look at circular motion, using Cartesian coordinates. Some simple features of the motion will be derived, which can be compared with more general results derived later.

Recall that the point $(\cos\theta, \sin\theta)$ lies on a circle of radius 1 and centre $(0,0)$. More generally, $(R\cos\theta, R\sin\theta)$ lies on a circle of radius R and centre $(0,0)$. If t denotes time and α is a constant, then a particle with position

$$(R\cos(\alpha t), R\sin(\alpha t))$$

at time t will move on a circle of radius R with centre $(0,0)$. The line joining the particle's position to $(0,0)$ makes an angle $\theta = \alpha t$ with the positive x-axis, and this angle is changing with time at a constant rate α, that is, $\dot\theta = \alpha$. The quantity $\omega = |\alpha| = |\dot\theta|$ is the **angular speed** of the particle. (The definition $\omega = |\dot\theta|$ applies even when $\dot\theta$ is not constant.)

Angular speed has units of radians per second.

The motion considered in this subsection, where $\dot\theta$ is constant, is called **uniform circular motion**. If the motion is circular and $\dot\theta$ is not constant, then the motion is called **non-uniform circular motion**.

To study the mechanics of a particle moving in a circle at constant angular speed, we need to know its velocity and acceleration.

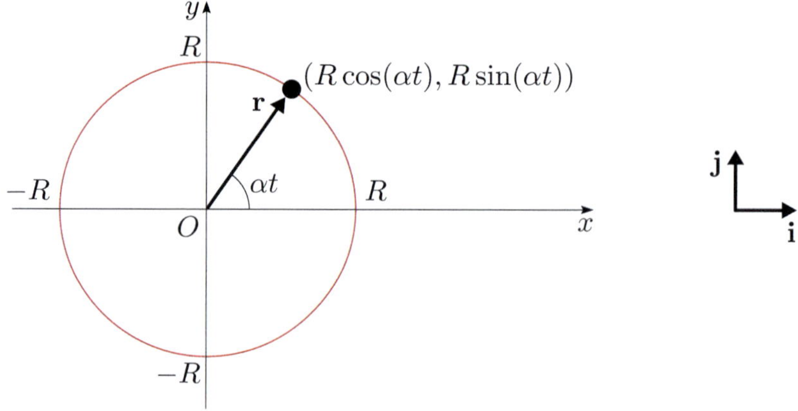

Figure 2 Position vector of a particle moving in a circle

Choosing Cartesian unit vectors $\mathbf{i}$ and $\mathbf{j}$ as shown in Figure 2, the position vector of the particle is given by

$$\mathbf{r} = R\cos(\alpha t)\mathbf{i} + R\sin(\alpha t)\mathbf{j}. \tag{1}$$

Before looking at the general case of a particle moving uniformly in a circle, try the following exercise.

We have

$$|\mathbf{r}| = \sqrt{R^2\cos^2(\alpha t) + R^2\sin^2(\alpha t)}$$
$$= R,$$

which is the radius of the circle.

Exercise 1

A particle has position vector $\mathbf{r}$ at time t given by

$$\mathbf{r}(t) = 3\cos(4t)\mathbf{i} + 3\sin(4t)\mathbf{j}.$$

(a) Find the velocity vector $\dot{\mathbf{r}}(t)$ and the acceleration vector $\ddot{\mathbf{r}}(t)$ of the particle at time t.

(b) What is the magnitude of the acceleration vector $\ddot{\mathbf{r}}(t)$?

(c) Show that $\ddot{\mathbf{r}}(t)$ is of the form $-c\,\mathbf{r}(t)$, where c is a positive constant.

Now consider again the case of a particle with position vector given by equation (1), and differentiate to obtain the particle's velocity

$$\mathbf{v} = \dot{\mathbf{r}} = -\alpha R\sin(\alpha t)\mathbf{i} + \alpha R\cos(\alpha t)\mathbf{j}$$
$$= \alpha R(-\sin(\alpha t)\mathbf{i} + \cos(\alpha t)\mathbf{j}). \tag{2}$$

Note that R is a constant (the radius of the circle).

What can we deduce about the velocity from this expression? First, $\mathbf{v}$ has magnitude

$$|\mathbf{v}| = \sqrt{\alpha^2 R^2\sin^2(\alpha t) + \alpha^2 R^2\cos^2(\alpha t)} = |\alpha|R = \omega R.$$

So the particle moves with speed ωR. Since both ω and R are independent of t, the speed of the particle is constant. (The *velocity* is not constant because its direction is changing.) Second, notice that

$$\mathbf{r}\cdot\mathbf{v} = \left(R\cos(\alpha t)\mathbf{i} + R\sin(\alpha t)\mathbf{j}\right)\cdot\left(-\alpha R\sin(\alpha t)\mathbf{i} + \alpha R\cos(\alpha t)\mathbf{j}\right)$$
$$= -\alpha R^2\cos(\alpha t)\sin(\alpha t) + \alpha R^2\cos(\alpha t)\sin(\alpha t)$$
$$= 0.$$

You saw in Unit 2 that if the dot product of two non-zero vectors is zero, then they are perpendicular. So the particle's velocity $\mathbf{v}$ is perpendicular to its position vector $\mathbf{r}$. This is illustrated in Figure 3, which shows that the velocity $\mathbf{v}$ is tangential to the circle that the particle is tracing.

Differentiating the velocity vector $\mathbf{v}$ (equation (2)) gives the acceleration

$$\mathbf{a} = \ddot{\mathbf{r}} = \alpha R(-\alpha\cos(\alpha t)\mathbf{i} - \alpha\sin(\alpha t)\mathbf{j})$$
$$= -\alpha^2 R(\cos(\alpha t)\mathbf{i} + \sin(\alpha t)\mathbf{j})$$
$$= -\alpha^2\mathbf{r}$$
$$= -\omega^2\mathbf{r}. \tag{3}$$

Since this vector is a negative multiple of $\mathbf{r}$, the acceleration is towards the centre of the circle (see Figure 3).

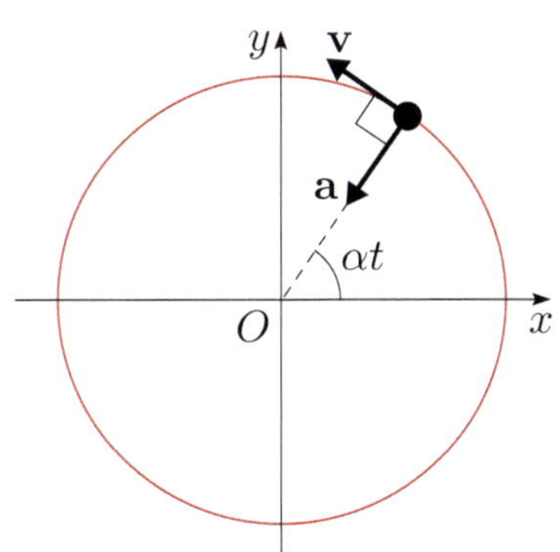

Figure 3 Velocity $\mathbf{v}$ and acceleration $\mathbf{a}$ of a particle moving in a circle at constant angular speed

The magnitude of the acceleration is $\omega^2 R$. Since $|\mathbf{v}| = \omega R$, the magnitude of the acceleration can also be written as $|\mathbf{v}|^2/R$. This form is useful if we know the speed of the particle, rather than its angular speed.

The main results derived above are summarised below.

If a particle follows a circle of radius R at a constant speed $|\mathbf{v}|$, then it has a constant angular speed $|\mathbf{v}|/R$, and these results apply.

Motion in a circle at constant angular speed

For a particle moving in a circle of radius R with constant angular speed ω:

- the speed (the magnitude of the velocity) is a constant, $|\mathbf{v}| = \omega R$
- the direction of the velocity is tangential to the circle
- the acceleration has constant magnitude, $|\mathbf{a}| = \omega^2 R = |\mathbf{v}|^2/R$
- the direction of the acceleration is towards the centre of the circle.

Since a particle moving in a circle in this way has a non-zero acceleration, it must be subject to some force, say $\mathbf{F}$. By Newton's second law, this force must act in the same direction as the acceleration $\mathbf{a}$, that is, towards the centre of the circle. Such a force might be provided by a string attaching the particle to the centre of the circle. Alternatively, if the particle is constrained to move in a circle by following a track, then the track must exert a sideways force on the particle (the force acting towards the centre of the circle). If the particle has mass m, then Newton's second law gives $\mathbf{F} = m\mathbf{a} = -m\omega^2\mathbf{r}$. The force is towards the centre of the circle and has magnitude $m\omega^2 R$, or $m|\mathbf{v}|^2/R$.

The results listed in the summary box above are sufficient to solve some problems involving circular motion, such as the following exercise.

Exercise 2

A miniature locomotive of mass $40\,\text{kg}$ is following a circular track of diameter $200\,\text{m}$ at constant speed. It completes one circuit in $50\,\text{s}$.

What is the magnitude of the sideways force exerted by the track on the locomotive?

In the form of motion discussed in this subsection, we have a particle subject to a force at right angles to its direction of motion. The effect of this force is not to change the magnitude of the particle's velocity (the speed remains constant), but to change the *direction* of the velocity. In general, forces in the direction of motion change the speed, while forces at right angles to the direction of motion change the direction.

In Exercise 2, a single force caused the circular motion. In order to solve more complicated problems involving circular motion, we need to be able to resolve forces radially and tangentially. For this we need appropriate unit vectors. The radial and tangential unit vectors are introduced in Subsection 1.3, but first we revise the differentiation of vector functions, which was introduced in Unit 3.

1.2 Differentiation of vector functions

In Unit 3 you saw that if $\mathbf{f}(t) = f_1(t)\,\mathbf{i} + f_2(t)\,\mathbf{j} + f_3(t)\,\mathbf{k}$ is a vector function of t, where $\mathbf{i}, \mathbf{j}, \mathbf{k}$ are the (constant) Cartesian unit vectors, then

$$\dot{\mathbf{f}}(t) = \dot{f}_1(t)\,\mathbf{i} + \dot{f}_2(t)\,\mathbf{j} + \dot{f}_3(t)\,\mathbf{k}. \tag{4}$$

We can use this result to obtain rules for differentiating sums and products of vectors. For example, if the scalar h and the vector $\mathbf{f}$ are both functions of the scalar variable t, then we can show that

$$\frac{d}{dt}(h\mathbf{f}) = \frac{dh}{dt}\mathbf{f} + h\frac{d\mathbf{f}}{dt}.$$

For ease of reading, we frequently write $\mathbf{f}$, f_i, etc., rather than $\mathbf{f}(t)$, $f_i(t)$, etc., i.e. omitting explicit reference to the variable t.

This follows by first writing the vector function $\mathbf{f}$ in Cartesian component form as

$$\mathbf{f} = f_1\mathbf{i} + f_2\mathbf{j} + f_3\mathbf{k},$$

so

$$h\mathbf{f} = (h f_1)\mathbf{i} + (h f_2)\mathbf{j} + (h f_3)\mathbf{k}.$$

To differentiate $h\mathbf{f}$, we apply equation (4) and differentiate the three components $h f_1$, $h f_2$ and $h f_3$. These are products of scalar functions, so we can use the rule for differentiating the product of two scalar functions, giving

$$\frac{d}{dt}(h f_i) = \frac{dh}{dt}f_i + h\frac{df_i}{dt} \quad (i = 1, 2, 3).$$

Therefore

$$\frac{d}{dt}(h\mathbf{f}) = \left(\frac{dh}{dt}f_1 + h\frac{df_1}{dt}\right)\mathbf{i} + \left(\frac{dh}{dt}f_2 + h\frac{df_2}{dt}\right)\mathbf{j} + \left(\frac{dh}{dt}f_3 + h\frac{df_3}{dt}\right)\mathbf{k}$$

$$= \frac{dh}{dt}(f_1\mathbf{i} + f_2\mathbf{j} + f_3\mathbf{k}) + h\left(\frac{df_1}{dt}\mathbf{i} + \frac{df_2}{dt}\mathbf{j} + \frac{df_3}{dt}\mathbf{k}\right)$$

$$= \frac{dh}{dt}\mathbf{f} + h\frac{d\mathbf{f}}{dt},$$

which is the required result.

Exercise 3

If the vectors $\mathbf{f}$ and $\mathbf{g}$ are both functions of the scalar variable t, show that

$$\frac{d}{dt}(\mathbf{f}\cdot\mathbf{g}) = \frac{d\mathbf{f}}{dt}\cdot\mathbf{g} + \mathbf{f}\cdot\frac{d\mathbf{g}}{dt}.$$

We can similarly derive the rules for differentiating the other possible combinations of scalar and vector functions.

The rules for differentiating products bear a strong resemblance to the rule for differentiating the product of two scalar functions. Note that in the rule for differentiating a cross product, the order of the factors in each term is important.

> **Rules for differentiating vector functions**
>
> If the scalar h and the vectors $\mathbf{f}$ and $\mathbf{g}$ are functions of the scalar variable t, then
>
> $$\frac{d}{dt}(\mathbf{f} + \mathbf{g}) = \frac{d\mathbf{f}}{dt} + \frac{d\mathbf{g}}{dt}, \tag{5}$$
>
> $$\frac{d}{dt}(h\mathbf{f}) = \frac{dh}{dt}\,\mathbf{f} + h\,\frac{d\mathbf{f}}{dt}, \tag{6}$$
>
> $$\frac{d}{dt}(\mathbf{f} \cdot \mathbf{g}) = \frac{d\mathbf{f}}{dt} \cdot \mathbf{g} + \mathbf{f} \cdot \frac{d\mathbf{g}}{dt}, \tag{7}$$
>
> $$\frac{d}{dt}(\mathbf{f} \times \mathbf{g}) = \frac{d\mathbf{f}}{dt} \times \mathbf{g} + \mathbf{f} \times \frac{d\mathbf{g}}{dt}. \tag{8}$$

These rules will prove useful in the next subsection.

Exercise 4

Differentiate $\mathbf{f} \cdot \mathbf{f}$, where $\mathbf{f}$ is a vector function of the scalar variable t.

In Exercise 4 you showed that

$$\frac{d}{dt}(\mathbf{f} \cdot \mathbf{f}) = 2\mathbf{f} \cdot \frac{d\mathbf{f}}{dt}.$$

This leads to an important result for a vector of *constant magnitude*. Suppose that $\mathbf{e}$ is such a vector, so $\mathbf{e} \cdot \mathbf{e} = |\mathbf{e}|^2 = \text{constant}$. Then

$$\frac{d}{dt}(\mathbf{e} \cdot \mathbf{e}) = 0.$$

Using the result of Exercise 4, this leads directly to

$$\mathbf{e} \cdot \frac{d\mathbf{e}}{dt} = 0.$$

Hence if a non-zero vector $\mathbf{e}$ has constant magnitude, then its derivative $d\mathbf{e}/dt$ is always either zero or perpendicular to $\mathbf{e}$. This can be a useful check on the derivatives of unit vectors, as will become clear in the next subsection.

1.3   Circular motion in polar coordinates

If position is given in Cartesian coordinates (x, y), then it is natural to write the position vector $\mathbf{r}$ in terms of the Cartesian unit vectors $\mathbf{i}$ and $\mathbf{j}$, that is,

$$\mathbf{r}(t) = x(t)\,\mathbf{i} + y(t)\,\mathbf{j}. \tag{9}$$

Then the velocity and acceleration vectors are

$$\mathbf{v}(t) = \dot{\mathbf{r}}(t) = \dot{x}(t)\,\mathbf{i} + \dot{y}(t)\,\mathbf{j}, \tag{10}$$

$$\mathbf{a}(t) = \dot{\mathbf{v}}(t) = \ddot{\mathbf{r}}(t) = \ddot{x}(t)\,\mathbf{i} + \ddot{y}(t)\,\mathbf{j}. \tag{11}$$

The case of derivation of equations (10) and (11) from equation (9) is due to the fact that $\mathbf{i}$ and $\mathbf{j}$ are constant vectors. Only the coordinates x and y, which are functions of time t, vary during the motion.

However, for a particle P moving in a circle (as shown in Figure 4(a)), it is convenient to use polar coordinates (r, θ), with an origin O at the centre of the circle, to describe the position vector $\mathbf{r}$ of the particle. With this choice of coordinate system, the coordinate r is a constant for circular motion (being the radius of the circle), and only the coordinate θ is a function of time t. However, in order to discuss the motion of a particle in a vector context using polar coordinates, we need the two orthogonal unit vectors $\mathbf{e}_r$ and $\mathbf{e}_\theta$ (see Figure 4(b)) that were introduced in Unit 15. Recall that $\mathbf{e}_r$ is a unit vector in the direction of increasing r (the **radial direction**), and $\mathbf{e}_\theta$ is a unit vector in the direction of increasing θ (the **tangential direction**).

In contexts where the motion is not circular – and sometimes even when it is – the direction of increasing θ is called the **transverse direction**. For non-circular motion, the transverse vector does not in general coincide with the tangent vector.

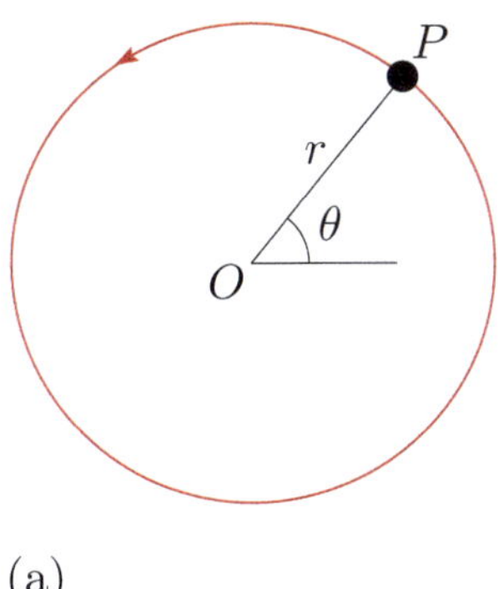
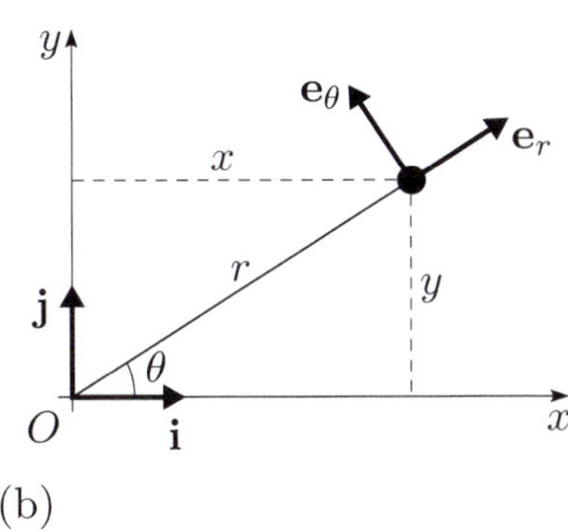

(a) (b)

Figure 4   (a) Particle P moving in a circle. (b) Unit vectors $\mathbf{e}_r$ and $\mathbf{e}_\theta$ for polar coordinates (r, θ), and their relationship to Cartesian coordinates (x, y) and unit vectors $\mathbf{i}$ and $\mathbf{j}$.

Before we obtain expressions for the position, velocity and acceleration of the particle P in terms of $\mathbf{e}_r$ and $\mathbf{e}_\theta$, we need to recall how each of $\mathbf{e}_r$ and $\mathbf{e}_\theta$ can be expressed as a linear combination of $\mathbf{i}$ and $\mathbf{j}$. In Unit 15 it was found that

$$\mathbf{e}_r = (\cos\theta)\mathbf{i} + (\sin\theta)\mathbf{j}, \tag{12}$$

$$\mathbf{e}_\theta = (-\sin\theta)\mathbf{i} + (\cos\theta)\mathbf{j}. \tag{13}$$

Since $\mathbf{e}_r$ and $\mathbf{e}_\theta$ are functions of the polar coordinate angle θ, which is a function of time t, it follows that $\mathbf{e}_r$ and $\mathbf{e}_\theta$ are not constant functions, like $\mathbf{i}$ and $\mathbf{j}$, but are functions of time. However, since their magnitudes are constant (they are unit vectors), it is only their directions that vary with time.

We also need to express the time derivatives $\dot{\mathbf{e}}_r$ and $\dot{\mathbf{e}}_\theta$ in terms of $\mathbf{e}_r$ and $\mathbf{e}_\theta$. Differentiating equation (12) with respect to time (since θ is a function of time) using the chain rule, we obtain

Recall that the chain rule is
$$\frac{df(\theta(t))}{dt} = \frac{df}{d\theta}\frac{d\theta}{dt} = \dot{\theta}\frac{df}{d\theta}$$

$$\dot{\mathbf{e}}_r = (-\sin\theta)\dot{\theta}\mathbf{i} + (\cos\theta)\dot{\theta}\mathbf{j} = \dot{\theta}((-\sin\theta)\mathbf{i} + (\cos\theta)\mathbf{j}) = \dot{\theta}\mathbf{e}_\theta,$$

where we have used the expression for $\mathbf{e}_\theta$ given by equation (13).

Exercise 5

By differentiating equation (13) with respect to time, and using equation (12), show that

$$\dot{\mathbf{e}}_\theta = -\dot\theta \mathbf{e}_r.$$

Since $\mathbf{e}_r$ and $\mathbf{e}_\theta$ are orthogonal, $\dot{\mathbf{e}}_r = \dot\theta \mathbf{e}_\theta$ is perpendicular to $\mathbf{e}_r$, and in addition, $\dot{\mathbf{e}}_\theta = -\dot\theta \mathbf{e}_r$ is perpendicular to $\mathbf{e}_\theta$. These are special cases of the result proved after Exercise 4 that the derivative of a non-zero vector of constant magnitude is always either zero or perpendicular to that vector.

> **Polar unit vectors $\mathbf{e}_r$ and $\mathbf{e}_\theta$**
>
> Given a polar coordinate system and a two-dimensional Cartesian coordinate system with the same origin O, where the positive x-axis corresponds to the axis from which the polar coordinate θ is measured, the polar unit vectors $\mathbf{e}_r$ and $\mathbf{e}_\theta$ are related to the Cartesian unit vectors $\mathbf{i}$ and $\mathbf{j}$ by
>
> $$\mathbf{e}_r = (\cos\theta)\mathbf{i} + (\sin\theta)\mathbf{j}, \tag{14}$$
> $$\mathbf{e}_\theta = (-\sin\theta)\mathbf{i} + (\cos\theta)\mathbf{j}. \tag{15}$$
>
> The derivatives with respect to time of the polar unit vectors are
>
> $$\dot{\mathbf{e}}_r = \dot\theta \mathbf{e}_\theta, \tag{16}$$
> $$\dot{\mathbf{e}}_\theta = -\dot\theta \mathbf{e}_r. \tag{17}$$

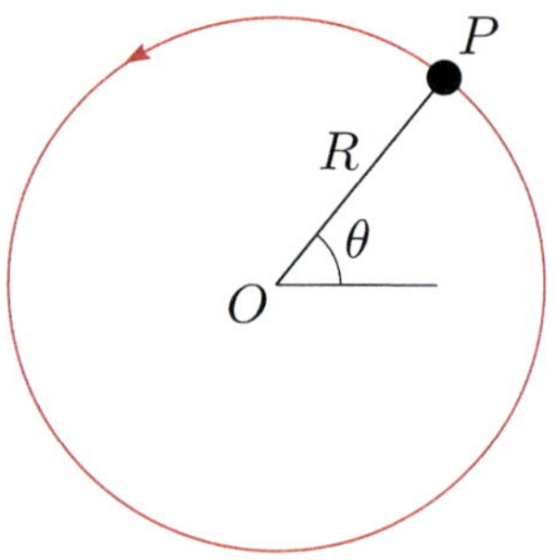

We are now in a position to obtain expressions for the position $\mathbf{r}$, the velocity $\mathbf{v} = \dot{\mathbf{r}}$ and the acceleration $\mathbf{a} = \dot{\mathbf{v}} = \ddot{\mathbf{r}}$ of a particle P moving in a circle of radius R, say (see Figure 5), in terms of the unit vectors $\mathbf{e}_r$ and $\mathbf{e}_\theta$. For any point on the circle $r = R$, the position vector of P is in the direction of $\mathbf{e}_r$ thus can be expressed as

$$\mathbf{r} = R\mathbf{e}_r.$$

Differentiating this equation with respect to time, and remembering that R is a constant, we have

$$\dot{\mathbf{r}} = R\dot{\mathbf{e}}_r.$$

Using equation (16), this may be written as

$$\dot{\mathbf{r}} = R\dot\theta \mathbf{e}_\theta. \tag{18}$$

Figure 5   Particle P moving in a circle of radius R

Exercise 6

By differentiating equation (18) with respect to time, remembering that θ and $\mathbf{e}_\theta$ are functions of time but that R is a constant, show that the acceleration of a particle moving in a circle of radius R is

$$\ddot{\mathbf{r}} = -R\dot\theta^2 \mathbf{e}_r + R\ddot\theta \mathbf{e}_\theta.$$

It is useful to collect together the results that apply whenever the motion is circular (for both uniform circular motion, considered in Section 2, and non-uniform circular motion, considered in Section 3).

Position, velocity and acceleration in polar coordinates for circular motion

In polar coordinates, the position, velocity and acceleration of a particle moving in the circle $r = R$ are

$$\mathbf{r} = R\mathbf{e}_r, \tag{19}$$

$$\dot{\mathbf{r}} = R\dot{\theta}\mathbf{e}_\theta, \tag{20}$$

$$\ddot{\mathbf{r}} = -R\dot{\theta}^2\mathbf{e}_r + R\ddot{\theta}\mathbf{e}_\theta. \tag{21}$$

It is geometrically clear (see Figure 6(a)) that the velocity of a particle moving in a circle is tangential to the circle, and equation (20) confirms this. On the other hand (see Figure 6(b)), the acceleration has a radial component $-R\dot{\theta}^2\mathbf{e}_r$ (arising from the rate of change of the direction of the velocity) as well as a tangential component $R\ddot{\theta}\mathbf{e}_\theta$ (arising from the rate of change of the speed). Furthermore, the component of the acceleration in the radial direction is *negative*, which indicates that the acceleration of a particle moving in a circle has a component of magnitude $R\dot{\theta}^2$ directed *inwards*, towards the centre of the circle. This inward component is commonly called the **centripetal acceleration**, but you must remember that the acceleration for circular motion in general also has a non-zero tangential component.

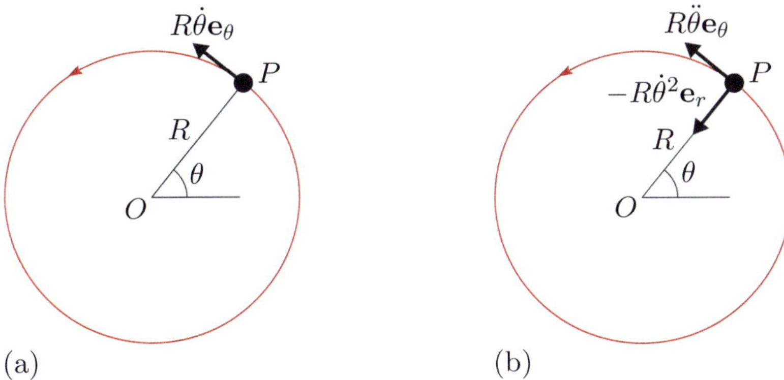

Figure 6   Particle P moving in a circle of radius R, showing (a) its velocity, and (b) its acceleration

Results (19) to (21) are the generalisation of the results of Subsection 1.1 to motion where the angular speed is not constant. If the angular speed is constant (i.e. the circular motion is *uniform*), then $\dot{\theta} = \text{constant}$ and $\ddot{\theta} = 0$, so equation (21) reduces to $\ddot{\mathbf{r}} = -R\dot{\theta}^2\mathbf{e}_r$ and the acceleration is solely towards the centre of the circle, as you saw in Subsection 1.1.

Exercise 7

A particle is moving on a unit circle centred on the origin O, that is, on the circle $r = 1$. Its polar coordinate angle θ at time t is given by $\theta(t) = t^2$.

Find the velocity and acceleration of the particle at time t in terms of the unit vectors $\mathbf{e}_r$ and $\mathbf{e}_\theta$.

---

2 Uniform circular motion

The previous section derived formulas that apply whenever an object is moving in a circle. This section considers the special case where the angular rate of rotation is constant, that is, **uniform circular motion**. Non-uniform circular motion is studied in Section 3.

2.1 Introductory examples

We begin with a summary of results from Section 1 specialised to the case of uniform circular motion, that is, the case where $\ddot{\theta} = 0$.

Kinematics of uniform circular motion

Consider a particle moving in a circle of radius R, with centre at the origin O, having a constant rate of rotation $\dot{\theta}$ and hence a constant angular speed $\omega = |\dot{\theta}|$.

- The velocity of the particle is

$$\dot{\mathbf{r}} = R\dot{\theta}\mathbf{e}_\theta, \tag{22}$$

  i.e. the velocity of the particle is purely tangential and has constant magnitude $v = |\dot{\mathbf{r}}| = R|\dot{\theta}| = R\omega$.

- The acceleration of the particle is

$$\ddot{\mathbf{r}} = -R\dot{\theta}^2\mathbf{e}_r = -R\omega^2\mathbf{e}_r = -\frac{v^2}{R}\mathbf{e}_r, \tag{23}$$

  i.e. the acceleration is directed towards the centre of the circle and has constant magnitude $|\ddot{\mathbf{r}}| = R\dot{\theta}^2 = R\omega^2 = v^2/R$.

- The time taken for one complete revolution of the circle is

$$T = \frac{2\pi}{\omega}. \tag{24}$$

This follows from ωt being the magnitude of the angle measured from the x-axis at time t, and the period being defined as the time for this angle to equal 2π.

Exercise 8

A fly sits at the tip of the minute hand of a clock. The hand has length
0.5 m and can be assumed to move with constant angular speed.

Find the angular speed, the speed and the magnitude of the acceleration of
the fly.

If there are no forces acting on a particle, then by Newton's first law, it
will either remain at rest or move with constant speed in a straight line.
Therefore if the particle moves on a circular path, there must be some
force or forces causing it to do so, even when the circular motion is
uniform. In Subsection 1.1 you saw that a particle moving on a circle of
radius R with constant angular speed ω and constant speed v is being
accelerated towards the centre of the circle with an acceleration of
magnitude $R\omega^2 = v^2/R$. If the particle has mass m, then Newton's second
law states that there must be a force

$$\mathbf{F} = m\ddot{\mathbf{r}} = -mR\omega^2\mathbf{e}_r = -\frac{mv^2}{R}\,\mathbf{e}_r$$

acting on the particle. This force is directed towards the centre of the
circle and has constant magnitude. There is no tangential component of
the acceleration, so there is no component of the total force (i.e. resultant
force) acting on the particle in this direction.

Similarly, there is no component of the resultant force acting at right
angles to the plane of the motion. In order to be able to use this fact to
solve problems, we need to introduce a third unit vector $\mathbf{k}$, which is
perpendicular to the plane of motion and whose direction is chosen so that
the unit vectors $\mathbf{e}_r$, $\mathbf{e}_\theta$ and $\mathbf{k}$ form a right-handed system. An example of
such a right-handed system is shown in Figure 7.

You met the term right-handed
system in Unit 2.

Example 1

Two particles A and B, of equal mass m, are connected by a model string
that passes through a small smooth hole in a smooth horizontal table.
Particle A is on the surface of the table, while particle B is suspended
beneath the table. Particle A describes circles of radius R with constant
angular speed ω, as a result of which particle B is static.

Find the angular speed of particle A.

Solution

Without loss of generality, we can assume that particle A moves in an
anticlockwise direction, when viewed from above the table. We will
describe the motion of particle A using a polar coordinate system in the
plane of the table surface with origin at O, where O is the centre of the
circle described by particle A.

The situation, together with the force diagrams for the two particles, is shown in Figure 7. $\mathbf{T}_1$ and $\mathbf{T}_2$ are the tension forces due to the string on particles A and B, respectively, $\mathbf{W}_1$ and $\mathbf{W}_2$ are the weights of particles A and B, respectively, and $\mathbf{N}$ is the normal reaction of the table on particle A (there is no friction force since the table is smooth).

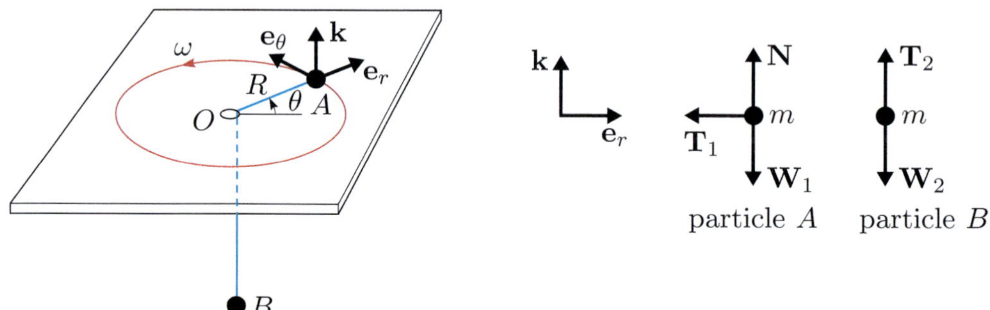

Figure 7 Two particles, A and B, connected by a string passing through the centre of a circle described by A

First consider particle B. There are two forces acting on this particle: the tension force $\mathbf{T}_2 = |\mathbf{T}_2|\mathbf{k}$ due to the string, and its weight $\mathbf{W}_2 = -mg\mathbf{k}$, where $\mathbf{k}$ is a unit vector pointing vertically upwards. Since particle B is static, the total force acting on it is zero, so

$$\mathbf{T}_2 + \mathbf{W}_2 = \mathbf{0},$$

that is,

$$|\mathbf{T}_2|\mathbf{k} - mg\mathbf{k} = \mathbf{0}.$$

Hence, resolving in the $\mathbf{k}$-direction, we obtain

$$|\mathbf{T}_2| = mg.$$

The vectors $\mathbf{e}_r$, $\mathbf{e}_\theta$ and $\mathbf{k}$ are mutually perpendicular, so we can resolve in these directions in the same way as for the Cartesian unit vectors $\mathbf{i}$, $\mathbf{j}$ and $\mathbf{k}$.

Now consider particle A. There are three forces acting on this particle: the tension force $\mathbf{T}_1 = -|\mathbf{T}_1|\mathbf{e}_r$ due to the string, the normal reaction $\mathbf{N} = |\mathbf{N}|\mathbf{k}$ of the table on the particle, and its weight $\mathbf{W}_1 = -mg\mathbf{k}$. Applying Newton's second law to particle A gives

$$m\ddot{\mathbf{r}} = \mathbf{T}_1 + \mathbf{N} + \mathbf{W}_1,$$

that is, using equation (23),

$$-mR\omega^2\mathbf{e}_r = -|\mathbf{T}_1|\mathbf{e}_r + |\mathbf{N}|\mathbf{k} - mg\mathbf{k}.$$

Resolving in the direction of $\mathbf{e}_r$ gives

$$|\mathbf{T}_1| = mR\omega^2.$$

Since we have a model string and a smooth hole, $|\mathbf{T}_1| = |\mathbf{T}_2|$ thus $mR\omega^2 = mg$. Hence $\omega^2 = g/R$ and $\omega = \sqrt{g/R}$.

So the angular speed of particle A is $\sqrt{g/R}$.

Exercise 9

A particle P of mass m moves in a circle with constant angular speed ω on a smooth horizontal table. It is attached to an end of a model string of length l. The other end of the string is fixed to a point on the table.

What is the tension in the string?

Exercise 10

A child places a small toy of mass m on a playground roundabout and spins the roundabout so that it describes one complete revolution every two seconds. The toy stays at the same point relative to the roundabout while the latter rotates. The coefficient of static friction between the toy and the roundabout is $\mu = \frac{1}{2}$.

By modelling the toy as a particle and the roundabout as a horizontal disc rotating uniformly about its centre, show that the greatest possible distance of the toy from the centre of the roundabout is about $0.5\,\mathrm{m}$.

(*Hint*: Recall from Unit 2 that the condition that an object on a rough surface does not slip is $|\mathbf{F}| \le \mu|\mathbf{N}|$, where $\mathbf{F}$ is the friction force and $\mathbf{N}$ is the normal reaction on the object.)

Exercise 11

A particle of mass m is joined by a model spring of stiffness k and natural length l_0 to a fixed point O on a smooth horizontal table. The particle is moving in a circle on the table, the circle being centred on O, with a constant angular speed ω (where $\omega < \sqrt{k/m}$).

Find the length of the model spring in terms of m, k, l_0 and ω.

(*Hint*: Recall Hooke's law from Unit 9, which states that the spring exerts a force $\mathbf{H} = k(l - l_0)\widehat{\mathbf{s}}$ on the particle, where l is the length of the spring and $\widehat{\mathbf{s}}$ is a unit vector from the particle to the centre of the spring.)

2.2 Geostationary satellites

An interesting application of uniform circular motion concerns communications satellites. Some of the most important satellites are those with *geostationary orbits*, that is, satellites that maintain a fixed position relative to the Earth's rotating surface. Signals can be sent to and from such a satellite without the need to adjust continually the orientations of the transmitter and the receiver. In order that the geostationary satellite should maintain its position, it must have a circular orbit in the equatorial plane and an orbital period of one sidereal day, so its angular speed is $\omega = 2\pi/T$, where

The sidereal day is the time required for a complete rotation of the Earth relative to a particular star. You will examine this in Example 3 in Section 4.

$$T = 1 \text{ day} = 23.9343 \times 60 \times 60\,\mathrm{s}.$$

Hence ω is approximately $7.292 \times 10^{-5}\,\mathrm{rad\,s^{-1}}$.

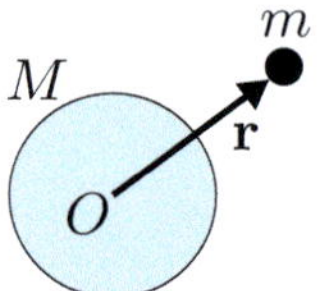

Figure 8 Position vector of a satellite

The force causing the circular motion is the gravitational force of attraction on the satellite due to the presence of the Earth. This force is attractive and directed towards the centre of the Earth. Empirical evidence has shown that it is well modelled by

$$\mathbf{F} = -\frac{GmM}{|\mathbf{r}|^3}\,\mathbf{r}, \tag{25}$$

where m is the mass of the satellite, M is the mass of the Earth, $\mathbf{r}$ is the position vector of the satellite with respect to the centre of the Earth (as shown in Figure 8), and G is a constant, known as the **gravitational constant**, whose value in SI units is $6.674 \times 10^{-11}\,\mathrm{N\,m^2\,kg^{-2}}$.

Equation (25) is a special case of *Newton's law of universal gravitation*, which models the gravitational force between any two particles. (We are assuming that all other forces can be neglected.)

Newton's law of universal gravitation

The gravitational force of attraction exerted on a particle of mass m_1 by a particle of mass m_2 is

$$\mathbf{F} = -\frac{Gm_1m_2}{|\mathbf{r}|^3}\,\mathbf{r}, \tag{26}$$

where $\mathbf{r}$ is the position vector of the particle of mass m_1, relative to the particle of mass m_2, and $G = 6.674 \times 10^{-11}$ (in $\mathrm{N\,m^2\,kg^{-2}}$) is the gravitational constant.

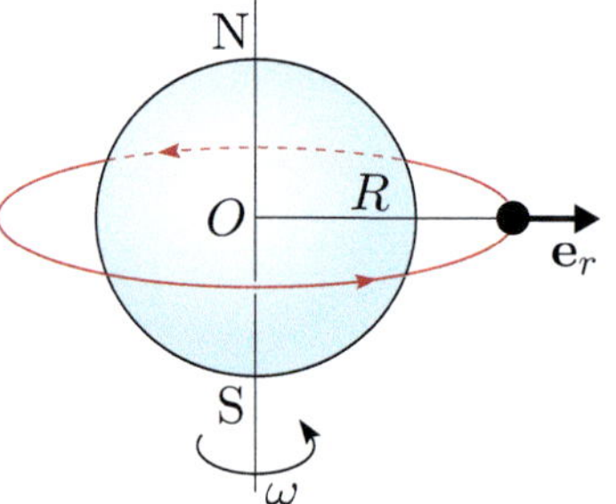

Figure 9 Satellite orbiting the Earth

In order to use Newton's law of universal gravitation to model the gravitational force on a geostationary satellite, we assume that we can model the Earth and the satellite as particles (evidence shows that this assumption is reasonable). Using polar coordinates in the equatorial plane, with origin O at the centre of the Earth, the position vector of the satellite can be written as $\mathbf{r} = R\mathbf{e}_r$, where R is the radius of the orbit (see Figure 9). So equation (25) can be written as

$$\mathbf{F} = -\frac{GmM}{R^2}\,\mathbf{e}_r.$$

Applying Newton's second law and equation (23) to the satellite, we obtain

$$-mR\omega^2\mathbf{e}_r = -\frac{GmM}{R^2}\,\mathbf{e}_r,$$

which leads to

$$R = \left(\frac{GM}{\omega^2}\right)^{1/3}.$$

Now the mass M of the Earth is $5.972 \times 10^{24}\,\mathrm{kg}$. Substituting the values of G, M and ω into the above equation, we obtain (in m)

$$R \simeq 4.216 \times 10^7.$$

The equatorial radius of the Earth is $6378.2\,\mathrm{km}$, so all geostationary satellites must orbit at approximately $35\,800\,\mathrm{km}$ above the Earth's surface.

Exercise 12

A good approximation for planetary motion is to assume that a planet orbits in uniform circular motion around the Sun under the action of their mutual gravitational attraction. Model the Sun and the planet as particles, take the origin at the centre of mass of the Sun, and take R to be the radius of the planet's orbit around this origin, so the position of the planet is $\mathbf{r} = R\mathbf{e}_r$. Newton's law of universal gravitation then tells us that a good model of the gravitational force on the planet is

$$\mathbf{F} = -\frac{GmM}{R^2}\,\mathbf{e}_r,$$

where m is the mass of the planet, and M is the mass of the Sun.

Verify that the square of the period T of the orbit is proportional to the cube of the radius of the orbit, that is,

$$T^2 \propto R^3.$$

This is Kepler's third law of planetary motion applied to circular orbits.

2.3 Conical pendulums

So far in this section we have considered examples of motion where the force causing the circular motion is in the plane of the motion. In Exercises 13–15 you are asked to consider examples where this is not the case.

Exercise 13

A *conical pendulum* is a pendulum whose bob traces out a horizontal circle rather than a vertical circle. The pendulum can be modelled as a particle of mass m joined by a light model rod of length l to a fixed point (the pivot). As the particle traces out a horizontal circle below the fixed point, the rod traces out the curved surface of a cone whose semi-vertical angle α is the inclination of the rod to the vertical (see Figure 10). Assume that for any given angle α, the bob performs uniform circular motion with angular speed ω.

(a) Determine the tension in the rod in terms of m, α and the magnitude of the acceleration due to gravity, g.

(b) Show that such motion is possible only if $\omega > \omega_0$, where the minimum angular speed is given by $\omega_0 = \sqrt{g/l}$, and that $0 < \alpha < \pi/2$. Show also that $\alpha \to \pi/2$ as $\omega \to \infty$.

(c) Show that for a given angular speed $\omega > \omega_0$, the vertical distance of the bob below the pivot is independent of l. In other words, regardless of the length of the rod, and with a given angular speed, the bob will always rotate at the same vertical distance below the pivot.

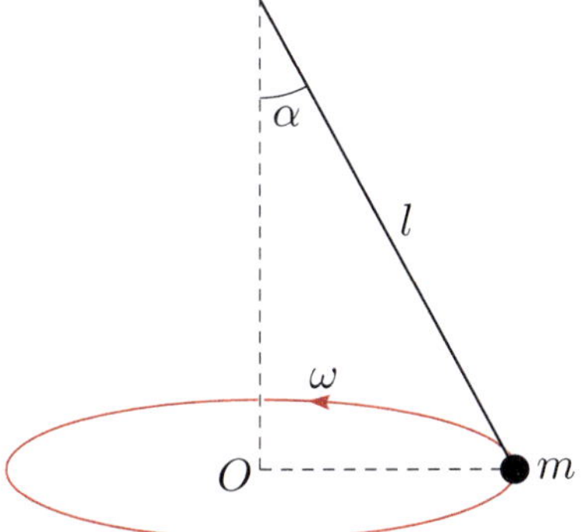

Figure 10   Conical pendulum

You saw in Exercise 13 that the angular speed of a conical pendulum is given by

$$\omega = \sqrt{g/(l \cos \alpha)},$$

where l is the length of the pendulum rod and α is the angle that the rod makes with the vertical. For a fixed l, this means that ω depends solely on α. As α increases, so does ω (since $\cos \alpha$ decreases), and $\omega \to \infty$ as $\alpha \to \frac{\pi}{2}$.

The fact that the angular speed of a conical pendulum depends on the angle made with the vertical is used in the design of a *governor* of a steam engine (see Figure 11), which regulates the supply of steam to the engine and thereby controls the speed at which the engine operates.

Figure 11   Steam engine with governor indicated

A model of a governor is illustrated in Figure 12.

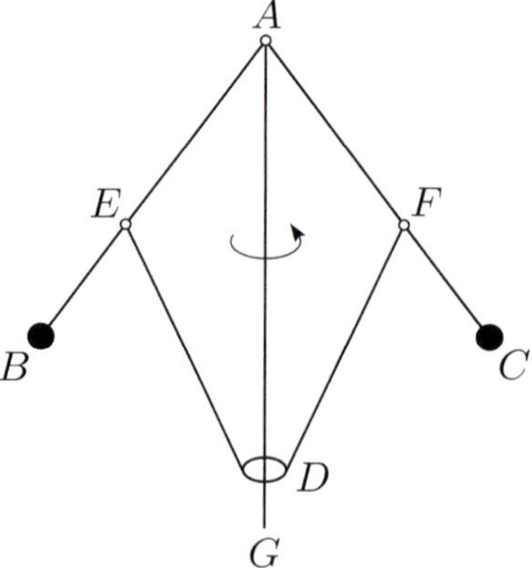

Figure 12   Model of a governor

The model consists of two light rods AB and AC, freely hinged at A to a vertical shaft AG that is rotated by the engine. At the other ends of the rods are two equal weights at B and C. Two other light rods DE and DF are freely hinged to the rods AB and AC, respectively, and to a collar at D that can slide freely up and down the shaft AG. A lever is attached to the collar at D and operates a valve that admits steam to the engine – as the collar rises, the amount of steam is decreased.

So when the angular speed of the shaft increases as more steam enters the engine, the weights at B and C rise and pull the collar at D up the shaft. This shuts off some of the steam entering the engine, and the speed of rotation is decreased. Similarly, when the speed of rotation decreases, the collar at D falls and more steam is admitted to the engine.

Exercise 14

A model of a steam governor assumes that in Figure 12, the points B and E coincide, and the points C and F coincide, that is, the rods DE and DF are freely hinged at B and C, respectively. From the symmetry of Figure 12, we need model only half of the governor, that is, two rods and one weight, and one half-collar at D. We model the rods and the shaft as light model rods, the weight as a particle P, and the half-collar as another particle. The model is shown in Figure 13.

Assume that the half-collar particle remains in smooth contact with the shaft, so there is no friction and the only force on the half-collar particle due to the shaft is the normal reaction. Also assume that there are no friction forces (or resistive torques) at any of the hinges, that both particles have the same mass m, and that both rods have the same length l.

The particle P rotates about the shaft in a horizontal circle with constant angular speed ω. Because both rods have the same length, each rod makes the same angle α with the vertical.

(a) Determine the tensions in the rods.

   (*Hint*: For fixed values of ω and α, the half-collar particle at D does not move.)

(b) Find an expression for the angular speed ω in terms of g, l and α, and find an expression for the angle α of inclination in terms of g, l and ω.

Exercise 15

A model string ACB has a particle of mass m attached at one end A and a particle of mass $2m$ attached at the other end B. The string is threaded through a smooth fixed ring C. The particle B is at rest vertically below the ring C, and the particle A is rotating in a horizontal circle of radius R with constant angular speed ω, as shown in Figure 14.

(a) Find the angle α of inclination of AC to the vertical.

(b) Find, in terms of R and g, the angular speed ω of particle A.

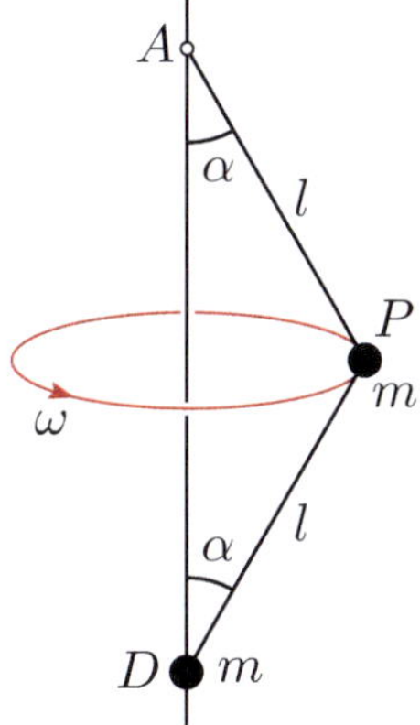

Figure 13   Simplified model of a governor

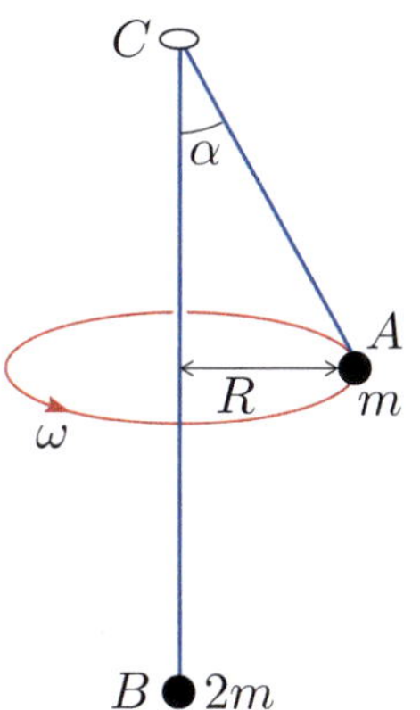

Figure 14

2.4 Cornering

In this subsection we look at the application of the formulas for circular motion to the real-world problem in the following example.

Example 2

Consider a vehicle of mass m moving at constant speed v around a circular bend of radius R. Suppose that the road has a horizontal surface, that is, it is not banked or cambered.

(a) If the coefficient of static friction between the vehicle and the road is μ, what is the maximum constant speed v at which the vehicle can go around the corner without slipping?

(b) Suppose that the vehicle is driven around a roundabout on a circle of radius $15\,\mathrm{m}$ at a constant speed of $10\,\mathrm{m\,s^{-1}}$ without skidding. What does this imply about the coefficient of static friction?

It may seem odd to use the coefficient of *static* friction here. However, there is no *sideways* motion between the tyres and the road.

Solution

(a) Model the vehicle as a particle, at position X. Choose the $\mathbf{e}_r$, $\mathbf{e}_\theta$ and $\mathbf{k}$ coordinate system shown in Figure 15.

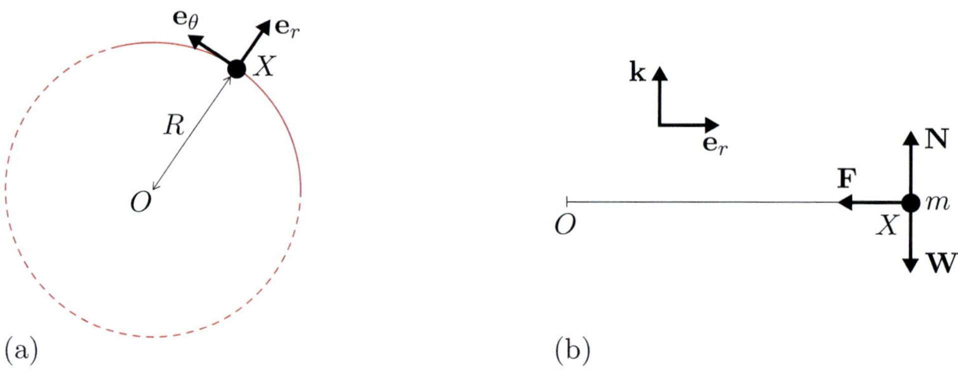

Figure 15 Vehicle following a circular bend at constant speed: (a) viewed from above; (b) viewed from the side, showing a vertical plane through OX and the forces acting in this plane

The forces acting on the vehicle are not all in one plane, so it is hard to draw a complete force diagram. Figure 15(b) shows the forces in a vertical plane through the vehicle and the centre of the circle, which are the weight $\mathbf{W}$, the normal reaction $\mathbf{N}$, and the sideways friction force $\mathbf{F}$. There are other forces acting on the vehicle, such as air resistance and the force propelling the vehicle. For the purposes of this model, we lump them together as one force $\mathbf{C}$, whose magnitude is unknown but whose direction is along the direction of travel, that is, tangential to the circle and perpendicular to the plane shown in Figure 15(b).

The vehicle is being treated as a particle, so $\mathbf{W}$ is the weight of the vehicle together with its occupants. The normal reaction $\mathbf{N}$ is the resultant of the separate forces acting at each tyre/road contact.

Applying Newton's second law to the vehicle gives

$$\mathbf{N} + \mathbf{W} + \mathbf{F} + \mathbf{C} = m\mathbf{a}. \tag{27}$$

Now we model the forces that appear in this equation. By inspection of Figure 15(b), we have

$$\mathbf{W} = -mg\mathbf{k}, \quad \mathbf{N} = |\mathbf{N}|\mathbf{k}, \quad \mathbf{F} = -|\mathbf{F}|\mathbf{e}_r.$$

By definition, the force $\mathbf{C}$ is tangential to the circle, so

$$\mathbf{C} = |\mathbf{C}|\mathbf{e}_\theta.$$

Since the vehicle is moving in a circular path at constant speed, the acceleration is $\mathbf{a} = (-v^2/R)\mathbf{e}_r$, using equation (23). Now we can resolve equation (27) in each of three orthogonal directions.

Resolving equation (27) in the $\mathbf{e}_\theta$-direction gives

$$|\mathbf{C}| = 0, \tag{28}$$

resolving equation (27) in the $\mathbf{k}$-direction gives

$$|\mathbf{N}| - mg = 0, \tag{29}$$

and resolving equation (27) in the $\mathbf{e}_r$-direction gives

$$-|\mathbf{F}| = -\frac{mv^2}{R}. \tag{30}$$

From equation (28) we see that the forces tangential to the circle must balance, since we have uniform circular motion. This justifies the modelling assumption that we can lump them all together.

From equation (29) we obtain $|\mathbf{N}| = mg$, so by the force model for static friction, we have

$$|\mathbf{F}| \leq \mu|\mathbf{N}| = \mu mg.$$

But from equation (30) we obtain $|\mathbf{F}| = mv^2/R$, which we can substitute into this inequality to obtain

$$\frac{mv^2}{R} \leq \mu mg,$$

that is,

$$v^2 \leq \mu Rg. \tag{31}$$

So the maximum (constant) speed at which the circular bend can be negotiated without skidding is $v = \sqrt{\mu Rg}$.

(b) We have $v = 10$ and $R = 15$. So from expression (31), we require that

$$\mu \geq \frac{v^2}{Rg} = \frac{100}{15 \times 9.81} \simeq 0.68.$$

So the coefficient of static friction must be at least 0.68 to avoid skidding.

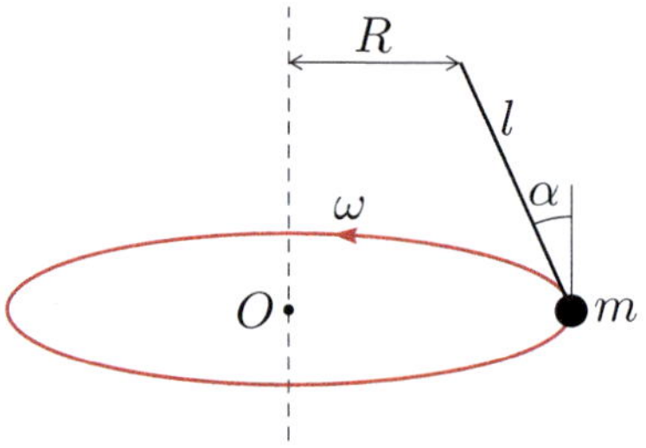

Figure 16 Model of a chair-o-plane roundabout

Exercise 16

Suppose that the coefficient of static friction between a particular car and the road in dry conditions is 1.3, and assume that the car follows a circular path at constant speed.

(a) At what speed could the car be driven around a roundabout on a circular path of radius $10\,\mathrm{m}$ in dry conditions without skidding?

(b) What would be the smallest radius of the roundabout that would enable the car to drive around it in dry conditions at $15\,\mathrm{m\,s^{-1}}$?

(c) Suppose that when conditions are wet, the coefficient of static friction between the car and the road is halved. What would be the smallest radius of the roundabout that would enable the car to be driven around it in wet conditions at $15\,\mathrm{m\,s^{-1}}$?

2.5 Chair-o-planes

A chair-o-plane, as mentioned in the Introduction (see Figure 1), relies on circular motion to supply its thrills. In this subsection we examine its motion by first working through an exercise.

Exercise 17

A chair-o-plane roundabout consists of seats that are suspended from chains. As the roundabout rotates, the chairs swing out at an angle. We will model a seat and passenger as a particle of mass m suspended from a model string of length l. The other end of the string is fixed to a point a horizontal distance R from the vertical axis of rotation, as shown in Figure 16. We assume that the roundabout is rotating at constant angular speed ω and that the system has reached a steady state with the string inclined to the vertical at an angle α.

Determine the equation of motion of the system, and hence show that

$$\frac{g}{\omega^2}\tan\alpha = R + l\sin\alpha. \tag{32}$$

It is not easy to solve equation (32) for α, but this is not necessary in order to answer the questions posed in the Introduction.

Question 1 Will a child swing out at a greater angle than a heavier adult?

Question 2 Will the people on the outside swing out at a greater angle than those on the inside? (In terms of the parameters in equation (32), this relates to how α varies with R, all other parameters being fixed.)

Question 3 How does the angle at which people swing out depend on the speed of rotation of the roundabout?

You may like to stop at this point and try to answer these questions yourself.

The answer to Question 1 is that the child and the adult will swing outwards at the same angle, because equation (32) is independent of m, the mass of the particle.

To answer Questions 2 and 3, the arguments can be made easier by looking at the graphs shown in Figure 17. Plotted on the same set of axes are the graphs of $y = (g/\omega^2)\tan\alpha$ and $y = R + l\sin\alpha$ (the two sides of equation (32)), for fixed ω, R and l, and for $0 \le \alpha \le \frac{\pi}{2}$. From equation (32), the intersection of these two graphs gives the angle α of inclination of the chain to the vertical.

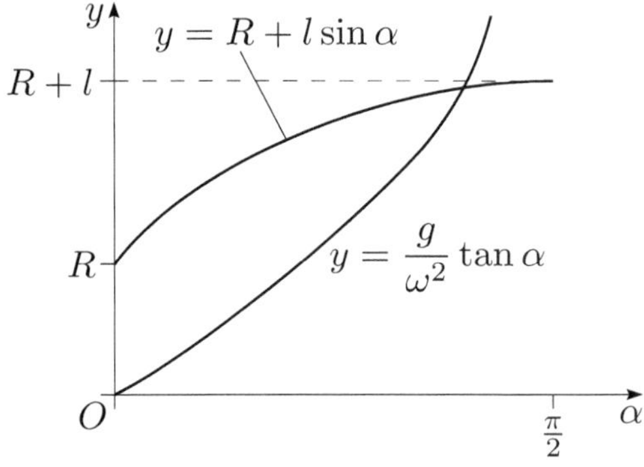

Figure 17   Graphical solution of $(g/\omega^2)\tan\alpha = R + l\sin\alpha$, at the intersection point of the two curves

The first point to note is that for any positive values of R, l and ω, the graphs intersect once for $0 \le \alpha < \frac{\pi}{2}$, so the motion is possible for all angular speeds ω.

To answer Question 2, note from Figure 17 that as R increases, the graph of $y = R + l\sin\alpha$ is translated upwards (assuming that all other parameters are fixed). This moves the point of intersection of the two graphs to the right. So a larger value of R corresponds to an angle α closer to $\alpha = \pi/2$. The answer to Question 2 is therefore that people on the outside do swing out further than those on the inside.

To answer Question 3, consider what happens to the point of intersection of the two graphs as ω increases. In this case, g/ω^2 decreases, and for a given α, the curve $y = (g/\omega^2)\tan\alpha$ becomes less steep. This has the effect of moving the point of intersection of the graphs to the right and hence increases the angle α of inclination. So the answer to Question 3 is that increasing the angular speed ω of the roundabout increases the angle of inclination α: as ω tends to infinity, the angle α tends to $\pi/2$, that is, the chains are almost horizontal if the roundabout has a very large angular speed.

(It can also be deduced from the graphs that the angle α of inclination increases as the length l of the chains increases.)

3 Non-uniform circular motion

In this section we look at **non-uniform circular motion**, that is, circular motion in which the angular rate of rotation is not constant. We will continue to make use of the polar unit vectors $\mathbf{e}_r$ and $\mathbf{e}_\theta$. Most of the section is concerned with the motion of a simple pendulum, which you examined in Unit 11. In that unit you modelled the motion using a second-order differential equation obtained by examining the mechanical energy of the system. You have modelled the motion using Newton's second law and Cartesian unit vectors. Here we use Newton's second law and polar unit vectors to obtain an equation for the varying angular speed of the pendulum.

We model the pendulum as a particle P of mass m attached to a light model rod of length l. The other end of the rod is attached to a fixed point O (a model pivot), and the particle moves in a vertical circle about this fixed point, as shown in Figure 18. The angle θ made by the rod with the vertical is measured anticlockwise from the downward vertical. We want to find expressions for the rate of rotation $\dot{\theta}$ and the tension in the rod in terms of θ.

We will use a polar coordinate system in the vertical plane of motion of the pendulum, with origin at the fixed point O, the centre of the circular motion. The forces on the particle are its weight $\mathbf{W}$ and the tension $\mathbf{T}$ due to the rod. The polar unit vectors $\mathbf{e}_r$ and $\mathbf{e}_\theta$, and a force diagram for the system, are shown in Figure 18.

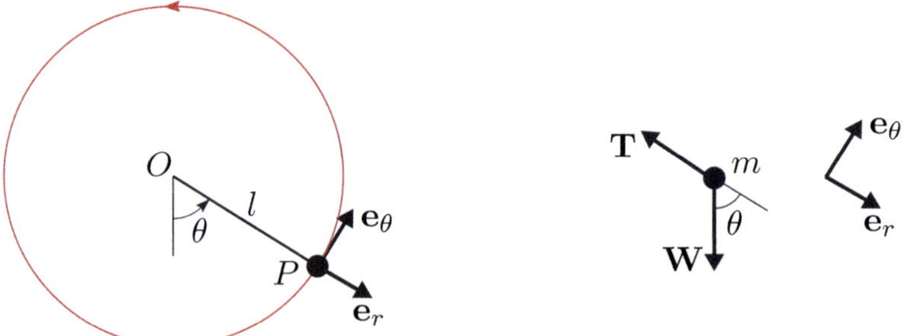

Figure 18 Vertical pendulum pivoted at O with a bob at P

With respect to this coordinate system,

$$\mathbf{W} = (mg\cos\theta)\mathbf{e}_r - (mg\sin\theta)\mathbf{e}_\theta, \tag{33}$$

$$\mathbf{T} = -T\mathbf{e}_r. \tag{34}$$

Note that we cannot write $\mathbf{T} = -|\mathbf{T}|\mathbf{e}_r$, as is often done elsewhere in this module, since the direction of $\mathbf{T}$ can point either towards or away from O depending on whether the rod is under tension or under compression. Instead, equation (34) uses T, which is defined as

$$T = \begin{cases} |\mathbf{T}| & \text{when the rod is under tension,} \\ -|\mathbf{T}| & \text{when the rod is under compression.} \end{cases}$$

Recall from Unit 2 that the modulus of a vector is a non-negative scalar, so $|\mathbf{T}| \geq 0$.

So if $T > 0$, then the force $\mathbf{T}$ is directed towards O so that the rod is under tension, as would be the case when the pendulum is hanging vertically downwards. On the other hand, if $T < 0$, then $\mathbf{T}$ is directed in the opposite direction, away from O so that the rod is under compression and exerts a normal reaction on the bob. Of course, in neither case does the rod's length change. The point where T vanishes can play an important role in further applications, as you will see later in this section.

The acceleration of the particle is non-zero and is given by equation (21):

$$\ddot{\mathbf{r}} = -l\dot{\theta}^2\mathbf{e}_r + l\ddot{\theta}\mathbf{e}_\theta. \tag{35}$$

Newton's second law applied to the particle gives

$$m\ddot{\mathbf{r}} = \mathbf{W} + \mathbf{T},$$

so, substituting from equations (33), (34) and (35), we obtain

$$-ml\dot{\theta}^2\mathbf{e}_r + ml\ddot{\theta}\mathbf{e}_\theta = (mg\cos\theta)\mathbf{e}_r - (mg\sin\theta)\mathbf{e}_\theta - T\mathbf{e}_r.$$

Resolving in the $\mathbf{e}_r$-direction gives

$$-ml\dot{\theta}^2 = mg\cos\theta - T. \tag{36}$$

Resolving in the $\mathbf{e}_\theta$-direction gives

$$ml\ddot{\theta} = -mg\sin\theta. \tag{37}$$

We concentrate first on integrating equation (37) to obtain an expression for the rate of rotation $\dot{\theta}$. By dividing equation (37) by ml it can be simplified to

$$\ddot{\theta} = -\frac{g}{l}\sin\theta, \tag{38}$$

which is an equation of the form

$$\ddot{\theta} = f(\theta) \tag{39}$$

(where in this case, $f(\theta) = -(g/l)\sin\theta$).

We then multiply equation (39) by $\dot{\theta}$ to obtain

$$\dot{\theta}\ddot{\theta} = \dot{\theta}\,f(\theta), \tag{40}$$

and apply the chain rule to give

$$\frac{d}{dt}(\dot{\theta}^2) = \frac{d(\dot{\theta}^2)}{d\dot{\theta}}\frac{d\dot{\theta}}{dt} = 2\dot{\theta}\ddot{\theta}.$$

Using this result after multiplying equation (40) by 2 leads to

$$\frac{d}{dt}(\dot{\theta}^2) = 2\dot{\theta}\,f(\theta),$$

which can be integrated to yield

$$\dot{\theta}^2 = 2\int f(\theta)\,\dot{\theta}\,dt = 2\int f(\theta)\frac{d\theta}{dt}\,dt = 2\int f(\theta)\,d\theta.$$

So with $f(\theta) = -(g/l)\sin\theta$, we arrive at the desired expression for $\dot\theta^2$, namely

$$\dot\theta^2 = 2\int f(\theta)\,d\theta = -\frac{2g}{l}\int \sin\theta\,d\theta = \frac{2g}{l}\cos\theta + C.$$

The constant of integration C can be found, for example, in terms of the rate of rotation $\dot\theta_0$ of the particle at the lowest point of the path, where $\theta = 0$. This initial condition leads to

$$C = \dot\theta_0^2 - \frac{2g}{l},$$

so

$$\dot\theta^2 = \dot\theta_0^2 - \frac{2g}{l}(1 - \cos\theta).$$

When $\theta = 0$, we have $v_0 = l|\dot\theta_0|$, so this equation becomes

$$\dot\theta^2 = \frac{v_0^2}{l^2} - \frac{2g}{l}(1 - \cos\theta). \tag{41}$$

In theory, this equation can now be integrated to find θ as a function of time t, but in practice, the integral involved cannot be evaluated in terms of elementary functions. However, equation (41) does permit us to calculate the speed $v = l|\dot\theta|$ of the particle at any point during its motion.

Equation (41) can also be derived more directly using the law of *conservation of energy*, without the need to resolve forces, let alone solve equation (38). This can be done by applying methods described in Units 9 and 16. Since gravity is a conservative force, the total mechanical energy of the pendulum is constant throughout its motion, and is given by

In addition to gravity there is also the tension $\mathbf{T}$, but since $\mathbf{T}$ is always perpendicular to the motion, it does not contribute to the work done on the particle or therefore to the energy (see Unit 16).

$$E = \tfrac{1}{2}mv^2 + U(\theta),$$

where $U(\theta)$ is the gravitational potential energy of the bob. This is given by $U = mgh$, where h is the height of the bob above the datum (see Figure 19 and Unit 9, Subsection 3.2). Taking the datum O to be the lowest point of the motion, that is, where $\theta = 0$, we have that $h = l - l\cos\theta$ and hence

$$E = \tfrac{1}{2}mv^2 + mgl(1 - \cos\theta). \tag{42}$$

Since E stays constant throughout the motion, its value can be determined by the condition that $v = v_0$ at $\theta = 0$, which implies that $E = \tfrac{1}{2}mv_0^2$. Substituting this into equation (42), together with $v^2 = l^2\dot\theta^2$, finally yields

$$\tfrac{1}{2}mv_0^2 = \tfrac{1}{2}ml^2\dot\theta^2 + mgl(1 - \cos\theta),$$

which can be rearranged to give equation (41).

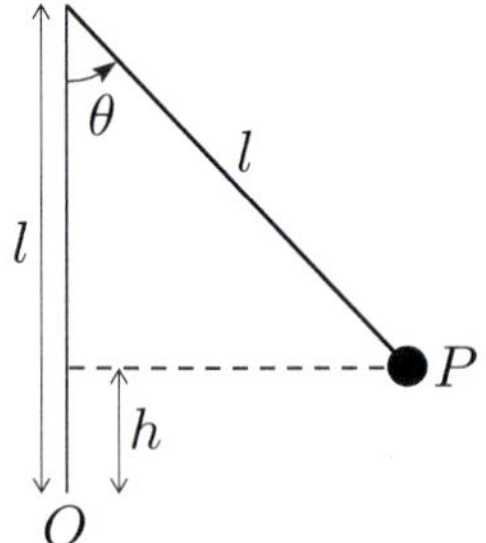

Figure 19 Vertical pendulum of length l with bob at P and datum at O

Exercise 18

Find the value of v_0 for which the bob just comes to rest at $\theta = \pi$, that is, at the highest point possible.

The solution to Exercise 18 implies that if $0 < v_0 < 2\sqrt{gl}$, then the bob oscillates back and forth with an amplitude of less than π, whereas for $v_0 > 2\sqrt{gl}$, the bob rotates in complete circles.

Let us now return to equation (36) and rewrite it as

$$T = mg\cos\theta + ml\dot\theta^2, \tag{43}$$

from which you can see that $T > 0$ (i.e. the rod is under tension) for all $|\theta| < \pi/2$ (i.e. whenever the bob is below the pivot), since $\cos\theta > 0$ on this interval, while $\dot\theta^2 \geq 0$ for any θ. We can use equation (41) to substitute for $\dot\theta^2$ in equation (43), to find the tension in the rod in terms of θ. Thus we have

$$T = mg\cos\theta + \frac{mv_0^2}{l} - 2mg(1 - \cos\theta),$$

so

$$T = \frac{mv_0^2}{l} + mg(3\cos\theta - 2). \tag{44}$$

Equation (44) can be used to determine the sign of T, but note that both $\dot\theta^2$ and T increase with increasing v_0.

Let us now consider a slightly different pendulum, where the light model rod is replaced with a model string.

Exercise 19

Consider the motion of a pendulum moving in a vertical plane, whose bob is attached to a fixed point by a model string.

For low, high and intermediate speeds, how is the motion of this new pendulum likely to differ from that of the original pendulum whose bob is attached to a fixed point by a light model rod?

The answer to Exercise 19 emphasises that in certain situations equations (41) and (44) apply only while the right-hand side of equation (44) is positive. Depending on the nature of the pendulum, other phenomena might occur if the right-hand side of equation (44) is negative. However, note that a pendulum consisting of a model rod and one consisting of a model string satisfy the same equations, (41) and (44), except that for a model string we require that $T > 0$.

Exercise 20

Consider the motion of a pendulum moving in a vertical plane, whose bob is attached to a fixed point by a model string.

(a) Show that if $v_0 > \sqrt{5gl}$, then the string never goes slack.

(b) If $v_0 = 2\sqrt{gl}$, find the angle θ for which the string goes slack.

In Exercise 20(a) you saw that the string never goes slack if the speed v_0 of the pendulum bob at its lowest point is greater than $\sqrt{5gl}$. In this case, the pendulum bob performs complete revolutions in the same direction. It was also predicted, in Exercise 19, that the string would not go slack and the pendulum would perform oscillatory motion for certain lower speeds. So what is the condition on v_0 for the string to remain taut at lower speeds?

To answer this question, look first at the tension in the string. Recall from equation (43), and the discussion immediately after it, that $T > 0$ whenever the bob is below the pivot. So the string can never go slack before the bob reaches the horizontal.

Above the horizontal, $\cos\theta$ is negative, in which case equation (43) implies that we have $T \leq ml\dot\theta^2$. So $T = 0$ before $\dot\theta = 0$, that is, the string becomes slack before the bob comes instantaneously to rest.

We know from Exercise 20(a) that the bob reaches the highest point of the circle if $v_0 > \sqrt{5gl}$. We have still to determine the initial speed so that the bob just reaches the horizontal.

Exercise 21

Find an initial speed v_0 so that the pendulum bob just reaches the horizontal when coming to rest.

We have found that the model of a pendulum consisting of a bob attached to a string predicts the following.

- For $0 < v_0 \leq \sqrt{2gl}$, the bob oscillates back and forth with an amplitude of less than or equal to $\pi/2$.

- For $\sqrt{2gl} < v_0 < \sqrt{5gl}$, the bob travels initially in a circular path with the string taut; then at some angle with $\pi/2 < |\theta| < \pi$, the string becomes slack and the bob falls under gravity as a projectile.

- For $v_0 \geq \sqrt{5gl}$, the bob travels continuously in a circle.

Note how this compares to the motion of a pendulum bob attached to a model rod, as described immediately below Exercise 18.

Exercise 22

A smooth hemispherical bowl of radius R, whose lowest point is at A, is fixed with its rim uppermost and horizontal. A particle of mass m is set in motion along the inner surface of the bowl towards A, so its subsequent motion is in a vertical plane through A. The particle is set in motion from a point at a vertical height $\frac{1}{2}R$ above A, with initial speed $\sqrt{gR}$.

(a) Show that the particle will just reach the top of the bowl.

(b) Find the magnitude of the normal reaction of the bowl on the particle when the particle is at a vertical height $\frac{2}{3}R$ above A.

The next two exercises involve noting a particular property of the normal reaction $\mathbf{N}$. Recall that when an object lies on a surface, the surface exerts a force $\mathbf{N}$ on the object, with $\mathbf{N}$ directed along the outward normal to the surface. You can think of $\mathbf{N}$ as the reaction of the surface to the object pressing down on it. Clearly, $\mathbf{N}$ cannot point in the opposite direction, towards the surface, and if the normal reaction vanishes, that is, $\mathbf{N} = \mathbf{0}$, then the object is no longer pressing down on the surface and is on the point of leaving the surface.

Exercise 23

Consider a particle P of mass m, sliding on the outside of a fixed smooth sphere of radius R. Assume that the particle has started from the highest point of the sphere with an initial speed v_0, and that the radius from the centre of the sphere to the particle makes an angle θ with the upward vertical, as shown in Figure 20. Assume also that the particle's path lies in a vertical plane.

If $v_0 = \sqrt{\frac{1}{2}gR}$, show that the particle leaves the surface of the sphere when $\theta = \arccos \frac{5}{6} \simeq 33.6°$.

(*Hint*: When considering the motion of the particle in a vertical plane from an equilibrium position at the highest point on the sphere, measuring θ anticlockwise from the upward vertical is similar to measuring θ anticlockwise from the downward vertical in the case of a pendulum.)

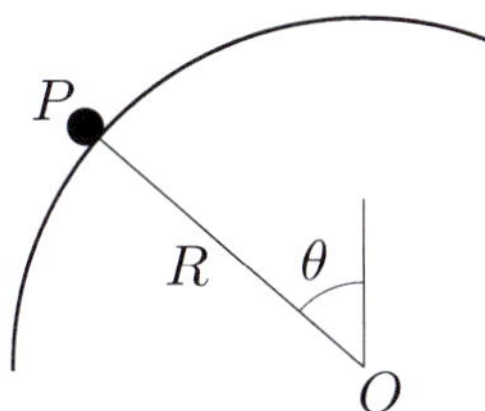

Figure 20   Particle sliding on the outside of a sphere

Exercise 24

The axis of a smooth fixed cylinder of radius R is horizontal. Two particles P_1, of mass $3m$, and P_2, of mass $5m$, are joined by a model string of length $\pi R/2$ and placed on the surface of the cylinder in a vertical plane perpendicular to the axis of the cylinder, as shown in Figure 21. Initially, the position of the particles is symmetrical about the vertical plane containing the axis of the cylinder and the origin O, with both OP_1 and OP_2 inclined at an angle of $\pi/4$ to the upward vertical through O. The particles are now released from rest, and θ is the angle between OP_1 and the horizontal t seconds after release.

For parts (a) and (b), assume that both particles are on the surface of the cylinder, and note that the angle that OP_2 makes with the vertical at time t is also θ.

(a) Show that $|\mathbf{N}|$, the magnitude of the normal reaction of the cylinder on P_2, and $\dot{\theta}$ are related by

$$|\mathbf{N}| = 5m(g\cos\theta - R\dot{\theta}^2).$$

(*Hint*: Consider the forces acting on particle P_2 at time t, and the acceleration of P_2.)

(b) By assuming that the total mechanical energy of the combined system is conserved, show that the speed v of the particles satisfies

$$v^2 = \tfrac{1}{4}gR(4\sqrt{2} - 5\cos\theta - 3\sin\theta).$$

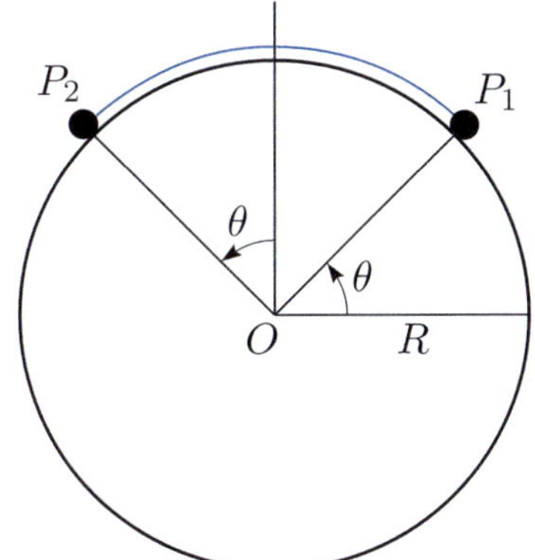

Figure 21   Particles P_1 and P_2 joined by a string and placed on the surface of a cylinder with axis passing through O. The points P_1, P_2 and O lie in the same vertical plane.

(c) Hence show that the angle θ at which P_2 leaves the cylinder is the solution of

$$9\cos\theta + 3\sin\theta = 4\sqrt{2} \quad (0 < 0 < \pi/2).$$

By re-expressing the left-hand side of this equation in terms of a single trigonometric function, show that this implies that

$$\theta = \arccos\left(\tfrac{4}{3\sqrt{5}}\right) + \arctan\left(\tfrac{1}{3}\right) \simeq 1.254 \ (\simeq 71.8^\circ).$$

4 Angular velocity and angular momentum

This section starts to look beyond circular motion to more general rotational motion. In this section we define two key quantities that will be useful in Unit 21, which looks at the motion of rigid bodies.

Subsection 4.1 defines the *angular velocity* of a particle. The magnitude of the angular velocity of a particle is the angular speed that we have been considering up to this point.

Linear momentum is defined both for a single particle and for an n-particle system in Unit 19.

In Unit 19 you were introduced to linear momentum and its conservation in the absence of external forces. You saw how the conservation of linear momentum could be used in the analysis of certain problems involving linear motion. For circular motion, and for rotational motion in general, the corresponding concept is *angular momentum*. Subsection 4.2 defines angular momentum for a particle and shows how it is linked to torques.

4.1 Angular velocity

The rotation of rigid bodies is considered in Unit 21.

If a particle is moving in a circle or if a rigid body is rotating about a fixed axis, then the (angular) rate of rotation and the direction of the axis of rotation are of primary importance. These two quantities can be encapsulated in the vector $\boldsymbol{\omega}$, known as the *angular velocity*. The magnitude of $\boldsymbol{\omega}$ is the angular speed, and the direction of $\boldsymbol{\omega}$ is defined to be along the axis of rotation in the sense given by the right-hand grip rule (or right-hand thumbs-up rule), that is, in the direction in which the thumb of a right hand would point if the fingers were curled in the direction of rotation (see Figure 22).

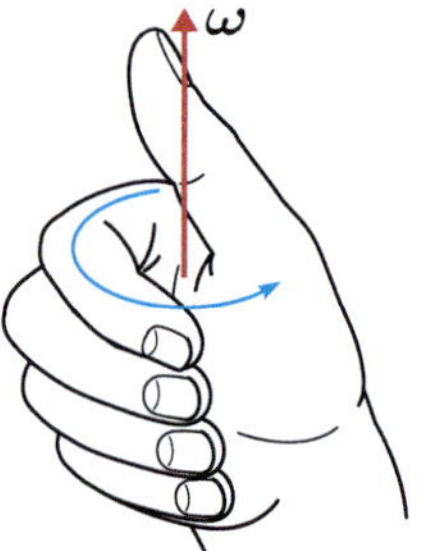

Figure 22 Right-hand grip rule

> **Angular velocity**
>
> The **angular velocity** $\boldsymbol{\omega}$ of a particle moving in a circle, or of a rigid body rotating about a fixed axis, is a vector with magnitude equal to the angular speed and with direction along the axis of rotation in the sense given by the right-hand grip rule.

Example 3

The Earth rotates in an easterly direction once in a sidereal day (the rotation period with respect to the fixed stars, which is less than that with respect to the Sun by one part in 365.25), which is 23.9343 hours.

Specify the angular velocity $\boldsymbol{\omega}$ of the rotation of the Earth.

Solution

The angular speed (in $\mathrm{rad\,s^{-1}}$) is

$$\omega = \frac{2\pi}{23.9343 \times 60 \times 60}$$
$$\simeq 7.292 \times 10^{-5}.$$

The direction of the angular velocity $\boldsymbol{\omega}$ is from the South Pole to the North Pole, as shown in Figure 23.

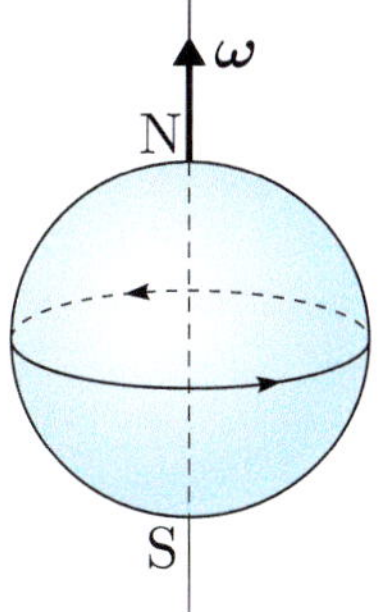

Figure 23   Rotation of the Earth

The direction of the angular velocity $\boldsymbol{\omega}$ is along the axis of rotation and is therefore perpendicular to the plane of motion. To be more specific about this direction, we use the usual right-handed system shown in Figure 24. Using the right-hand grip rule, the direction of the angular velocity vector $\boldsymbol{\omega}$ will be in the positive $\mathbf{k}$-direction if $\dot{\theta} > 0$ and in the negative $\mathbf{k}$-direction if $\dot{\theta} < 0$, as shown in Figure 25(a) and Figure 25(b), respectively.

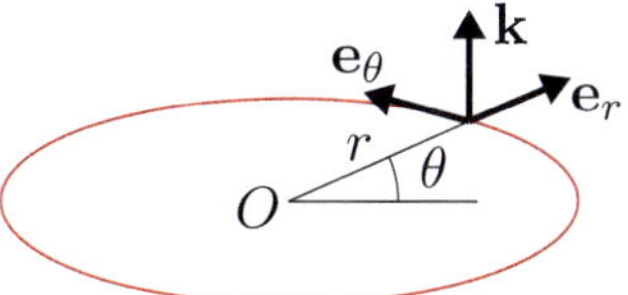

Figure 24   Right-handed rotating system

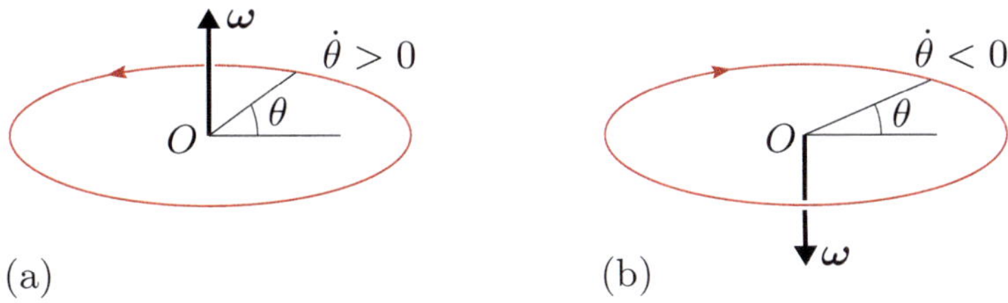

(a)   (b)

Figure 25   Rotating system with (a) $\dot{\theta} > 0$ and (b) $\dot{\theta} < 0$

Therefore

$$\boldsymbol{\omega} = \begin{cases} \omega\mathbf{k} & (\dot{\theta} > 0) \\ -\omega\mathbf{k} & (\dot{\theta} < 0) \end{cases} = \begin{cases} |\dot{\theta}|\mathbf{k} & (\dot{\theta} > 0) \\ -|\dot{\theta}|\mathbf{k} & (\dot{\theta} < 0) \end{cases} = \dot{\theta}\mathbf{k}.$$

So the relationship $\boldsymbol{\omega} = \dot{\theta}\mathbf{k}$ holds, regardless of the direction of rotation.

The relationship also holds if there is no rotation, when $\dot{\theta} = 0$ and $\boldsymbol{\omega} = \mathbf{0}$.

Relationship between angular velocity and the axis and rate of rotation

The angular velocity $\boldsymbol{\omega}$ of a particle moving in a circle about an axis in the $\mathbf{k}$-direction, with angular rate of rotation given by $\dot{\theta}$, is

$$\boldsymbol{\omega} = \dot{\theta}\mathbf{k}. \tag{45}$$

We complete this subsection by establishing the link between the angular velocity $\boldsymbol{\omega}$ of a particle in circular motion and its velocity $\mathbf{v} = \dot{\mathbf{r}}$. For circular motion of radius R, we know from equation (19) that $\mathbf{r} = R\mathbf{e}_r$, relative to the centre of the circle, and from equation (20) that $\mathbf{v} = \dot{\mathbf{r}} = R\dot{\theta}\mathbf{e}_\theta$. We also know that $\boldsymbol{\omega} = \dot{\theta}\mathbf{k}$. Therefore

You can check that $\mathbf{k} \times \mathbf{e}_r = \mathbf{e}_\theta$ by looking at Figure 24.

$$\boldsymbol{\omega} \times \mathbf{r} = \dot{\theta}\mathbf{k} \times R\mathbf{e}_r = R\dot{\theta}(\mathbf{k} \times \mathbf{e}_r) = R\dot{\theta}\mathbf{e}_\theta = \mathbf{v}.$$

This result also holds if we take the origin to be anywhere on the axis of rotation. For then $\mathbf{r} = R\mathbf{e}_r + z\mathbf{k}$ for some z, and

Recall from Unit 2 that $\mathbf{k} \times \mathbf{k} = \mathbf{0}$.

$$\begin{aligned}
\boldsymbol{\omega} \times \mathbf{r} &= (\dot{\theta}\mathbf{k}) \times (R\mathbf{e}_r + z\mathbf{k}) \\
&= \dot{\theta}\mathbf{k} \times R\mathbf{e}_r + \dot{\theta}\mathbf{k} \times z\mathbf{k} \\
&= R\dot{\theta}(\mathbf{k} \times \mathbf{e}_r) \\
&= R\dot{\theta}\mathbf{e}_\theta \\
&= \mathbf{v}.
\end{aligned}$$

> **Relationship between velocity and angular velocity**
>
> The velocity $\mathbf{v}$ of a particle moving in a circle with angular velocity $\boldsymbol{\omega}$ is
>
> $$\mathbf{v} = \boldsymbol{\omega} \times \mathbf{r}, \tag{46}$$
>
> where $\mathbf{r}$ is the position vector of the particle relative to an origin on the axis of rotation.

Exercise 25

A particle moves clockwise in a circle at a constant angular speed of $2\,\mathrm{rad\,s^{-1}}$.

(a) Find the angular velocity of the particle.

(b) If the position vector of the particle, relative to the centre of the circle, at a certain instant is $3\mathbf{i} + 4\mathbf{j}$, find its velocity at that instant both in terms of the Cartesian unit vectors $\mathbf{i}$ and $\mathbf{j}$, and in terms of the corresponding polar unit vectors $\mathbf{e}_r$ and $\mathbf{e}_\theta$.

4.2 Angular momentum

Consider a particle of mass m moving along a curve in space. This is illustrated in Figure 26.

At time t the particle has position vector $\mathbf{r} = \mathbf{r}(t)$, velocity $\dot{\mathbf{r}} = \dot{\mathbf{r}}(t)$ and acceleration $\ddot{\mathbf{r}} = \ddot{\mathbf{r}}(t)$ relative to a fixed Cartesian coordinate system. The linear momentum of the particle is

$$\mathbf{p} = m\dot{\mathbf{r}}.$$

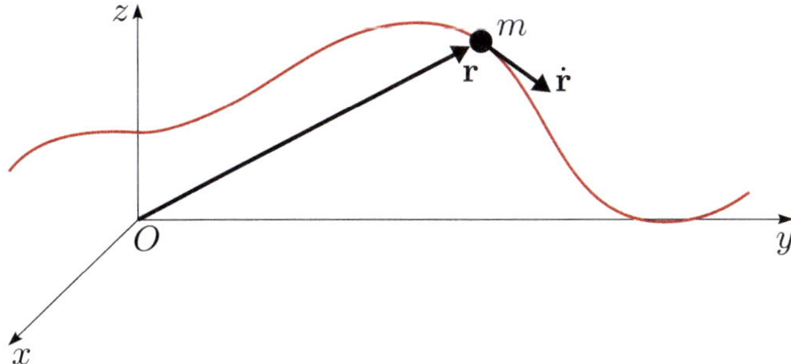

Figure 26   Particle with position $\mathbf{r}$ and velocity $\dot{\mathbf{r}}$, moving along a curve

You saw in Unit 19 that the linear momentum is related to the force $\mathbf{F}$ acting on the particle according to the equation

$$\mathbf{F} = \dot{\mathbf{p}}, \tag{47}$$

which says that the rate of change of linear momentum of the particle is equal to the force acting on the particle.

In Unit 2 we quantified the turning effect of a force about a fixed point O by defining the torque $\boldsymbol{\Gamma}$ of the force about O as

$$\boldsymbol{\Gamma} = \mathbf{r} \times \mathbf{F}.$$

By analogy with the case of linear momentum, when analysing rotational motion it would be useful to have a quantity that plays the same role in relation to torques as linear momentum does in relation to forces. So by analogy with equation (47), we would like the derivative of this quantity to equal the torque on the particle. A suitable quantity is

$$\mathbf{L} = \mathbf{r} \times \mathbf{p} = \mathbf{r} \times m\dot{\mathbf{r}},$$

since, using the rule for differentiating a cross product, we have

$$\begin{aligned}
\dot{\mathbf{L}} &= \frac{d}{dt}(\mathbf{r} \times m\dot{\mathbf{r}}) \\
&= (\dot{\mathbf{r}} \times m\dot{\mathbf{r}}) + (\mathbf{r} \times m\ddot{\mathbf{r}}) \\
&= \mathbf{r} \times m\ddot{\mathbf{r}} \\
&= \mathbf{r} \times \mathbf{F} \\
&= \boldsymbol{\Gamma}.
\end{aligned}$$

We call $\mathbf{L} = \mathbf{L}(t)$ the *angular momentum* of the particle about the fixed point.

> **Angular momentum**
>
> For a particle that has linear momentum $\mathbf{p} = m\dot{\mathbf{r}}$ and position vector $\mathbf{r}$ relative to an origin O, its **angular momentum** $\mathbf{L}$ about O is
>
> $$\mathbf{L} = \mathbf{r} \times \mathbf{p} = \mathbf{r} \times m\dot{\mathbf{r}}. \tag{48}$$

This equation is Newton's second law $\mathbf{F} = m\ddot{\mathbf{r}}$ expressed in terms of linear momentum.

The cross product of parallel vectors is zero, so $\dot{\mathbf{r}} \times m\dot{\mathbf{r}} = \mathbf{0}$.

The SI units for the magnitude of angular momentum are $\mathrm{kg\,m^2\,s^{-1}}$.

As with torques, the angular momentum of a particle is dependent on the choice of origin O. Figure 27 illustrates this, where the vector $\mathbf{L} = \mathbf{r} \times m\dot{\mathbf{r}}$ represents the angular momentum relative to the origin O, while the vector $\mathbf{L}' = \mathbf{r}' \times m\dot{\mathbf{r}}'$ represents the angular momentum relative to the origin O'. These vectors may differ in both magnitude and direction. (However, since both coordinate systems are static, the velocity of the particle is the same in each case, i.e. $\dot{\mathbf{r}} = \dot{\mathbf{r}}'$.)

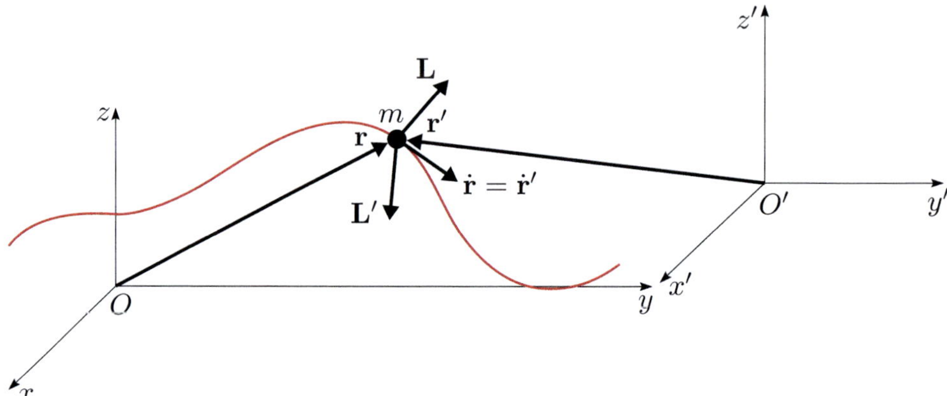

Figure 27   Angular momentum relative to different static origins O and O'

The fact that $\dot{\mathbf{L}} = \boldsymbol{\Gamma}$ is given a special name.

The torque law is fundamental to the study of rotating bodies.

Torque law for a particle

The rate of change of a particle's angular momentum about a fixed point is equal to the applied torque about that point, that is,

$$\dot{\mathbf{L}} = \boldsymbol{\Gamma}.$$

Example 4

A particle of mass m, on which no forces act, moves with constant speed u parallel to the y-axis in the positive y-direction in the (x, y)-plane. At time $t = t_0$, its path intersects the x-axis at $x = b$ (see Figure 28).

Calculate the magnitude and direction of the angular momentum of the particle with respect to the origin O.

Solution

The position of the particle in the coordinate system of Figure 28 is

$$\mathbf{r} = b\mathbf{i} + u(t - t_0)\mathbf{j},$$

while the velocity is $\dot{\mathbf{r}} = u\mathbf{j}$.

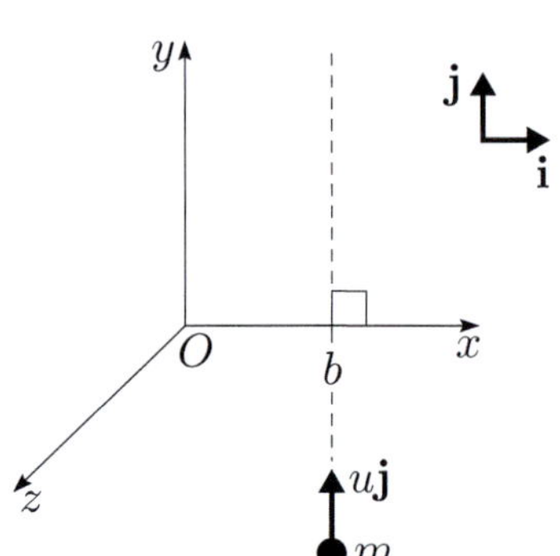

Figure 28   Particle moving parallel to the y-axis at constant speed

So using $\mathbf{i} \times \mathbf{j} = \mathbf{k}$ and $\mathbf{j} \times \mathbf{j} = \mathbf{0}$, the angular momentum is

$$\mathbf{L} = \mathbf{r} \times m\dot{\mathbf{r}}$$
$$= (b\mathbf{i} + u(t - t_0)\mathbf{j}) \times mu\mathbf{j}$$
$$= bmu\mathbf{k}.$$

Thus even though the vector $\mathbf{r}$ is changing continuously, the cross product $\mathbf{r} \times m\dot{\mathbf{r}}$ is a constant vector.

Hence the angular momentum points along the positive z-axis and has magnitude bmu.

Exercise 26

At a given time, a particle of mass 3 has position $\mathbf{i} + 2\mathbf{j}$ and velocity $2\mathbf{i} - \mathbf{j}$. All quantities are measured in SI units.

What is the particle's angular momentum about the origin of the coordinate system at this time?

Example 5

A particle of mass m at position $\mathbf{r} = x\mathbf{i} + y\mathbf{j} + z\mathbf{k}$ moves in a circle, whose centre lies on the z-axis, with angular velocity $\boldsymbol{\omega} = \dot{\theta}\mathbf{k}$.

Find the angular momentum $\mathbf{L}$ of the particle about the origin O in terms of x, y, z, $\dot{\theta}$, $\mathbf{i}$, $\mathbf{j}$ and $\mathbf{k}$.

Solution

The motion of the particle is shown in Figure 29. By definition, we have $\mathbf{L} = \mathbf{r} \times m\dot{\mathbf{r}}$. From equation (46), the velocity is

$$\dot{\mathbf{r}} = \boldsymbol{\omega} \times \mathbf{r} = \dot{\theta}\mathbf{k} \times (x\mathbf{i} + y\mathbf{j} + z\mathbf{k})$$
$$= \dot{\theta}(x\mathbf{j} - y\mathbf{i}).$$

It follows that

$$\mathbf{L} = \mathbf{r} \times m\dot{\mathbf{r}} = (x\mathbf{i} + y\mathbf{j} + z\mathbf{k}) \times m\dot{\theta}(x\mathbf{j} - y\mathbf{i})$$
$$= m\dot{\theta}(-xz\mathbf{i} - yz\mathbf{j} + (x^2 + y^2)\mathbf{k}).$$

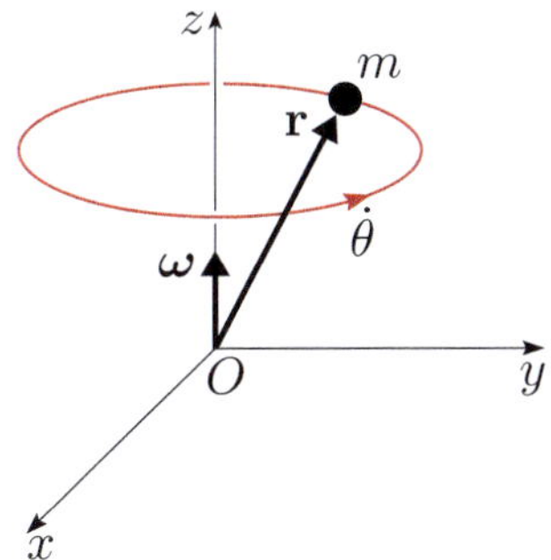

Figure 29 Particle moving in a circle

Example 5 demonstrates that even for circular motion, the direction of the angular momentum $\mathbf{L}$ is *not*, in general, the same as the direction of the angular velocity $\boldsymbol{\omega}$. However, if the origin is at the centre of the circular motion, then the angular velocity and the angular momentum of the particle are in the same direction, since then $z = 0$ in Example 5 and $\mathbf{L} = m\dot{\theta}(x^2 + y^2)\mathbf{k} = mR^2\dot{\theta}\mathbf{k}$, where R is the radius of the circular motion. Note that we can write $\mathbf{L} = mR^2\boldsymbol{\omega}$.

These results hold even when $\boldsymbol{\omega}$ is not constant.

The relationship between angular velocity and angular momentum will be explored further in Unit 21.

This result can be extended to general motion confined to the (x, y)-plane, not just circular motion. If the particle has position vector

$$\mathbf{r} = x\mathbf{i} + y\mathbf{j} = r\mathbf{e}_r,$$

where r, the distance to the origin, is not necessarily constant, then using the results of Subsection 1.2 we have for the velocity

$$\dot{\mathbf{r}} = \frac{d}{dt}(r\mathbf{e}_r) = \dot{r}\mathbf{e}_r + r\dot{\mathbf{e}}_r = \dot{r}\mathbf{e}_r + r\dot{\theta}\mathbf{e}_\theta,$$

where we have used $\dot{\mathbf{e}}_r = \dot{\theta}\mathbf{e}_\theta$ from equation (16). Hence the angular momentum about the origin is

$$\begin{aligned} \mathbf{L} = \mathbf{r} \times m\dot{\mathbf{r}} &= mr\mathbf{e}_r \times (\dot{r}\mathbf{e}_r + r\dot{\theta}\mathbf{e}_\theta) \\ &= mr(\dot{r}\mathbf{e}_r \times \mathbf{e}_r + r\dot{\theta}\mathbf{e}_r \times \mathbf{e}_\theta) \\ &= mr^2\dot{\theta}\mathbf{k}. \end{aligned}$$

We have used $\mathbf{e}_r \times \mathbf{e}_r = \mathbf{0}$ and $\mathbf{e}_r \times \mathbf{e}_\theta = \mathbf{k}$.

Angular momentum for general planar motion

The angular momentum $\mathbf{L}$ about the origin of a particle of mass m with motion (not necessarily circular) confined to the (x, y)-plane is

$$\mathbf{L} = mr^2\dot{\theta}\mathbf{k}. \tag{49}$$

Exercise 27

A particle of mass m moves in the (x, y)-plane in a circle of radius R, whose centre is at the origin, with angular velocity $\boldsymbol{\omega} = \dot{\theta}\mathbf{k}$.

Show that $|\mathbf{L}| = m|\mathbf{r}|\,|\dot{\mathbf{r}}|$, where $\mathbf{L}$ is the angular momentum of the particle about the origin, $\mathbf{r}$ is its position relative to the origin, and $\dot{\mathbf{r}}$ is its velocity.

Example 6

A pendulum bob of mass m is fixed to an origin O by a taut light string of constant length l. The bob swings to and fro in a vertical (x, y)-plane.

Use the torque law to derive the equation of motion of the bob.

Solution

Model the bob as a particle and the string as a model string. Use a polar coordinate system with origin O at the centre of the motion, and let $\mathbf{k}$ be a unit vector pointing out of the page for the set-up shown in Figure 30.

The total force $\mathbf{F}$ acting on the particle is the sum of its weight $\mathbf{W}$ and the tension force $\mathbf{T}$ due to the string, that is,

$$\mathbf{F} = \mathbf{W} + \mathbf{T},$$

where $\mathbf{W} = (mg\cos\theta)\mathbf{e}_r - (mg\sin\theta)\mathbf{e}_\theta$ and $\mathbf{T} = -|\mathbf{T}|\mathbf{e}_r$. The corresponding torque about O is

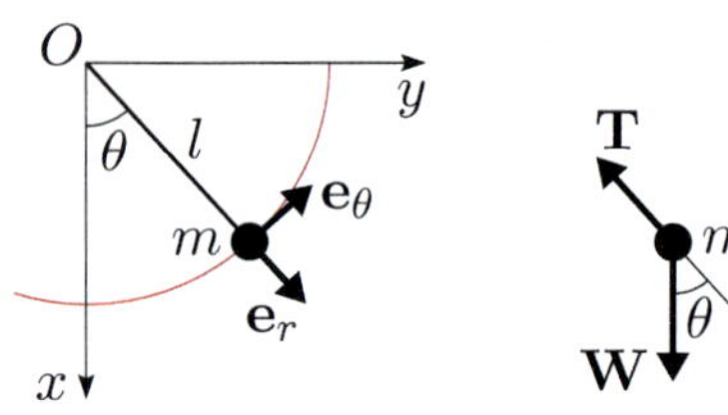

Figure 30 Pendulum swinging in a vertical plane

$$\boldsymbol{\Gamma} = \mathbf{r} \times \mathbf{F} = \mathbf{r} \times (\mathbf{W} + \mathbf{T}) = l\mathbf{e}_r \times ((mg\cos\theta)\mathbf{e}_r - (mg\sin\theta)\mathbf{e}_\theta - |\mathbf{T}|\mathbf{e}_r)$$
$$= -(mgl\sin\theta)\mathbf{k}.$$

From equation (49), with $r = l$, the angular momentum of the particle about O is

$$\mathbf{L} = ml^2\dot{\theta}\mathbf{k}.$$

Hence, by the torque law, we have

$$\frac{d}{dt}(ml^2\dot{\theta}\mathbf{k}) = -(mgl\sin\theta)\mathbf{k}.$$

Differentiating, resolving in the **k**-direction and rearranging, we obtain the equation of motion

$$\ddot{\theta} = -\frac{g}{l}\sin\theta.$$

This result is the same as that obtained by different means in Unit 11 and in Section 3.

Exercise 28

The position of a particle of mass m at time t is given by

$$\mathbf{r}(t) = a\cos(\alpha t)\mathbf{i} + a\sin(\alpha t)\mathbf{j} + ut\mathbf{k},$$

where a, α and u are constants. The particle is moving in a spiral path as shown in Figure 31.

Write down expressions for the linear momentum of the particle, the force acting on the particle, the angular momentum of the particle about the origin O, and the torque about O acting on the particle. Verify the torque law for the motion of the particle.

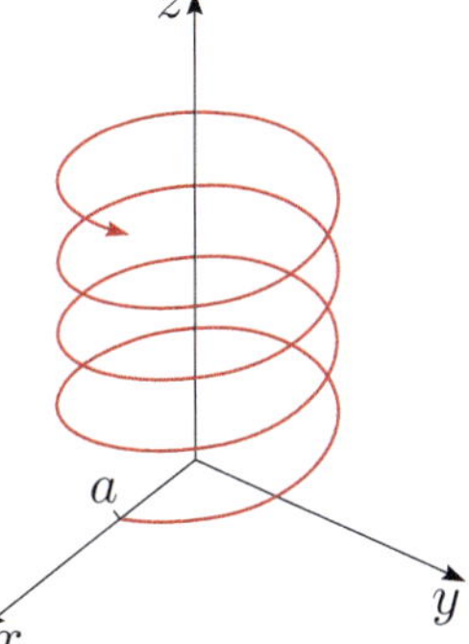

Figure 31   Particle moving in a spiral path

Exercise 29

In the film *The Wizard of Oz*, a house is swept up in a whirlwind. After a while, the whirlwind begins to dissipate and the house is subject to a decreasing torque. Although the house moves slowly towards the axis of the whirlwind, its motion is almost circular around this axis.

Model the house as a particle of mass m performing anticlockwise circular motion of radius R around a vertical z-axis in a horizontal (x, y)-plane, as shown in Figure 32. Model the decreasing torque by $\boldsymbol{\Gamma} = \Gamma_0 e^{-\alpha t}\mathbf{k}$, where Γ_0 and α are positive constants, t represents time, and $\mathbf{k}$ is a unit vector in the positive z-direction.

(a) Find an expression for the angular speed of the particle in terms of the other variables and parameters. To what value does this angular speed tend as t increases?

(b) Suppose that we no longer ignore the motion towards the axis of the whirlwind, and allow R to become smaller and smaller. How does this affect the angular speed? As R becomes smaller, does it remain valid to model the house as a particle?

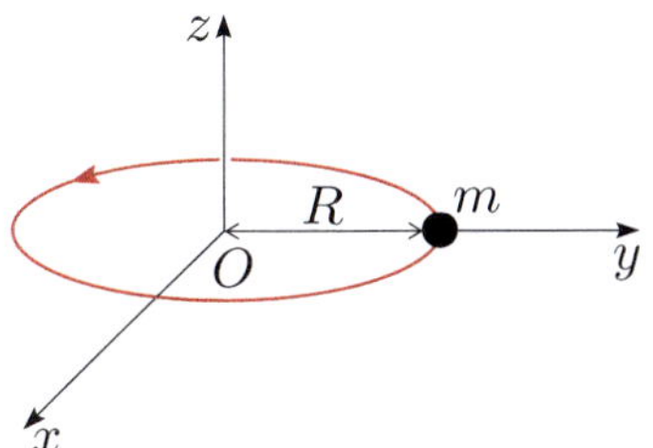

Figure 32   House swept up by a whirlwind

The most important special case for the torque law is the case when no resultant torque acts on a particle. In this case the torque law $\dot{\mathbf{L}} = \boldsymbol{\Gamma}$ reduces to $\dot{\mathbf{L}} = \mathbf{0}$, so $\mathbf{L}$ is a constant vector. So if there is no resultant torque, then the angular momentum is constant. This is usually stated in the following form.

> **Law of conservation of angular momentum for a particle**
>
> If the total torque acting on a particle about a fixed point is zero, then the angular momentum of the particle about that point is constant. In other words, if $\boldsymbol{\Gamma} = \mathbf{0}$, then $\mathbf{L}$ is constant.

Central force fields were introduced in Unit 16 and were shown to be conservative.

The total torque acting on a particle will be zero if the total force acting on the particle is zero, but this is not an especially interesting situation. More noteworthy is the case when the total force acting on a particle is directed towards a fixed point, which we can choose to be the origin. A force of this type is called a **central force**, and in the case where it is also spherically symmetric, it can be expressed as $\mathbf{F} = f(r)\,\widehat{\mathbf{r}}$, where $\mathbf{r} = x\mathbf{i} + y\mathbf{j} + z\mathbf{k}$, $r = |\mathbf{r}|$ and $\widehat{\mathbf{r}} = \mathbf{r}/|\mathbf{r}|$. The gravitational force given in equation (26) is an important example of a central force.

Exercise 30

Show that the total torque of the central force $\mathbf{F} = f(r)\,\widehat{\mathbf{r}}$ about the origin is zero.

Exercise 31

Consider a particle moving around a circle at constant speed.

Show that the total torque about the centre of the circle is zero.

Learning outcomes

After studying this unit, you should be able to:

- differentiate products of vector functions
- understand and use the polar unit vectors $\mathbf{e}_r$ and $\mathbf{e}_\theta$
- understand and use the expressions for position, velocity and acceleration for the circular motion of a particle in polar coordinates
- understand and use Newton's law of universal gravitation
- understand and use angular velocity in the description of circular motion
- relate the angular velocity to the velocity of a particle in circular motion
- use polar unit vectors to solve mechanics problems involving uniform and non-uniform circular motion
- understand and use the concept of angular momentum for the motion of a particle
- use the torque law to solve simple mechanics problems involving a single particle.

Solutions to exercises

Solution to Exercise 1

(a) Differentiating and noting that $\mathbf{i}$ and $\mathbf{j}$ are constant vectors, we find that

$$\dot{\mathbf{r}}(t) = -12\sin(4t)\mathbf{i} + 12\cos(4t)\mathbf{j},$$
$$\ddot{\mathbf{r}}(t) = -48\cos(4t)\mathbf{i} - 48\sin(4t)\mathbf{j}.$$

(b) The magnitude of the acceleration vector is

$$|\ddot{\mathbf{r}}(t)| = \sqrt{(-48\cos(4t))^2 + (-48\sin(4t))^2}$$
$$= \sqrt{48^2(\cos^2(4t) + \sin^2(4t))} = 48.$$

(c) The acceleration vector can be written as

$$-16(3\cos(4t)\mathbf{i} + 3\sin(4t)\mathbf{j}) = -16\mathbf{r}(t).$$

So $\ddot{\mathbf{r}}(t) = -c\,\mathbf{r}(t)$, with $c = 16$.

Solution to Exercise 2

The magnitude of the sideways force, towards the centre of the circle, exerted on the locomotive by the track is $|\mathbf{F}| = m|\mathbf{a}| = m\omega^2 R$. We have $m = 40$ and $R = 100$. If one circuit is completed in $50\,\mathrm{s}$, then the angular speed is $\omega = 2\pi/50$. So the magnitude of the force is

$$40\left(\frac{2\pi}{50}\right)^2 100 \simeq 63.17.$$

So the magnitude of the sideways force exerted by the track on the locomotive is $63.17\,\mathrm{N}$.

Solution to Exercise 3

We write $\mathbf{f}$ and $\mathbf{g}$ in component form as

$$\mathbf{f} = f_1\mathbf{i} + f_2\mathbf{j} + f_3\mathbf{k}, \quad \mathbf{g} = g_1\mathbf{i} + g_2\mathbf{j} + g_3\mathbf{k},$$

so

$$\mathbf{f} \cdot \mathbf{g} = f_1 g_1 + f_2 g_2 + f_3 g_3.$$

Using the formula for differentiating the product of two scalar functions, we obtain

$$\frac{d}{dt}(\mathbf{f} \cdot \mathbf{g}) = \frac{d}{dt}(f_1 g_1) + \frac{d}{dt}(f_2 g_2) + \frac{d}{dt}(f_3 g_3)$$

$$= \left(\frac{df_1}{dt} g_1 + f_1 \frac{dg_1}{dt}\right) + \left(\frac{df_2}{dt} g_2 + f_2 \frac{dg_2}{dt}\right) + \left(\frac{df_3}{dt} g_3 + f_3 \frac{dg_3}{dt}\right)$$

$$= \left(\frac{df_1}{dt} g_1 + \frac{df_2}{dt} g_2 + \frac{df_3}{dt} g_3\right) + \left(f_1 \frac{dg_1}{dt} + f_2 \frac{dg_2}{dt} + f_3 \frac{dg_3}{dt}\right)$$

$$= \left(\frac{df_1}{dt}\mathbf{i} + \frac{df_2}{dt}\mathbf{j} + \frac{df_3}{dt}\mathbf{k}\right) \cdot (g_1\mathbf{i} + g_2\mathbf{j} + g_3\mathbf{k})$$

$$+ (f_1\mathbf{i} + f_2\mathbf{j} + f_3\mathbf{k}) \cdot \left(\frac{dg_1}{dt}\mathbf{i} + \frac{dg_2}{dt}\mathbf{j} + \frac{dg_3}{dt}\mathbf{k}\right)$$

$$= \frac{d\mathbf{f}}{dt} \cdot \mathbf{g} + \mathbf{f} \cdot \frac{d\mathbf{g}}{dt},$$

as required.

Solution to Exercise 4

Using equation (7),

$$\frac{d}{dt}(\mathbf{f} \cdot \mathbf{f}) = \frac{d\mathbf{f}}{dt} \cdot \mathbf{f} + \mathbf{f} \cdot \frac{d\mathbf{f}}{dt} = 2\mathbf{f} \cdot \frac{d\mathbf{f}}{dt}.$$

Solution to Exercise 5

Differentiating equation (13) with respect to time using the chain rule, we have

$$\dot{\mathbf{e}}_\theta = (-\cos\theta)\dot{\theta}\mathbf{i} + (-\sin\theta)\dot{\theta}\mathbf{j} = -\dot{\theta}((\cos\theta)\mathbf{i} + (\sin\theta)\mathbf{j}) = -\dot{\theta}\mathbf{e}_r,$$

using equation (12).

Solution to Exercise 6

Differentiating equation (18) with respect to time (where θ and $\mathbf{e}_\theta$ are functions of time) using the product rule, we obtain

$$\ddot{\mathbf{r}} = R\ddot{\theta}\mathbf{e}_\theta + R\dot{\theta}\dot{\mathbf{e}}_\theta,$$

which, by equation (17), may be written as

$$\ddot{\mathbf{r}} = R\ddot{\theta}\mathbf{e}_\theta + (R\dot{\theta})(-\dot{\theta}\mathbf{e}_r) = -R\dot{\theta}^2\mathbf{e}_r + R\ddot{\theta}\mathbf{e}_\theta.$$

Solution to Exercise 7

The radius of the circle is $R = 1$, and $\theta(t) = t^2$. Hence $\dot{\theta} = 2t$ and $\ddot{\theta} = 2$. Therefore the velocity of the particle is (from equation (20))

$$\dot{\mathbf{r}} = R\dot{\theta}\mathbf{e}_\theta = 2t\mathbf{e}_\theta,$$

and the acceleration is (from equation (21))

$$\ddot{\mathbf{r}} = -R\dot{\theta}^2\mathbf{e}_r + R\ddot{\theta}\mathbf{e}_\theta = -4t^2\mathbf{e}_r + 2\mathbf{e}_\theta.$$

Solution to Exercise 8

The minute hand of a clock makes one complete revolution (clockwise when viewed from the front of the clock) in 1 hour $= 3600$ seconds. So the fly's angular speed (in $\mathrm{rad\,s^{-1}}$) is

$$\omega = \frac{2\pi}{3600} \simeq 1.745 \times 10^{-3}.$$

The fly's speed (in $\mathrm{m\,s^{-1}}$) is

$$v = R\omega = (0.5)\frac{2\pi}{3600} \simeq 8.727 \times 10^{-4}.$$

The acceleration has magnitude (in $\mathrm{m\,s^{-2}}$)

$$R\omega^2 = 0.5\left(\frac{2\pi}{3600}\right)^2 \simeq 1.523 \times 10^{-6}.$$

Solution to Exercise 9

Use a polar coordinate system with origin O at the fixed end of the string, and let $\mathbf{k}$ be a unit vector pointing vertically upwards, as shown in the figure below.

There are three forces acting on the particle: the tension force $\mathbf{T} = -|\mathbf{T}|\mathbf{e}_r$ due to the string, the normal reaction $\mathbf{N} = |\mathbf{N}|\mathbf{k}$ of the table on the particle, and its weight $\mathbf{W} = -mg\mathbf{k}$. As the particle is in uniform circular motion with radius l and angular speed ω, the acceleration is, using equation (23),

$$\ddot{\mathbf{r}} = -l\omega^2\mathbf{e}_r.$$

Applying Newton's second law gives

$$m\ddot{\mathbf{r}} = \mathbf{T} + \mathbf{N} + \mathbf{W},$$

so we have

$$-ml\omega^2\mathbf{e}_r = -|\mathbf{T}|\mathbf{e}_r + |\mathbf{N}|\mathbf{k} - mg\mathbf{k}.$$

Resolving in the $\mathbf{e}_r$-direction gives the required result,

$$|\mathbf{T}| = ml\omega^2.$$

Solution to Exercise 10

Use a polar coordinate system in the plane of the surface of the disc, with origin O at the centre of the disc, and let $\mathbf{k}$ be a unit vector pointing vertically upwards, as shown in the figure below.

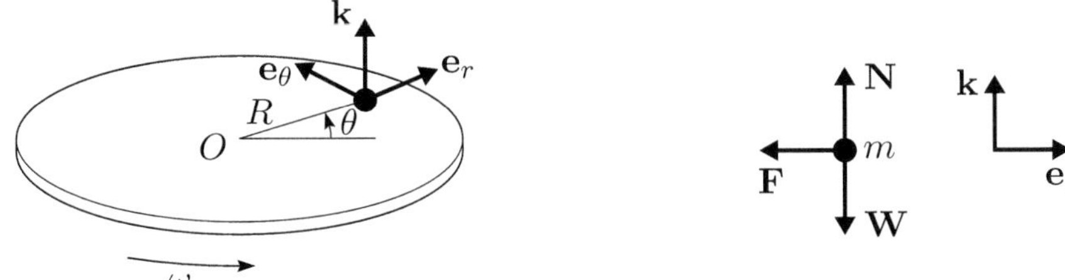

If the disc rotates through one complete revolution every two seconds, then its angular speed is

$$\omega = 2\pi/2 = \pi.$$

The forces acting on the particle are its weight $\mathbf{W} = -mg\mathbf{k}$, the normal reaction $\mathbf{N} = |\mathbf{N}|\mathbf{k}$ from the disc, and the friction force $\mathbf{F}$. The direction of $\mathbf{F}$ is in the horizontal plane of the disc.

Since the particle stays at the same point on the disc, at a distance R, say, from the centre, it performs uniform circular motion. Hence its acceleration is, using equation (23),

$$\ddot{\mathbf{r}} = -R\omega^2\mathbf{e}_r.$$

Applying Newton's second law to the particle, we obtain

$$m\ddot{\mathbf{r}} = \mathbf{W} + \mathbf{N} + \mathbf{F},$$

so we have

$$-mR\omega^2\mathbf{e}_r = -mg\mathbf{k} + |\mathbf{N}|\mathbf{k} + \mathbf{F}.$$

Since the friction force is in the horizontal plane, resolving in the $\mathbf{k}$-direction gives

$$|\mathbf{N}| = mg.$$

Hence we can deduce that

$$\mathbf{F} = -mR\omega^2\mathbf{e}_r,$$

so the friction force is directed towards the centre of the disc and has magnitude

$$|\mathbf{F}| = mR\omega^2.$$

If the particle is to remain static (relative to the disc), we must have

$$|\mathbf{F}| \leq \mu|\mathbf{N}|,$$

which, from the expressions above for $|\mathbf{F}|$ and $|\mathbf{N}|$, and from $\omega = \pi$ and $\mu = \frac{1}{2}$, leads to

$$R \leq \frac{\mu g}{\omega^2} = \frac{g}{2\pi^2} \simeq 0.497.$$

Thus the maximum possible distance of the toy from the centre of the roundabout is about $0.5\,\mathrm{m}$.

Solution to Exercise 11

Use a polar coordinate system in the horizontal plane of the surface of the table, with origin at O, and take $\mathbf{k}$ to be a unit vector pointing vertically upwards. The forces on the particle are its weight $\mathbf{W}$, the normal reaction $\mathbf{N}$ of the table, and the spring force $\mathbf{H}$. Let R be the length of the spring. The situation is shown in the figure below.

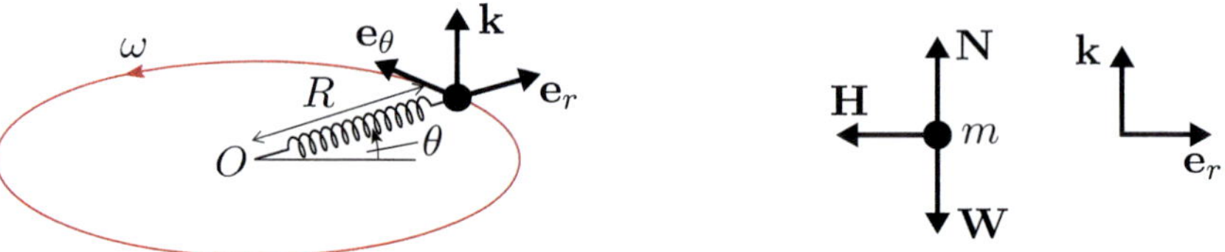

Applying Newton's second law to the particle, we have

$$m\ddot{\mathbf{r}} = \mathbf{W} + \mathbf{N} + \mathbf{H},$$

where $\ddot{\mathbf{r}} = -R\omega^2 \mathbf{e}_r$ (using equation (23)), $\mathbf{W} = -mg\mathbf{k}$, $\mathbf{N} = |\mathbf{N}|\mathbf{k}$ and $\mathbf{H} = k(R - l_0)(-\mathbf{e}_r)$.

Resolving in the $\mathbf{e}_r$-direction gives

$$-mR\omega^2 = -k(R - l_0),$$

so the length of the model spring is

$$R = \frac{kl_0}{k - m\omega^2}.$$

(Since $R > 0$, we must have $k - m\omega^2 > 0$, that is, $\omega < \sqrt{k/m}$.)

Solution to Exercise 12

Applying Newton's second law and equation (23) to the planet, we obtain

$$-mR\omega^2 \mathbf{e}_r = -\frac{GmM}{R^2}\mathbf{e}_r.$$

Hence

$$\omega = \sqrt{\frac{GM}{R^3}}.$$

So the period of the planet in its orbit is

$$T = \frac{2\pi}{\omega}$$
$$= 2\pi\sqrt{\frac{R^3}{GM}}.$$

It follows that

$$T^2 = \frac{4\pi^2}{GM}R^3,$$

that is,

$$T^2 \propto R^3.$$

Solution to Exercise 13

(a) Use a polar coordinate system in the horizontal plane of the particle's motion, with origin O at the centre of its circular path, and take $\mathbf{k}$ to be a unit vector pointing vertically upwards. Let R be the radius of the circular path so $R = l \sin \alpha$. The only forces acting on the particle are its weight $\mathbf{W}$ and the tension force $\mathbf{T}$ due to the rod. The situation is shown in the figure below.

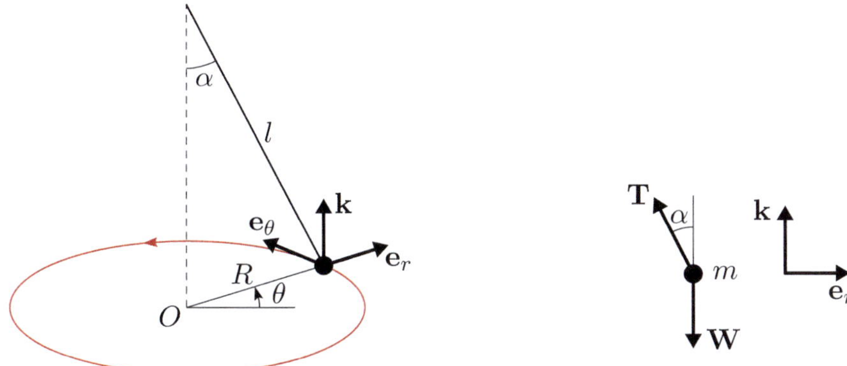

Applying Newton's second law and equation (23) to the particle gives

$$-mR\omega^2 \mathbf{e}_r = \mathbf{W} + \mathbf{T},$$

where $\mathbf{T} = -(|\mathbf{T}| \sin \alpha)\mathbf{e}_r + (|\mathbf{T}| \cos \alpha)\mathbf{k}$ and $\mathbf{W} = -mg\mathbf{k}$.

Resolving in the $\mathbf{k}$-direction gives

$$0 = -mg + |\mathbf{T}| \cos \alpha,$$

so the tension in the rod is $|\mathbf{T}| = mg/\cos \alpha$.

(b) Resolving in the $\mathbf{e}_r$-direction gives

$$-mR\omega^2 = -|\mathbf{T}| \sin \alpha.$$

Substituting for $|\mathbf{T}|$ from the result in part (a), dividing through by m and rearranging, we obtain

$$\omega = \sqrt{\frac{g \sin \alpha}{R \cos \alpha}},$$

and using $R = l \sin \alpha$ we finally have

$$\omega = \sqrt{\frac{g}{l \cos \alpha}}.$$

Since ω is real, $\cos \alpha$ must be non-negative, and therefore $0 < \alpha < \pi/2$. Also, the minimal value of ω is determined by the maximal value of $\cos \alpha$, which occurs when $\alpha = 0$, in which case $\cos \alpha = 1$. Hence $\omega > \omega_0 = \sqrt{g/l}$.

Rearranging the equation above gives

$$\alpha = \arccos\left(\frac{g}{l\omega^2}\right),$$

from which it follows that $\alpha \to \arccos(0) = \pi/2$ as $\omega \to \infty$.

(c) The vertical distance h of the bob below the pivot is given by $h = l\cos\alpha$, but from above we have $\omega = \sqrt{g/h}$, so $h = g/\omega^2$, which is independent of l for a given ω.

See Unit 9.

Note that this means that the potential energy of the conical pendulum, which is given by the height above the datum times mg, does not change as l is varied if ω is kept fixed. Thus modifying l while keeping ω fixed changes the system's kinetic energy but not its potential energy.

Solution to Exercise 14

(a) Use a polar coordinate system in the horizontal plane of the motion of particle P, with origin O at the centre of its circular path, and take $\mathbf{k}$ to be a unit vector pointing vertically upwards. Let R be the radius of the circular path. The forces acting on particle P are its weight $\mathbf{W}_1$, and the tension forces $\mathbf{T}_1$ and $\mathbf{T}_2$ due to the rods PA and PD, respectively. The forces acting on particle D are its weight $\mathbf{W}_2$, the tension force $\mathbf{T}_3$ due to the rod DP, and the normal reaction $\mathbf{N}$ of the shaft on the particle. The situation is shown in the figure below.

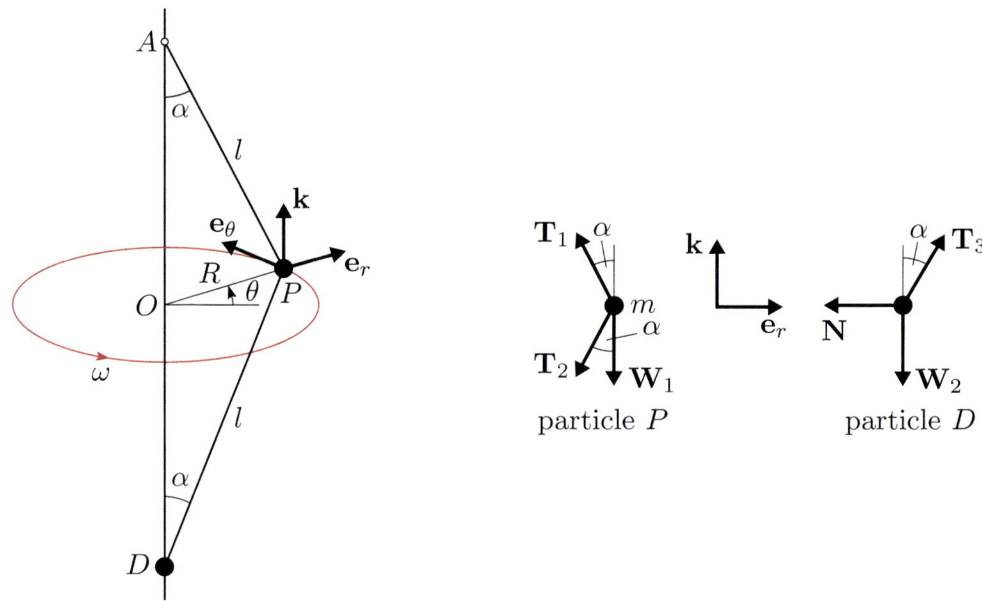

Applying Newton's second law and equation (23) to particle P gives

$$-mR\omega^2\mathbf{e}_r = \mathbf{W}_1 + \mathbf{T}_1 + \mathbf{T}_2,$$

where $\mathbf{T}_1 = -(|\mathbf{T}_1|\sin\alpha)\mathbf{e}_r + (|\mathbf{T}_1|\cos\alpha)\mathbf{k}$, $\mathbf{W}_1 = -mg\mathbf{k}$ and $\mathbf{T}_2 = -(|\mathbf{T}_2|\sin\alpha)\mathbf{e}_r - (|\mathbf{T}_2|\cos\alpha)\mathbf{k}$.

For fixed ω and α, particle D is in equilibrium, so

$$\mathbf{W}_2 + \mathbf{T}_3 + \mathbf{N} = \mathbf{0},$$

where $\mathbf{W}_2 = -mg\mathbf{k}$, $\mathbf{T}_3 = -\mathbf{T}_2$ (by Newton's third law) and $\mathbf{N} = -|\mathbf{N}|\mathbf{e}_r$.

We want to find $|\mathbf{T}_1|$ and $|\mathbf{T}_2| = |\mathbf{T}_3|$. Resolving the above equations in the **k**-direction gives, respectively,

$$0 = -mg + |\mathbf{T}_1|\cos\alpha - |\mathbf{T}_2|\cos\alpha,$$
$$-mg + |\mathbf{T}_2|\cos\alpha = 0$$

(since **N** is horizontal, it has no **k**-component).

Hence $|\mathbf{T}_2| = |\mathbf{T}_3| = mg/\cos\alpha$ and $|\mathbf{T}_1| = 2mg/\cos\alpha$.

(b) Resolving the particle P equation in the $\mathbf{e}_r$-direction gives

$$-mR\omega^2 = -|\mathbf{T}_1|\sin\alpha - |\mathbf{T}_2|\sin\alpha.$$

Substituting the values found in part (a) for $|\mathbf{T}_1|$ and $|\mathbf{T}_2|$, and using $R = l\sin\alpha$, we obtain

$$-m(l\sin\alpha)\omega^2 = -\frac{3mg\sin\alpha}{\cos\alpha},$$

so $\omega = \sqrt{3g/(l\cos\alpha)}$ and $\alpha = \arccos(3g/(l\omega^2))$.

Solution to Exercise 15

(a) Use a polar coordinate system in the horizontal plane of the motion of particle A, with origin O at the centre of the circular motion, and take **k** to be a unit vector pointing vertically upwards. The forces on particle A are its weight $\mathbf{W}_1$ and the tension force $\mathbf{T}_1$ due to the string. The forces on particle B are its weight $\mathbf{W}_2$ and the tension force $\mathbf{T}_2$ due to the string. The situation is shown in the figure below.

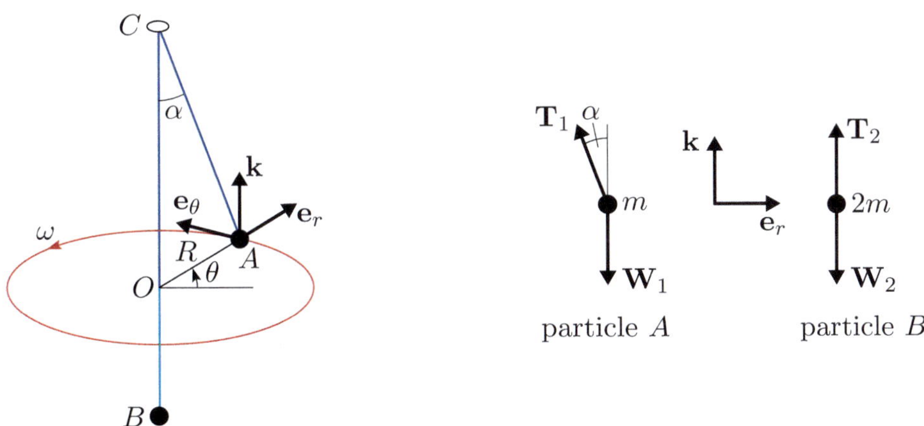

Applying Newton's second law and equation (23) to particle A gives

$$-mR\omega^2\mathbf{e}_r = \mathbf{W}_1 + \mathbf{T}_1,$$

where $\mathbf{T}_1 = -(|\mathbf{T}_1|\sin\alpha)\mathbf{e}_r + (|\mathbf{T}_1|\cos\alpha)\mathbf{k}$ and $\mathbf{W}_1 = -mg\mathbf{k}$.

Since particle B is in equilibrium, we have

$$\mathbf{W}_2 + \mathbf{T}_2 = \mathbf{0},$$

where $\mathbf{W}_2 = -2mg\mathbf{k}$ and $\mathbf{T}_2 = |\mathbf{T}_2|\mathbf{k}$.

Resolving the above equations in the **k**-direction gives, respectively,

$$0 = -mg + |\mathbf{T}_1| \cos \alpha,$$
$$-2mg + |\mathbf{T}_2| = 0,$$

so $|\mathbf{T}_1| = mg/\cos \alpha$ and $|\mathbf{T}_2| = 2mg$.

But $|\mathbf{T}_1| = |\mathbf{T}_2|$, so $mg/\cos \alpha = 2mg$, giving $\cos \alpha = \frac{1}{2}$ and $\alpha = \frac{\pi}{3}$.

(b) Resolving the particle A equation in the $\mathbf{e}_r$-direction gives

$$-mR\omega^2 = -|\mathbf{T}_1| \sin \alpha,$$

so, using $|\mathbf{T}_1| = mg/\cos \alpha$ and $\alpha = \frac{\pi}{3}$, we have

$$\omega = \sqrt{\sqrt{3}g/R}.$$

Solution to Exercise 16

(a) Using equation (31) with $\mu = 1.3$, the maximum speed with which the car can follow a circular path of radius $R = 10$ is

$$v = \sqrt{\mu R g} = \sqrt{1.3 \times 10 \times 9.81} \simeq 11.29.$$

So the maximum speed here is about $11.29\,\mathrm{m\,s^{-1}}$.

(b) Rearranging equation (31) with $v = 15$, the condition to avoid slipping is

$$R \geq \frac{v^2}{\mu g} = \frac{15^2}{1.3 \times 9.81} \simeq 17.64.$$

So the radius of the roundabout needs to be at least $18\,\mathrm{m}$.

(c) We use the same condition as in part (b), but with $\mu = \frac{1}{2} \times 1.3 = 0.65$. So we need $R \geq 35.29$.

In this case, the roundabout needs a minimum radius of about $36\,\mathrm{m}$.

Solution to Exercise 17

Use a polar coordinate system in the horizontal plane of the particle's motion, with origin at the centre of the circular motion, and take **k** to be a unit vector pointing vertically upwards. The radius of the uniform circular motion of the particle is $R + l \sin \alpha$. The forces on the particle are its weight **W** and the tension force **T** due to the string. The situation is shown in the figure below.

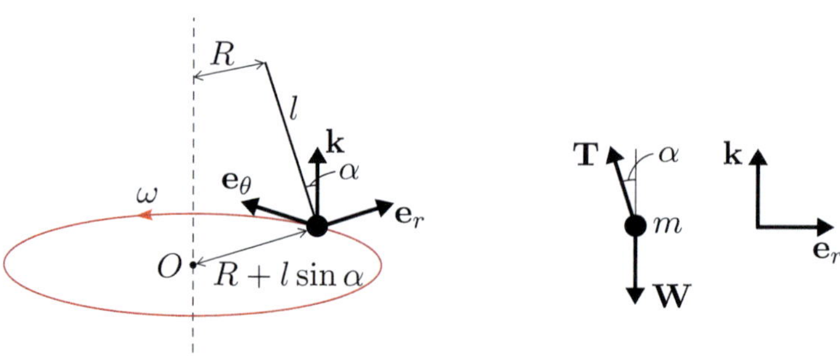

Applying Newton's second law and equation (23) to the particle gives

$$-m(R + l\sin\alpha)\omega^2\mathbf{e}_r = \mathbf{W} + \mathbf{T},$$

where $\mathbf{T} = -|\mathbf{T}|\sin\alpha\,\mathbf{e}_r + |\mathbf{T}|\cos\alpha\,\mathbf{k}$ and $\mathbf{W} = -mg\mathbf{k}$.

Resolving in the $\mathbf{e}_r$- and $\mathbf{k}$-directions gives, respectively,

$$-m(R + l\sin\alpha)\omega^2 = -|\mathbf{T}|\sin\alpha,$$
$$0 = -mg + |\mathbf{T}|\cos\alpha,$$

so $|\mathbf{T}| = mg/\cos\alpha$.

Substituting for $|\mathbf{T}|$ gives the equation of motion

$$m(R + l\sin\alpha)\omega^2 = mg\tan\alpha.$$

Hence, dividing through by $m\omega^2$, we have

$$\frac{g}{\omega^2}\tan\alpha = R + l\sin\alpha.$$

Solution to Exercise 18

This is obtained by substituting $\dot\theta = 0$ and $\theta = \pi$ into equation (41), which yields

$$\frac{v_0^2}{l^2} - \frac{4g}{l} = 0,$$

or

$$v_0^2 = 4gl,$$

so

$$v_0 = 2\sqrt{gl}.$$

Alternatively, conservation of energy can be used so that the result is obtained by equating loss of kinetic energy, $\frac{1}{2}mv_0^2$, with gain in potential energy, $2mgl$, so that $\frac{1}{2}v_0^2 = 2gl$ and hence the result.

Solution to Exercise 19

The essential difference between a model rod and model string is that a string needs to be kept under tension (i.e. with $T > 0$) in order for it to stay taut. For low speeds (at the lowest point), both pendulums will oscillate gently back and forth about the stable equilibrium position. For very high speeds, both will move in a circle around the fixed point. For intermediate speeds, the original pendulum will perform large oscillations or will circle around the fixed point, but the new pendulum will behave in this way only while the string stays taut (when the bob is above the level of the fixed point). If the speed is not high enough to keep the string taut, then the only force acting on the bob will be its weight, and the bob will behave like a projectile until the string becomes taut once more.

Solution to Exercise 20

(a) From equation (44), the string will not go slack provided that

$$T = \frac{mv_0^2}{l} + mg(3\cos\theta - 2) > 0.$$

Dividing by m/l and rearranging, we obtain

$$v_0^2 > gl(2 - 3\cos\theta).$$

The greatest value of $gl(2 - 3\cos\theta)$ occurs when $\cos\theta = -1$ (corresponding to $\theta = \pm\pi$), giving $v_0^2 > 5gl$ and $v_0 > \sqrt{5gl}$. So the string never goes slack if the speed v_0 of the pendulum bob when $\theta = 0$ is greater than $\sqrt{5gl}$ (in which case the bob performs complete revolutions).

(b) If $v_0 = 2\sqrt{gl}$, then from equation (41) we have

$$\dot{\theta}^2 = \frac{4g}{l} - \frac{2g}{l}(1 - \cos\theta)$$

$$= \frac{2g}{l}(1 + \cos\theta),$$

so $\dot{\theta} = 0$ only when $\cos\theta = -1$, that is, when $\theta = \pi$. So the bob will reach the top of the circle unless the string goes slack first. We know from part (a) that for the string never to go slack, we must have $v_0 > \sqrt{5gl}$. Therefore since $2\sqrt{gl} < \sqrt{5gl}$, we know that the string must go slack before the bob comes momentarily to rest.

To see when the string goes slack, substitute $v_0 = 2\sqrt{gl}$ into equation (44) to obtain

$$T = 4mg + mg(3\cos\theta - 2)$$

$$= mg(3\cos\theta + 2)$$

$$= 0.$$

So the string goes slack when $3\cos\theta + 2 = 0$, which gives $\cos\theta = -2/3$. In the range $-\pi \le \theta \le \pi$, this gives θ as approximately ± 2.3 radians or $\pm 132°$. So the string goes slack when the pendulum makes an angle of approximately $42°$ above the horizontal.

Solution to Exercise 21

The condition for the pendulum bob to just reach the horizontal is $\dot{\theta} = 0$ at $\theta = \pm\frac{\pi}{2}$. So from equation (41), we find that

$$0 = \frac{v_0^2}{l^2} - \frac{2g}{l}.$$

Hence $v_0 = \sqrt{2gl}$.

Alternatively, this result could be obtained by equating loss of kinetic energy, $\frac{1}{2}mv_0^2$, with gain in potential energy, mgl, so that $\frac{1}{2}v_0^2 = gl$ and hence the result.

Solution to Exercise 22

(a) Use a polar coordinate system in the plane of motion of the particle, with origin O at the centre of the circular motion. The forces acting on the particle are its weight $\mathbf{W}$ and the normal reaction $\mathbf{N}$ of the bowl. Let θ be the angle between the radius to the particle and the downward vertical, measured anticlockwise from the vertical, and let h be the vertical height of the particle above A. The situation is shown in the figure below.

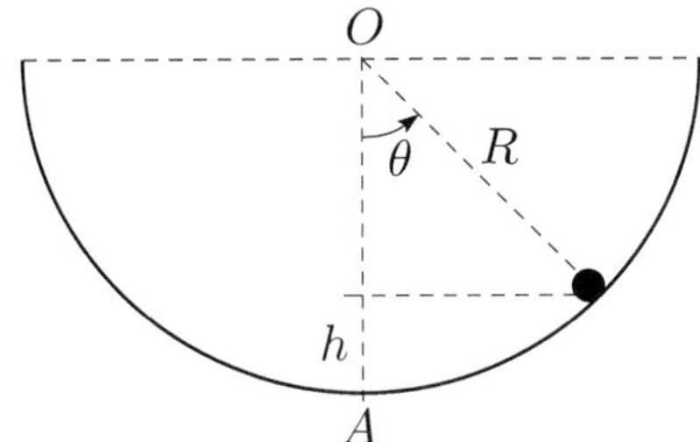
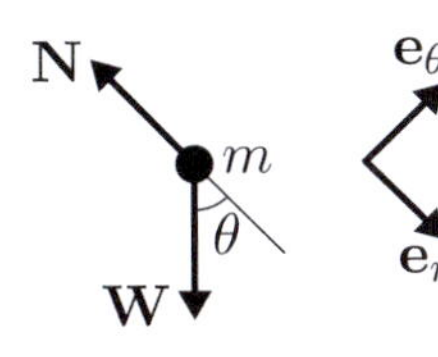

The analysis of this situation is identical to that for the pendulum at the beginning of Section 3, with $\mathbf{T}$ replaced by $\mathbf{N}$ and l by R.

From equation (41), we have

$$v^2 = v_0^2 - 2gR(1 - \cos\theta),$$

where v_0 is the speed of the particle at A, and $v = R|\dot{\theta}|$ is its speed at angle θ. Initially, we have $h = \frac{1}{2}R$, $v = \sqrt{gR}$ and $\cos\theta = (R - h)/R = \frac{1}{2}$, giving

$$gR = v_0^2 - 2gR \times \tfrac{1}{2},$$

so $v_0^2 = 2gR$. Hence the first equation becomes

$$v^2 = 2gR\cos\theta.$$

The speed at the top of the bowl, where $\theta = \pm\frac{\pi}{2}$, is thus zero, showing that the particle just reaches this level.

(b) From equation (36), we have

$$|\mathbf{N}| = mg\cos\theta + mR\dot{\theta}^2.$$

From the equation at the end of the solution to part (a), we have

$$\dot{\theta}^2 = v^2/R^2 = (2g\cos\theta)/R,$$

so

$$|\mathbf{N}| = mg\cos\theta + 2mg\cos\theta = 3mg\cos\theta.$$

(We could, alternatively, have obtained this expression for $|\mathbf{N}|$ by use of equation (44) with $v_0^2 = 2gR$ and $l = R$.)

With $h = \frac{2}{3}R$, we have $\cos\theta = (R - h)/R = \frac{1}{3}$, so $|\mathbf{N}| = mg$, that is, the magnitude of the normal reaction when $h = \frac{2}{3}R$ is mg.

Solution to Exercise 23

Use a polar coordinate system in the vertical plane of the particle's motion, with origin O at the centre of the sphere. The forces acting on the particle are its weight $\mathbf{W}$ and the normal reaction $\mathbf{N}$ of the sphere. The situation is shown in the figure below.

The analysis of this situation is similar to that for the pendulum, with $\mathbf{T}$ replaced by $-\mathbf{N}$, $\mathbf{W}$ replaced by $-\mathbf{W}$, and l replaced by R, so we can make use of the results obtained for the pendulum at the beginning of the section.

With respect to the chosen coordinate system, we have

$$\mathbf{W} = -(mg\cos\theta)\mathbf{e}_r + (mg\sin\theta)\mathbf{e}_\theta,$$
$$\mathbf{N} = |\mathbf{N}|\mathbf{e}_r.$$

The acceleration of the particle is

$$\ddot{\mathbf{r}} = -R\dot{\theta}^2\mathbf{e}_r + R\ddot{\theta}\mathbf{e}_\theta.$$

Applying Newton's second law to the particle gives

$$m\ddot{\mathbf{r}} = \mathbf{W} + \mathbf{N},$$

so we have

$$-mR\dot{\theta}^2\mathbf{e}_r + mR\ddot{\theta}\mathbf{e}_\theta = -(mg\cos\theta)\mathbf{e}_r + (mg\sin\theta)\mathbf{e}_\theta + |\mathbf{N}|\mathbf{e}_r.$$

Resolving in the $\mathbf{e}_r$-direction gives

$$-mR\dot{\theta}^2 = |\mathbf{N}| - mg\cos\theta,$$

and rearranging this equation yields

$$|\mathbf{N}| = mg\cos\theta - mR\dot{\theta}^2.$$

We could use the appropriately amended version of equation (41), with $l = R$, to substitute for $\dot{\theta}^2$ in this equation. Alternatively, we can use conservation of energy to obtain an expression for $\dot{\theta}^2$ as follows. The potential energy is $U = mgR(\cos\theta - 1)$, taking the highest point of the sphere ($\theta = 0$) as the datum. So the total conserved mechanical energy is

$$E = \tfrac{1}{2}mv^2 + mgR(\cos\theta - 1),$$

where $v = R|\dot{\theta}|$.

The value of E can be determined from the initial condition where $v = v_0$ at $\theta = 0$. Hence we have

$$\tfrac{1}{2}mR^2\dot\theta^2 + mgR(\cos\theta - 1) = \tfrac{1}{2}mv_0^2,$$

or

$$mR\dot\theta^2 = \frac{mv_0^2}{R} + 2mg(1 - \cos\theta),$$

and substituting into the equation for $|\mathbf{N}|$ gives

$$|\mathbf{N}| = mg\cos\theta - \frac{mv_0^2}{R} - 2mg(1 - \cos\theta)$$
$$= mg(3\cos\theta - 2) - \frac{mv_0^2}{R}.$$

For $v_0 = \sqrt{\tfrac{1}{2}gR}$, we have

$$|\mathbf{N}| = mg(3\cos\theta - 2) - \tfrac{1}{2}mg = mg\left(3\cos\theta - \tfrac{5}{2}\right).$$

The particle will leave the surface when $\mathbf{N} = \mathbf{0}$, that is, when $3\cos\theta - \tfrac{5}{2} = 0$, which gives

$$\theta = \arccos\tfrac{5}{6} \simeq 0.59 \ (\simeq 33.6°).$$

Solution to Exercise 24

(a) Use a polar coordinate system with θ being the angle that OP_2 makes with the upward vertical, similar to the set-up in Exercise 23, with the polar unit vectors $\mathbf{e}_r$ and $\mathbf{e}_\theta$ defined with respect to P_2 as indicated in the figure below.

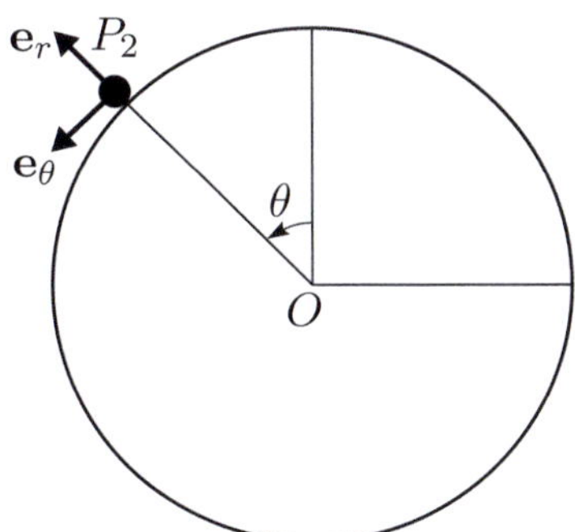
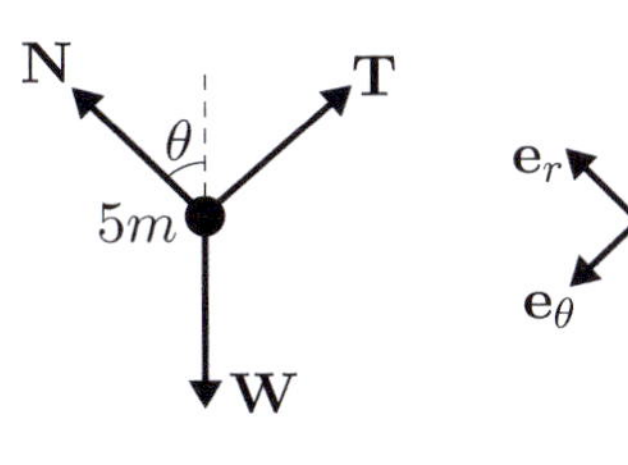

The forces acting on particle P_2 are its weight $\mathbf{W}$, the normal reaction of the cylinder $\mathbf{N}$, and the tension of the string $\mathbf{T}$. These forces are indicated in the figure and are expressed in terms of $\mathbf{e}_r$ and $\mathbf{e}_\theta$ as

$$\mathbf{N} = |\mathbf{N}|\mathbf{e}_r,$$
$$\mathbf{T} = -|\mathbf{T}|\mathbf{e}_\theta,$$
$$\mathbf{W} = 5mg(\mathbf{e}_\theta\sin\theta - \mathbf{e}_r\cos\theta).$$

The acceleration of P_2 is

$$\ddot{\mathbf{r}} = R\ddot{\theta}\mathbf{e}_\theta - R\dot{\theta}^2\mathbf{e}_r$$

(as in Exercise 23). Newton's second law applied at P_2 gives

$$5m\ddot{\mathbf{r}} = \mathbf{W} + \mathbf{N} + \mathbf{T},$$

so we have

$$5mR(\ddot{\theta}\mathbf{e}_\theta - \dot{\theta}^2\mathbf{e}_r) = 5mg(\mathbf{e}_\theta\sin\theta - \mathbf{e}_r\cos\theta) + |\mathbf{N}|\mathbf{e}_r - |\mathbf{T}|\mathbf{e}_\theta,$$

and resolving in the $\mathbf{e}_r$-direction yields

$$-5mR\dot{\theta}^2 = |\mathbf{N}| - 5mg\cos\theta,$$

which can be rearranged to give the required result.

(b) The total mechanical energy of the combined system is conserved since the gravitational forces are conservative forces and the sum of the contributions of the string tensions to work done on the particles vanishes as a result of Newton's third law.

The total mechanical energy E is the sum of the energies of both particles, P_1 and P_2. The kinetic energy of P_1 is $\frac{1}{2} \times 3m \times v^2$, and taking the horizontal diameter through O as the datum, its potential energy is $3mgR\sin\theta$. Similarly, the kinetic energy of P_2 is $\frac{1}{2} \times 5m \times v^2$, and its potential energy is $5mgR\cos\theta$. Thus the total mechanical energy of both particles is

$$E = \tfrac{3}{2}mv^2 + 3mgR\sin\theta + \tfrac{5}{2}mv^2 + 5mgR\cos\theta$$
$$= 4mv^2 + mgR(3\sin\theta + 5\cos\theta),$$

which stays constant throughout the motion (provided that both particles stay on the cylinder).

The value of E can be determined by the initial condition where $v = 0$ when $\theta = \pi/4$. Given that $\sin(\pi/4) = \cos(\pi/4) = 1/\sqrt{2}$, we have

$$E = mgR(8/\sqrt{2}) = 4\sqrt{2}mgR,$$

and substituting this into the equation above and dividing through by m gives

$$4v^2 + gR(3\sin\theta + 5\cos\theta) = 4\sqrt{2}gR,$$

which can be rearranged to give the required result.

(c) Recalling that $v = R|\dot{\theta}|$, the result of part (b) gives

$$R\dot{\theta}^2 = \frac{v^2}{R} = \tfrac{1}{4}g(4\sqrt{2} - 5\cos\theta - 3\sin\theta),$$

which can be substituted into the expression for $|\mathbf{N}|$ obtained in part (a) to give

$$|\mathbf{N}| = 5m\left(g\cos\theta - \tfrac{1}{4}g(4\sqrt{2} - 5\cos\theta - 3\sin\theta)\right)$$
$$= \tfrac{5}{4}mg(9\cos\theta + 3\sin\theta - 4\sqrt{2}).$$

Particle P_2 will leave the cylinder when $|\mathbf{N}| = 0$, that is, when

$$9 \cos \theta + 3 \sin \theta = 4\sqrt{2},$$

as required. The left-hand side of this equation can be expressed as

$$9 \cos \theta + 3 \sin \theta = A \cos(\theta - \phi)$$

with

$$A \cos \phi = 9 \quad \text{and} \quad A \sin \phi = 3,$$

so $A^2 = 9^2 + 3^2 = 90$ or $A = 3\sqrt{10}$, and $\tan \phi = 3/9 = 1/3$ or $\phi = \arctan(1/3)$. Thus we have

$$3\sqrt{10} \cos(\theta - \phi) = 4\sqrt{2},$$

or

$$\theta - \phi = \arccos\left(\tfrac{4}{3\sqrt{5}}\right),$$

and with $\phi = \arctan(1/3)$ we obtain the result $\theta \simeq 1.254$ (in radians), which is equivalent to $71.8°$ to three significant figures.

Solution to Exercise 25

(a) Since the motion is clockwise, θ is decreasing during the motion, so $\dot{\theta} = -2$ and the particle's angular velocity is

$$\boldsymbol{\omega} = \dot{\theta}\mathbf{k} = -2\mathbf{k}.$$

(b) From equation (46), the particle's velocity is

$$\mathbf{v} = \boldsymbol{\omega} \times \mathbf{r} = -2\mathbf{k} \times (3\mathbf{i} + 4\mathbf{j}) = 8\mathbf{i} - 6\mathbf{j}.$$

Since the radius of the circular path is $\sqrt{3^2 + 4^2} = 5$, the velocity is also given by

$$\mathbf{v} = \dot{\mathbf{r}} = R\dot{\theta}\mathbf{e}_\theta = 5 \times (-2)\mathbf{e}_\theta = -10\mathbf{e}_\theta.$$

Alternatively, we have

$$\mathbf{v} = \boldsymbol{\omega} \times \mathbf{r} = -2\mathbf{k} \times 5\mathbf{e}_r = -10\mathbf{e}_\theta.$$

Solution to Exercise 26

By definition, the angular momentum is

$$\mathbf{L} = \mathbf{r} \times m\dot{\mathbf{r}} = (\mathbf{i} + 2\mathbf{j}) \times 3(2\mathbf{i} - \mathbf{j}) = (-3\mathbf{i} \times \mathbf{j}) + (12\mathbf{j} \times \mathbf{i}) = -15\mathbf{k},$$

where we have used $\mathbf{i} \times \mathbf{i} = \mathbf{j} \times \mathbf{j} = \mathbf{0}$ and $\mathbf{j} \times \mathbf{i} = -\mathbf{i} \times \mathbf{j} = -\mathbf{k}$.

Solution to Exercise 27

We know, from the text just before the exercise, that $\mathbf{L} = mR^2\dot{\theta}\mathbf{k}$, so $|\mathbf{L}| = mR^2|\dot{\theta}|$. We also know, from Section 1, that $\mathbf{r} = R\mathbf{e}_r$ and $\dot{\mathbf{r}} = R\dot{\theta}\mathbf{e}_\theta$, so $|\mathbf{r}| = R$ and $|\dot{\mathbf{r}}| = R|\dot{\theta}|$. Hence we have

$$m|\mathbf{r}|\,|\dot{\mathbf{r}}| = mR^2|\dot{\theta}| = |\mathbf{L}|.$$

Solution to Exercise 28

The position of the particle is

$$\mathbf{r} = a\cos(\alpha t)\mathbf{i} + a\sin(\alpha t)\mathbf{j} + ut\mathbf{k}.$$

Hence the linear momentum of the particle is

$$\mathbf{p} = m\dot{\mathbf{r}} = m(-a\alpha\sin(\alpha t)\mathbf{i} + a\alpha\cos(\alpha t)\mathbf{j} + u\mathbf{k}).$$

Therefore the force acting on the particle is

$$\mathbf{F} = \dot{\mathbf{p}} = -ma\alpha^2(\cos(\alpha t)\mathbf{i} + \sin(\alpha t)\mathbf{j}).$$

The angular momentum of the particle about the origin O is

$$\begin{aligned}
\mathbf{L} &= \mathbf{r} \times m\dot{\mathbf{r}} \\
&= (a\cos(\alpha t)\mathbf{i} + a\sin(\alpha t)\mathbf{j} + ut\mathbf{k}) \\
&\quad \times m(-a\alpha\sin(\alpha t)\mathbf{i} + a\alpha\cos(\alpha t)\mathbf{j} + u\mathbf{k}) \\
&= mau(\sin(\alpha t) - \alpha t\cos(\alpha t))\mathbf{i} - mau(\alpha t\sin(\alpha t) + \cos(\alpha t))\mathbf{j} \\
&\quad + ma^2\alpha(\cos^2(\alpha t) + \sin^2(\alpha t))\mathbf{k} \\
&= mau(\sin(\alpha t) - \alpha t\cos(\alpha t))\mathbf{i} - mau(\cos(\alpha t) + \alpha t\sin(\alpha t))\mathbf{j} \\
&\quad + ma^2\alpha\,\mathbf{k}.
\end{aligned}$$

The torque about O acting on the particle is

$$\begin{aligned}
\boldsymbol{\Gamma} &= \mathbf{r} \times \mathbf{F} \\
&= -(a\cos(\alpha t)\mathbf{i} + a\sin(\alpha t)\mathbf{j} + ut\mathbf{k}) \times ma\alpha^2(\cos(\alpha t)\mathbf{i} + \sin(\alpha t)\mathbf{j}) \\
&= ma\alpha^2(ut\sin(\alpha t)\mathbf{i} - ut\cos(\alpha t)\mathbf{j}) \\
&= mau\alpha^2 t(\sin(\alpha t)\mathbf{i} - \cos(\alpha t)\mathbf{j}).
\end{aligned}$$

To verify the torque law, we need to check that $\dot{\mathbf{L}} = \boldsymbol{\Gamma}$. We have

$$\begin{aligned}
\dot{\mathbf{L}} &= mau(\alpha\cos(\alpha t) - (\alpha\cos(\alpha t) - \alpha^2 t\sin(\alpha t)))\mathbf{i} \\
&\quad - mau(-\alpha\sin(\alpha t) + (\alpha\sin(\alpha t) + \alpha^2 t\cos(\alpha t)))\mathbf{j} \\
&= mau\alpha^2 t(\sin(\alpha t)\mathbf{i} - \cos(\alpha t)\mathbf{j}) \\
&= \boldsymbol{\Gamma},
\end{aligned}$$

as required.

Solution to Exercise 29

(a) Let $\boldsymbol{\omega} = \dot{\theta}\mathbf{k}$ be the angular velocity of the particle. From Example 5 and the discussion that follows it, the angular momentum is $\mathbf{L} = mR^2\dot{\theta}\mathbf{k}$. The torque law gives $\dot{\mathbf{L}} = \boldsymbol{\Gamma}$, so we have

$$mR^2\ddot{\theta}\mathbf{k} = \Gamma_0 e^{-\alpha t}\mathbf{k}.$$

Resolving in the $\mathbf{k}$-direction and rearranging, we obtain

$$\ddot{\theta} = \frac{\Gamma_0}{mR^2}e^{-\alpha t}.$$

Integrating, we obtain

$$\dot{\theta} = -\frac{\Gamma_0}{\alpha mR^2}e^{-\alpha t} + C.$$

Taking $t = 0$ when the house is first picked up by the whirlwind, so $\dot{\theta}(0) = 0$, we have $C = \Gamma_0/(\alpha m R^2)$ and hence

$$\dot{\theta} = \frac{\Gamma_0}{\alpha m R^2}\left(1 - e^{-\alpha t}\right).$$

So the angular speed is

$$\omega = |\dot{\theta}| = \frac{\Gamma_0}{\alpha m R^2}\left(1 - e^{-\alpha t}\right).$$

As $t \to \infty$, $\omega \to \Gamma_0/(\alpha m R^2)$.

(b) Assuming that the model remains valid for decreasing R, as R becomes smaller, ω becomes larger and larger without limit. However, as R becomes smaller, the distance from the centre of mass of the house to the axis of the whirlwind gets smaller and smaller. Eventually, some parts of the house will coincide with this axis and the particle model will no longer be valid because the dimensions of the house can no longer be neglected.

Solution to Exercise 30

The total torque about the origin is given by

$$\boldsymbol{\Gamma} = \mathbf{r} \times \mathbf{F} = \mathbf{r} \times f(r)\,\widehat{\mathbf{r}} = \mathbf{0},$$

since $\mathbf{r}$ and $\widehat{\mathbf{r}}$ are parallel. So the total torque about the origin is zero.

Solution to Exercise 31

Choose the origin to be the centre of the circle. Let ω be the (constant) angular speed of the particle, and let m be its mass. Let R be the radius of the circular motion. Using equation (23), the total force acting on the particle is $-mR\omega^2 \mathbf{e}_r$, so it is a central force. Hence, from the result of Exercise 30, the total torque about the centre of the circle is zero.

Acknowledgements

Grateful acknowledgement is made to the following sources:

Figure 1: © Niels Van Gijn / JAI / Corbis.

Figure 11: © Science Museum / Science & Society Picture Library.

Every effort has been made to contact copyright holders. If any have been inadvertently overlooked, the publishers will be pleased to make the necessary arrangements at the first opportunity.

Rotating bodies and angular momentum

Introduction

When an ice skater is performing a spin, if she brings her arms in and folds them across her chest, her rate of rotation will increase. Why is this? Take another example. Suppose that two cylindrical objects have equal size and mass, but one is hollow and the other is solid. If they are released together from the top of a slope, will they roll down at the same rate? If not, which cylinder will reach the bottom of the slope first?

To answer such questions, we need to bring together a number of ideas that have been introduced earlier in the module. In earlier mechanics units, such as Units 3 and 9, we modelled moving objects as particles. However, this approach is inadequate for dealing with the questions above, because we are now concerned with extended bodies, that is, objects that have size, and we are interested in aspects of their motion where that size is important. In Unit 19, you saw that the motion of an extended body can often be modelled by the motion of a representative particle located at the centre of mass of the body. But in the case of the spinning skater, the centre of mass may well be more or less stationary – it is the skater's rotation about the centre of mass that is of interest. For the rolling cylinders, it is perhaps less obvious that the particle model is inappropriate, but again, rotation about the centre of mass is a crucial part of the motion.

An *extended* body has one or more of length, breadth and depth.

In this unit we deal with the motion of extended bodies, and in particular with their rotational motion. In Unit 2, we considered such bodies when stationary: you learned that for a rigid body in equilibrium, the sum of all the external forces must be zero, and that the sum of the external *torques* must be zero. In Unit 20, you saw that if a non-zero torque is applied to a particle, then this changes the rotational motion of the particle. Now we combine these ideas and consider the motion of an *extended body* subject to a *non-zero torque*.

A *rigid* body is an extended body whose shape does not change.

In Section 1 we give an overview of the unit and freely apply principles that are not formally established until Section 4. Section 2 begins by reviewing concepts from earlier units, and then goes on to develop a theoretical basis for modelling the motion of extended bodies. Section 3 looks at the rotation of extended bodies about an axis that is fixed, such as the spinning ice skater. In Section 4 we explore situations where rotational motion is combined with other types of motion. For example, consider a diver in flight after leaving a high diving board (see Figure 1). From Unit 19 we would expect the diver's centre of mass to follow a parabolic trajectory (if we ignore air resistance), but the diver's rotation about her centre of mass is also a major factor in the success of the dive (as is the change of shape of the body during the dive).

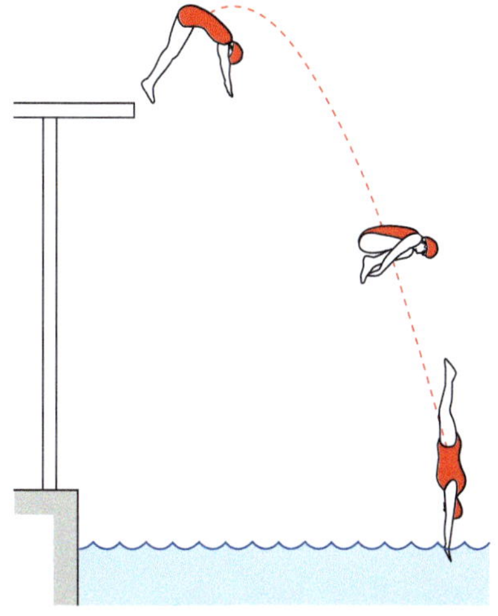

Figure 1 Diver with centre of mass following a parabolic path. Note that the diver is an extended body but *not* a rigid body.

1 Rotating bodies

Rotation is an important aspect of motion in many situations. Understanding such motion involves various mechanical concepts that you have met earlier in the module: these include torque (Unit 2), moment of inertia (Unit 17) and angular momentum (Unit 20).

To get an idea of some of the factors involved in analysing rotational motion, consider someone pushing a roundabout in a playground so as to make it move (see Figure 2). Initially the roundabout is stationary, but when it is pushed, it rotates with increasing rotational speed. Even after the person stops pushing, the roundabout will continue to rotate.

While the roundabout is being pushed, the total force on it is zero: the force supplied by the pusher is balanced by a force exerted by the support at the centre of the roundabout. The roundabout 'as a whole' is not going anywhere, that is, its centre of mass is not moving. However, although the two horizontal forces shown in Figure 2 are equal in magnitude and opposite in direction, they have different lines of action. As a result, there is a torque on the roundabout. This torque initiates the rotation of the roundabout and gives it angular momentum. In Unit 20, you saw that for a particle, if there is no torque being applied, then the angular momentum is constant. As you will see later, this result can be generalised to extended bodies. This means that even when the pushing stops, the roundabout will maintain its angular momentum and will continue to rotate; indeed, in the absence of resistive forces, it would go on rotating forever without the need for further pushing (but in practice resistive forces are always present).

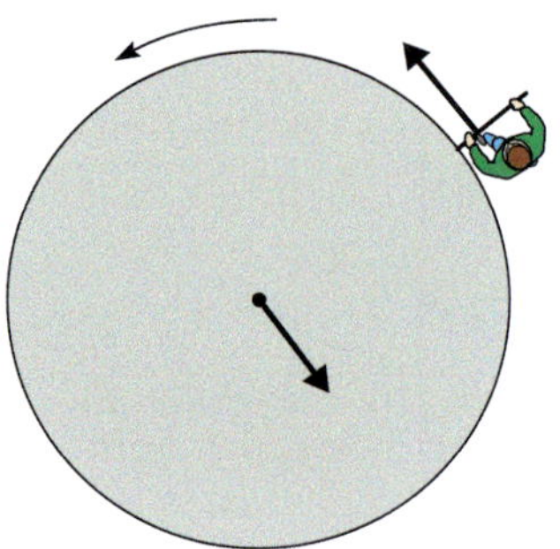

Figure 2　Playground roundabout being pushed

Later in this unit we will develop quantitative models of rotational motion. But at this stage you should note the following general aspects of rotational motion.

- The motion of an extended body can be treated in two parts: the motion of an equivalent particle located at the centre of mass, and rotation about the centre of mass. This result will be established in Section 4.

- A torque applied to an extended body that is initially stationary will initiate rotation and supply angular momentum.

- Once an extended body has angular momentum, that momentum will remain constant provided that no further torque is applied to the body.

- The component L of the angular momentum along the axis of rotation of an extended body that is rotating about a fixed axis is the product of the angular speed ω and the moment of inertia I about that axis, that is,

You will see later, in Subsection 3.1, that this scalar equation can be obtained from a vector equation by resolving in the direction of the (fixed) axis of rotation.

$$L = I\omega. \tag{1}$$

This scalar equation is sufficient for the needs of this section.

- An ice skater can vary her speed of rotation during a spin by changing body shape. This is in accord with equation (1) since although the angular momentum L is constant, the moment of inertia is changing. When modelling *rigid* bodies, the moment of inertia is a constant, but the human body is flexible and its moment of inertia can change.

In the following examples and exercises we will analyse some sporting situations, thus exploring further the relationships between angular momentum, moment of inertia and rotational motion.

Example 1

(a) A diver is executing a simple dive in which her body shape remains constant, as illustrated in Figure 3(a). She starts in a handstand position. The subsequent motion can be divided into two phases: first, rotation about the point O while the diver remains in contact with the diving board; second, motion in flight after the diver lets go of the board, but before she enters the water.

   (i)   In the first phase, how is the diver's angular momentum about O changing?

   (ii)   In the second phase, what would you expect to happen to the angular momentum about the diver's centre of mass? Assume that resistive forces are negligible.

(b) Suppose that the diver goes into a tuck position (see Figure 3(b)) in the second phase. What aspect of the motion will be different from that in part (a)?

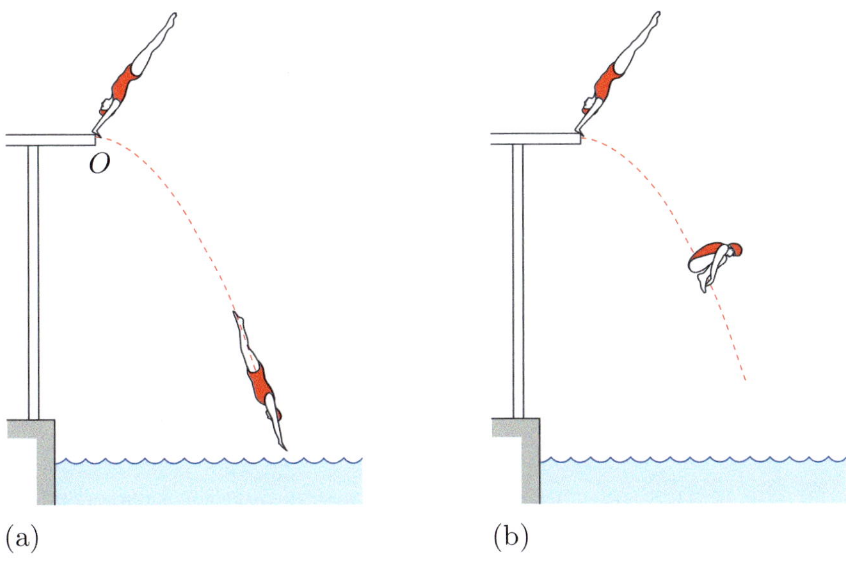

Figure 3   Diver performing a dive: (a) with body shape constant, (b) taking a tuck position

Solution

(a) (i)   In the first phase, the diver's weight provides a torque about O that will act to increase the angular momentum about O. During this phase of the motion, the angular momentum (and angular speed) about O are increasing.

   (ii)   In the second phase, the only force on the diver is her weight, which acts through her centre of mass. This means that there is no torque about the centre of mass, and consequently the angular momentum about the centre of mass will be constant.

(b) In the situation in part (a), the diver's body shape, and hence her moment of inertia, remain constant throughout the dive, so the angular speed of rotation about the diver's centre of mass does not change after she has let go of the board. However, if the diver adopts a tuck position after letting go of the board, this will reduce her moment of inertia, and will increase her angular speed of rotation about her centre of mass.

In Example 2 and Exercise 1, we will make use of the idea that is implicit in Example 1(a)(ii): that angular momentum about the centre of mass is constant for an (effectively rigid) extended body in flight (assuming that the body is subject only to gravity, that is, that resistive forces are negligible). This result will be established in Section 4.

Example 2

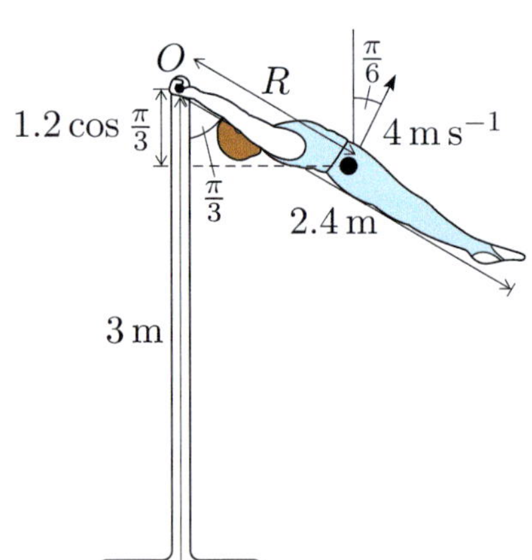

Figure 4 Gymnast rotating anticlockwise around a bar

Figure 4 shows a gymnast rotating anticlockwise around a bar that is 3 m above the ground, at the point of letting go of the bar and dismounting. Assume that the gymnast does not change body shape, so he can be modelled as a rigid rod of length 2.4 m (with his arms extended). Also assume that his centre of mass is at his midpoint, that is, a distance R of 1.2 m from O, while he is contact with the bar. Just before he lets go of the bar, his centre of mass is moving in a circle at a speed of $4\,\mathrm{m\,s^{-1}}$. At the moment of release, his body makes an angle of $\frac{\pi}{3}$ with the vertical.

(a) What is the gymnast's angular speed about his centre of mass just after he releases the bar?

(b) If the gymnast is to land successfully, that is, on his feet with his body in a vertical position, through what angle must he rotate while in the air? How long will it take him to rotate through this angle?

(c) Consider the vertical movement of an equivalent particle located at the centre of mass of the gymnast. What is the vertical component of the velocity of his centre of mass just after the gymnast releases the bar? Through what distance will his centre of mass fall during the time calculated in part (b)?

(d) Will the gymnast complete a dismount successfully using this approach? If not, what can he do to achieve a successful dismount?

Solution

(a) Just after releasing the bar, the gymnast's hands are stationary, and his centre of mass has the same velocity (say $\mathbf{u}$) as immediately before leaving the bar. So, relative to his centre of mass, his hands have velocity $-\mathbf{u}$. (To determine motion relative to the centre of mass, add $-\mathbf{u}$ to the velocities throughout the body, as shown in Figure 5, so that the point of view is taken in which the centre of mass is 'fixed'.) At that moment, the gymnast is rotating about his centre of mass with angular speed $|\mathbf{u}|/R$ in the direction in which he was circling the bar before letting go. (Recall from Unit 20 the relationship between speed and angular speed.) Since $|\mathbf{u}|$ is $4\,\mathrm{m\,s}^{-1}$ and R is $1.2\,\mathrm{m}$, the angular speed of rotation is $4/1.2 = \frac{10}{3}$ (in $\mathrm{rad\,s}^{-1}$).

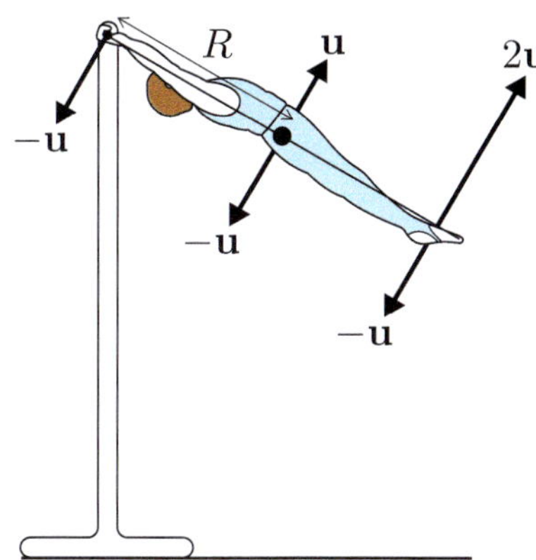

Figure 5   Velocities of parts of the gymnast

(b) At the time of release, the gymnast's body makes an angle of $\frac{\pi}{3}$ with the vertical, so the gymnast must rotate through $2\pi - \frac{\pi}{3} = \frac{5\pi}{3}$ in order to achieve a vertical landing. At the angular speed calculated in part (a), this will take $\frac{5\pi}{3} / \frac{10}{3} = \frac{\pi}{2}$ (in seconds).

(c) At the moment of release, the gymnast's centre of mass has an upward vertical component of velocity of $4\cos\frac{\pi}{6}$, which is $2\sqrt{3}\,\mathrm{m\,s}^{-1}$.

Now, taking into account the constant downward acceleration of magnitude g, and substituting $x_0 = 0$, $v_0 = 2\sqrt{3}$, $a_0 = -g$ and $t = \frac{\pi}{2}$ into the constant acceleration equation $x = x_0 + v_0 t + \frac{1}{2} a_0 t^2$, we find that the vertical component of position will increase in $\frac{\pi}{2}$ seconds by

The constant acceleration equation was derived in Unit 3.

$$2\sqrt{3}\tfrac{\pi}{2} - \tfrac{1}{2}g\left(\tfrac{\pi}{2}\right)^2 \simeq -6.66.$$

Therefore in the time that it takes for the gymnast to attain a vertical body position for landing, his centre of mass will have fallen through a distance of about $6.66\,\mathrm{m}$.

(d) The gymnast cannot complete a dismount successfully in this way. His centre of mass starts at $3 - 1.2\cos\frac{\pi}{3}$, which is $2.4\,\mathrm{m}$ above the ground. But as he would have to fall through $6.66\,\mathrm{m}$ before his body was vertical, he would hit the ground before he had completed the necessary rotation.

If the gymnast were to adjust his body position while in the air, so as to reduce his moment of inertia, he could increase his angular speed of rotation about his centre of mass. This might allow him to complete the necessary rotation before reaching the ground.

Figure 6   Gymnast in the tuck position

Exercise 1

Suppose that the gymnast in Example 2 adjusts his body position while in the air so as to achieve a successful dismount, landing on his feet in a vertical position. When he lands, his body is extended, with his centre of mass at a height of 1.2 m above the ground.

(a) How long is the gymnast in flight between letting go of the bar and landing?

(b) Assume that the gymnast is able to adjust his body position instantaneously, so that he is in a tuck position (see Figure 6) for *all* the time that he is in flight. What is the ratio of the moments of inertia of his body in the tuck and extended positions?

We end this section by discussing qualitatively, and in some detail, the Highland sport of tossing the caber. Tossing the caber can be divided into five phases (see Figure 7). Note that for the toss to be successful, the caber must land with the end originally held by the competitor pointing away from him. The five phases are as follows.

(i)   The competitor runs forward with the caber held in a vertical position, and then stops.

(ii) The caber rotates forwards about the competitor's hands, which are stationary.

(iii) The competitor pushes upwards on the bottom of the caber, and releases it.

(iv) After the competitor releases the caber, it is in flight prior to hitting the ground.

(v) The end of the caber strikes the ground and remains stationary, but the motion of the rest of the caber continues until the caber falls down flat.

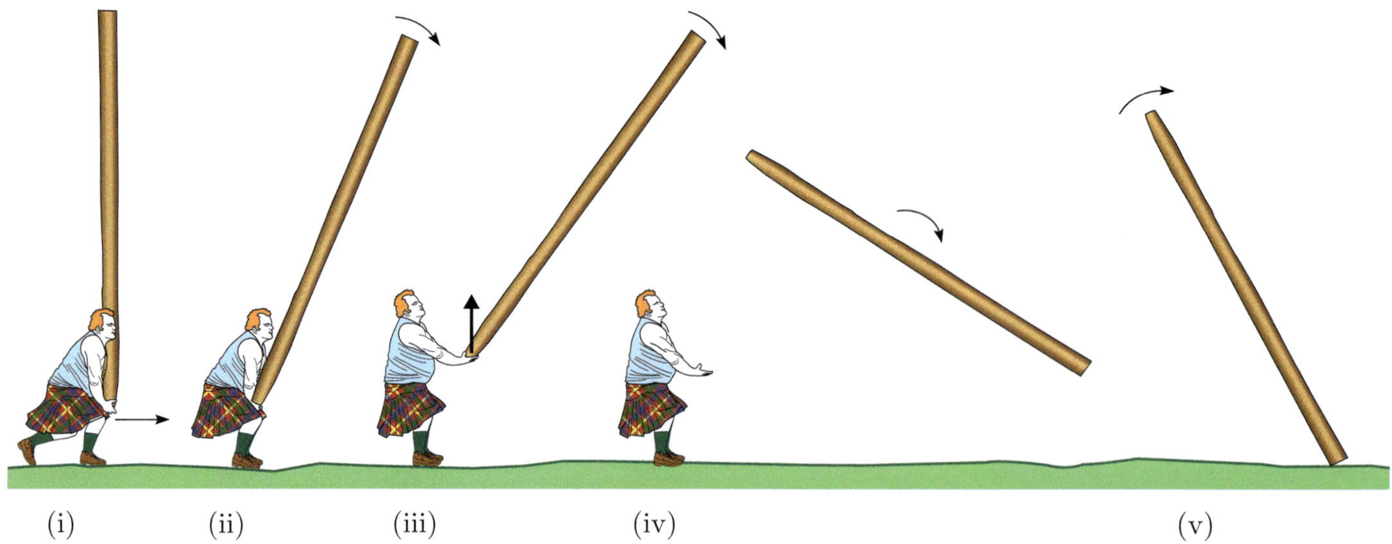

Figure 7   Tossing the caber in five phases

In the first phase, the whole caber gains a uniform horizontal speed. There is no rotation.

In the second phase, the bottom end of the caber is stationary. The upper part of the caber retains forward momentum. The result is to turn the forward motion into rotation about the bottom end of the caber. As the caber topples forwards, gravity supplies a torque about the hands, increasing the rate of rotation.

In the third phase, as the competitor pushes upwards, the resulting upward force applies a torque about the centre of mass, further increasing the rotation of the caber until the moment of release. (By waiting for the caber to rotate forward in the second phase before applying this force, the competitor increases the distance of the centre of mass from the line of action of the upward force that he applies, so increasing the torque.)

In the fourth phase, the centre of mass moves like a projectile. If resistive forces are ignored, the centre of mass follows a parabolic path and the angular speed about it is constant.

In the final phase, the end of the caber strikes the ground and becomes stationary (assuming that it does not skid or bounce). The caber will, however, retain some forward rotation about the end that hits the ground. If it lands as illustrated in Figure 7(v), with the upper end yet to reach the vertical, then the weight due to gravity will provide a torque about the end in contact with the ground that slows this rotation. Whether or not the caber will rotate past the vertical, ensuring that the toss is successful, will depend on the angle at which the caber lands, and on how much angular momentum it has from the preceding phase.

Typically the caber is tapered, with the thinner end held by the competitor. There are several good reasons for this. As the centre of mass of a tapered caber is nearer to the thicker end of the caber than to the thinner end being held by the competitor, in the third phase the centre of mass is further from the line of action of the force (applied by the competitor) than if the caber were not tapered, so increasing the applied torque about the centre of mass. Moreover, for a tapered caber, the centre of mass has further to fall during the fourth phase. This increases the time of flight, and hence increases the angle through which the caber rotates while in flight. Finally, suppose that the caber strikes the ground, thicker end first, before it has rotated past the vertical. Then, in the fifth phase, the centre of mass is closer to the ground than if the caber were symmetric, thereby reducing the torque that is slowing down the caber's rotation. Consequently the chances of a tapered caber reaching the vertical are greater.

A more quantitative mathematical treatment of various aspects of tossing the caber is given in Section 4.

2 Angular momentum

In Unit 20 we obtained the torque law for a particle,

$$\dot{\mathbf{L}} = \boldsymbol{\Gamma},$$

where $\mathbf{L}$ is the particle's angular momentum, and $\boldsymbol{\Gamma}$ is the torque applied to the particle (each taken about the same point). In this section, you will see that this result can be extended to a system of particles; in that case, $\mathbf{L}$ is the total angular momentum of all the particles in the system, and $\boldsymbol{\Gamma}$ is the total torque exerted by all the external forces on the particles in the system. These results can be applied to extended bodies by modelling them as systems of particles.

This section starts by reviewing a number of ideas that you have met earlier: in Subsection 2.1 we review Newton's third law of motion, and in Subsection 2.2 we review angular momentum and the torque law for a particle. Then in Subsection 2.3, we extend the torque law to a system of particles.

2.1 Newton's third law of motion revisited

In order to extend the results of Unit 20 from particles to extended bodies, we need to look again at Newton's third law, which deals with the interaction between particles. In the Introduction to Unit 2, Newton's third law was stated as follows:

> **Law III**   To every action (i.e. force) by one body on another there is always opposed an equal reaction (i.e. force) – that is, the actions of two bodies on each other are always equal in magnitude and opposite in direction.

In symbols this can be expressed by considering two particles, particle 1 and particle 2. Let $\mathbf{I}_{12}$ be the force exerted on particle 1 by particle 2, and let $\mathbf{I}_{21}$ be the force exerted on particle 2 by particle 1. Then the statement of Newton's third law above can be written as $\mathbf{I}_{12} = -\mathbf{I}_{21}$. Alternatively,

$$\mathbf{I}_{12} + \mathbf{I}_{21} = \mathbf{0}. \tag{2}$$

Thus the sum of the forces is zero. Recall that these forces between the particles in a system were called *internal* forces in Unit 19.

Exercise 2 considers how pairs of inter-particle forces that satisfy equation (2) might be represented diagrammatically.

Exercise 2

Figure 8 shows examples of two forces of equal magnitude acting on particles A and B. For each pair of forces, state whether equation (2) is satisfied.

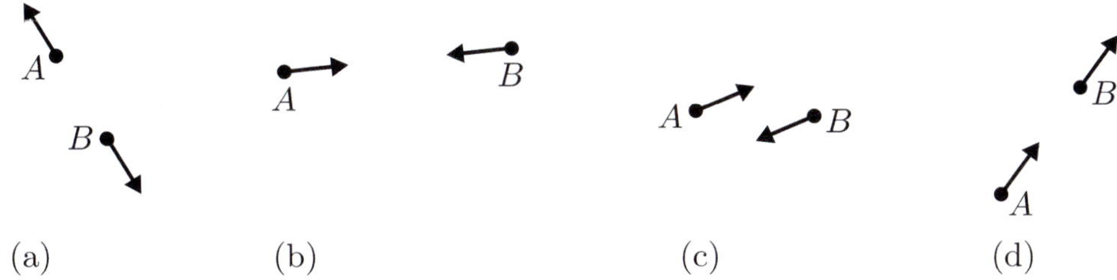

Figure 8   Pairs of forces acting on particles A and B

What we did not mention in Unit 2 was that Newton's third law also states that the pair of equal and opposite forces should have *the same line of action* (this was because the concept of line of action had yet to be introduced in Unit 2). Try the following exercise to see what this means.

Exercise 3

Look back at Figure 8 and find an example of a pair of forces that satisfy equation (2) but where the forces do *not* have the same line of action.

So there are pairs of forces that satisfy equation (2) but where the forces do *not* act in the same straight line. We therefore need a mathematical condition to test whether forces are acting in the same straight line, and hence whether Newton's third law is applicable. We will now develop such a condition.

Figure 9 illustrates a situation where Newton's third law applies: the force $\mathbf{I}_{12}$ exerted on particle 1 by particle 2 is equal and opposite to the force $\mathbf{I}_{21}$ exerted on particle 2 by particle 1, with $\mathbf{I}_{12}$ and $\mathbf{I}_{21}$ having the same line of action. Let $\mathbf{r}_1$ and $\mathbf{r}_2$ be the position vectors of particles 1 and 2, respectively, with respect to some origin O. The vectors $\mathbf{I}_{12}$ and $\mathbf{r}_1 - \mathbf{r}_2$ have the same direction, so their cross product must be $\mathbf{0}$, that is,

$$(\mathbf{r}_1 - \mathbf{r}_2) \times \mathbf{I}_{12} = \mathbf{0}.$$

By equation (2) we have $\mathbf{I}_{12} = -\mathbf{I}_{21}$, so this can be rewritten as

$$(\mathbf{r}_1 - \mathbf{r}_2) \times \mathbf{I}_{12} = \mathbf{r}_1 \times \mathbf{I}_{12} - \mathbf{r}_2 \times \mathbf{I}_{12}$$
$$= \mathbf{r}_1 \times \mathbf{I}_{12} + \mathbf{r}_2 \times \mathbf{I}_{21} = \mathbf{0}. \tag{3}$$

Let $\boldsymbol{\Gamma}_{12} = \mathbf{r}_1 \times \mathbf{I}_{12}$ be the torque about O exerted on particle 1 by the force from particle 2, and similarly let $\boldsymbol{\Gamma}_{21} = \mathbf{r}_2 \times \mathbf{I}_{21}$ be the torque about O exerted on particle 2 by the force from particle 1. Then from equation (3) we have

$$\boldsymbol{\Gamma}_{12} + \boldsymbol{\Gamma}_{21} = \mathbf{0},$$

that is, the sum of the inter-particle torques is zero when equal and opposite inter-particle forces have the same line of action. In other words, when Newton's third law applies, the sum of the inter-particle torques is zero.

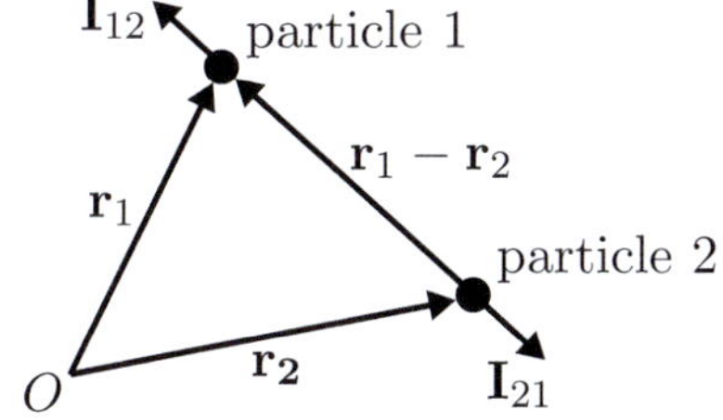

Figure 9   Application of Newton's third law

Recall from Unit 2 that the cross product of parallel vectors is zero.

Recall from Unit 2 that a force $\mathbf{F}$ whose point of action has position vector $\mathbf{r}$ exerts a torque $\boldsymbol{\Gamma} = \mathbf{r} \times \mathbf{F}$ about the origin O.

Conversely, if the sum of the inter-particle torques is not zero, then the line of action of the forces is not the same (by reversing the above argument). Newton's third law can therefore be expressed in terms of the properties of the forces and torques between two interacting particles, as follows.

> ### Newton's third law re-stated
>
> The force $\mathbf{I}_{12}$ exerted on particle 1 by particle 2 is equal in magnitude but opposite in direction to the force $\mathbf{I}_{21}$ exerted on particle 2 by particle 1, with both forces acting along the line joining the two particles. An equivalent condition in symbols is
>
> $$\mathbf{I}_{12} + \mathbf{I}_{21} = \mathbf{0} \quad \text{and} \quad \mathbf{\Gamma}_{12} + \mathbf{\Gamma}_{21} = \mathbf{0},$$
>
> where $\mathbf{\Gamma}_{12} = \mathbf{r}_1 \times \mathbf{I}_{12}$ and $\mathbf{\Gamma}_{21} = \mathbf{r}_2 \times \mathbf{I}_{21}$, with $\mathbf{r}_1$ and $\mathbf{r}_2$ being the position vectors of particles 1 and 2, respectively, relative to the origin.

These equations state that the sum of the inter-particle forces is zero, and the sum of the torques exerted (about the origin) by those forces is also zero.

In electromagnetism there is a type of inter-particle force that does not obey Newton's third law, but this exception need not concern you here.

Almost all inter-particle forces conform to Newton's third law. The gravitational and electrostatic forces between particles obey this law, as do the forces in model springs or model rods joining two particles. All the systems that you will meet in this module conform to Newton's third law.

2.2 Torque law for a particle

Recall, from Unit 20, the definition of angular momentum for a particle, which is as follows: for a particle of mass m that has linear momentum $m\dot{\mathbf{r}}$ and position vector $\mathbf{r}$ relative to an origin O, its angular momentum $\mathbf{L}$ about O is

$$\mathbf{L} = \mathbf{r} \times m\dot{\mathbf{r}}, \tag{4}$$

that is, the angular momentum is the cross product of the particle's position vector and its linear momentum.

Since the position vector is part of the definition, the angular momentum is dependent on the choice of origin (which is not the case for linear momentum). So the angular momentum will usually not be the same relative to different choices of origin.

Exercise 4

A particle of mass 20 has position vector $\mathbf{r} = 3\cos(2t)\mathbf{i} + 4\sin(2t)\mathbf{j} + 5\mathbf{k}$ with origin O (working in SI units), where $\mathbf{i}$, $\mathbf{j}$ and $\mathbf{k}$ are Cartesian unit vectors.

What is the angular momentum of the particle about O at time $t = 0$?

Exercise 5

A particle of mass m moves anticlockwise in a circle of radius R at constant speed v. The circle has its centre at the origin O and lies in the (x, y)-plane. Let $\mathbf{i}$, $\mathbf{j}$ and $\mathbf{k}$ be the Cartesian unit vectors.

(a) What is the angular velocity $\boldsymbol{\omega}$ of the particle?

(b) What is the angular momentum $\mathbf{L}$ of the particle about O?

> This result is also derived in Unit 20.

    (*Hint*: Work from the definition of angular momentum, and assume that the particle has position $R(\cos(\omega t)\mathbf{i} + \sin(\omega t)\mathbf{j})$, where ω is the angular speed.)

(c) What is the moment of inertia I of the particle about an axis through O in the $\mathbf{k}$-direction?

(d) Show that $\mathbf{L} = I\boldsymbol{\omega}$.

In Unit 20 we showed that if a particle is subject to a torque, then the rate of change of the particle's angular momentum $\mathbf{L}$ about a fixed point is equal to the applied torque $\boldsymbol{\Gamma}$ about that point, that is, $\dot{\mathbf{L}} = \boldsymbol{\Gamma}$. The derivation of this important result from Unit 20, known as the *torque law for a particle*, is repeated below.

Using equation (4), we find

$$\frac{d}{dt}\mathbf{L} = \frac{d}{dt}(\mathbf{r} \times m\dot{\mathbf{r}})$$
$$= (\dot{\mathbf{r}} \times m\dot{\mathbf{r}}) + (\mathbf{r} \times m\ddot{\mathbf{r}})$$
$$= \mathbf{r} \times m\ddot{\mathbf{r}}. \tag{5}$$

> Since $\dot{\mathbf{r}}$ and $m\dot{\mathbf{r}}$ are parallel vectors, $\dot{\mathbf{r}} \times m\dot{\mathbf{r}} = \mathbf{0}$.

Now, by Newton's second law, $\mathbf{F} = m\ddot{\mathbf{r}}$, where $\mathbf{F}$ is the total force on the particle. Substituting for $m\ddot{\mathbf{r}}$ in equation (5), we have

$$\dot{\mathbf{L}} = \mathbf{r} \times \mathbf{F}.$$

But the total torque on the particle (relative to the chosen origin) is $\mathbf{r} \times \mathbf{F} = \boldsymbol{\Gamma}$. So we obtain the torque law for a particle:

$$\dot{\mathbf{L}} = \boldsymbol{\Gamma}. \tag{6}$$

An important special case of this law is when the total external torque is zero. In this case equation (6) becomes $\dot{\mathbf{L}} = \mathbf{0}$, so $\mathbf{L}$ is a constant vector. This is called the *law of conservation of angular momentum for a particle*.

As an example of a case where angular momentum is conserved, consider any particle moving at constant velocity. By Newton's first law, the total force acting on the particle must be zero, so the total torque acting on the particle will also be zero. So by the torque law, the angular momentum of the particle about *any* point will be constant. However, there is an alternative way of understanding this conservation of angular momentum that comes directly from definition (4) and proceeds as follows.

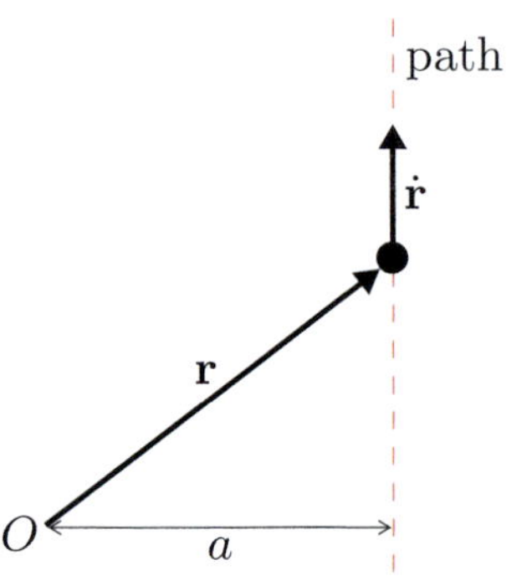

Figure 10   Particle moving with constant velocity $\dot{\mathbf{r}}$

This result was also derived in Unit 20.

By definition, the particle has angular momentum $\mathbf{L} = \mathbf{r} \times m\dot{\mathbf{r}}$, where $\mathbf{r}$ gives the particle's position relative to the origin. Since the velocity $\dot{\mathbf{r}}$ of the particle is constant, the particle will follow a straight-line path, as illustrated in Figure 10. The position vector $\mathbf{r}$ is varying, but wherever the particle is on the path, the vector $\mathbf{r} \times \dot{\mathbf{r}}$ will have magnitude $a|\dot{\mathbf{r}}|$, where a is the perpendicular distance from O to the particle's path, and direction normal to the plane shown in Figure 10 (and out of the page in the case illustrated).

Thus the angular momentum is constant, since both a and $|\dot{\mathbf{r}}|$ are constant and its direction is constant.

Exercise 6

(a) If $\mathbf{i}$, $\mathbf{j}$ and $\mathbf{k}$ are Cartesian unit vectors and $\mathbf{r} = x\mathbf{i} + y\mathbf{j}$, show that

$$\mathbf{r} \times \dot{\mathbf{r}} = (x\dot{y} - y\dot{x})\mathbf{k}.$$

(b) Suppose that a particle of mass m is moving in the (x, y)-plane. By expressing x and y in terms of polar coordinates r and θ, show that the angular momentum $\mathbf{L}$ of the particle about the origin O is

$$\mathbf{L} = mr^2\dot{\theta}\mathbf{k}. \tag{7}$$

Exercise 6(b) provides an expression for the angular momentum about the origin of any particle moving in the (x, y)-plane. It generalises the result obtained in Exercise 5 for the angular momentum of a particle travelling in a circle at constant speed. In Exercise 5 the particle followed a circle whose centre was at the origin. The next example concerns motion in a circle whose centre is not the origin (indeed, the circle is not in the (x, y)-plane).

Example 3

A particle of mass m moves with a constant angular velocity $\boldsymbol{\omega} = \omega\mathbf{k}$ in a circular path of radius R, centred on a fixed point with Cartesian coordinates $(0, 0, h)$. At $t = 0$, the particle is at the point $(R, 0, h)$. Let $\mathbf{i}$, $\mathbf{j}$ and $\mathbf{k}$ be Cartesian unit vectors in the x-, y- and z-directions, respectively, and let O be the origin.

(a) Express the particle's position vector $\mathbf{r}$ as a function of t.

(b) (i)   Calculate the particle's velocity $\dot{\mathbf{r}}$ by differentiation.

(ii)   Recall from Unit 20 that a particle executing circular motion with angular velocity $\boldsymbol{\omega}$ has velocity $\dot{\mathbf{r}} = \boldsymbol{\omega} \times \mathbf{r}$. Verify that this is consistent with your result in part (b)(i).

(c) Calculate the angular momentum of the particle about O.

(d) Show that the torque about O acting to sustain the motion of the particle is $\boldsymbol{\Gamma} = -m\omega h\dot{\mathbf{r}}$.

Solution

(a) The angular velocity is in the $\mathbf{k}$-direction, so the particle must be moving parallel to the (x, y)-plane. The plane of motion is $z = h$. The particle is travelling with constant angular velocity $\omega\mathbf{k}$ in a circle of radius R, and is at $x = R$, $y = 0$ when $t = 0$. Therefore its position vector is

$$\mathbf{r} = R\cos(\omega t)\mathbf{i} + R\sin(\omega t)\mathbf{j} + h\mathbf{k}. \tag{8}$$

In Unit 20, you saw that the angular velocity is perpendicular to the plane of motion.

See Unit 20.

(b) (i) Differentiating equation (8) gives the particle's velocity as

$$\dot{\mathbf{r}} = -R\omega\sin(\omega t)\mathbf{i} + R\omega\cos(\omega t)\mathbf{j}. \tag{9}$$

(ii) Using $\mathbf{r}$ from equation (8) and $\boldsymbol{\omega} = \omega\mathbf{k}$, we substitute into $\dot{\mathbf{r}} = \boldsymbol{\omega} \times \mathbf{r}$ and obtain

$$\dot{\mathbf{r}} = \omega\mathbf{k} \times \left(R\cos(\omega t)\mathbf{i} + R\sin(\omega t)\mathbf{j} + h\mathbf{k}\right)$$
$$= R\omega\cos(\omega t)\mathbf{j} - R\omega\sin(\omega t)\mathbf{i},$$

which is consistent with the result in part (b)(i).

(c) By definition (equation (4)), the angular momentum of the particle about O is $\mathbf{L} = \mathbf{r} \times m\dot{\mathbf{r}}$. On substituting for $\mathbf{r}$ and $\dot{\mathbf{r}}$ from equations (8) and (9), respectively, we obtain

$$\mathbf{L} = \left(R\cos(\omega t)\mathbf{i} + R\sin(\omega t)\mathbf{j} + h\mathbf{k}\right)$$
$$\times\, m\left(-R\omega\sin(\omega t)\mathbf{i} + R\omega\cos(\omega t)\mathbf{j}\right)$$
$$= m\left(R^2\omega\cos^2(\omega t)\mathbf{k} + R^2\omega\sin^2(\omega t)\mathbf{k}\right.$$
$$\left. -\, hR\omega\sin(\omega t)\mathbf{j} - hR\omega\cos(\omega t)\mathbf{i}\right)$$
$$= mR^2\omega\mathbf{k} - mhR\omega\left(\cos(\omega t)\mathbf{i} + \sin(\omega t)\mathbf{j}\right). \tag{10}$$

(d) From the torque law for a particle, we have $\boldsymbol{\Gamma} = \dot{\mathbf{L}}$. Now, by differentiating equation (10) we get

$$\dot{\mathbf{L}} = -mhR\omega^2\left(\cos(\omega t)\mathbf{j} - \sin(\omega t)\mathbf{i}\right). \tag{11}$$

Comparing this with equation (9), we see that

$$\dot{\mathbf{L}} = -mh\omega\dot{\mathbf{r}},$$

hence $\boldsymbol{\Gamma} = -mh\omega\dot{\mathbf{r}}$.

To sum up: in both Exercise 5 and Example 3 we looked at a particle moving at constant speed in a circle. In Exercise 5 you saw that, relative to the centre of the circle, angular momentum is constant and the total torque acting on the particle is zero. In Example 3, we considered angular momentum and torque relative to a point on the axis of rotation but *not* at the centre of the circle. Relative to that point, the angular momentum is not constant (although its component in the direction of the angular velocity is constant); consequently, there must be a non-zero total torque on the particle about that point if such motion is to be sustained.

Equation (10) shows that the component of $\mathbf{L}$ in the $\mathbf{k}$-direction is constant.

For a particle moving as described in Example 3, calculate the acceleration $\ddot{\mathbf{r}}$. Use Newton's second law to find the total force $\mathbf{F}$ acting on the particle. Then calculate the total torque $\boldsymbol{\Gamma}$ on the particle by using $\boldsymbol{\Gamma} = \mathbf{r} \times \mathbf{F}$, and show that you obtain the same result as in Example 3(d).

2.3   Torque law for an n-particle system

We now move on to consider angular momentum and torque for systems involving more than one particle. First, we look at systems that have two particles, then we go on to examine a general system containing an arbitrary number of particles.

Two-particle systems

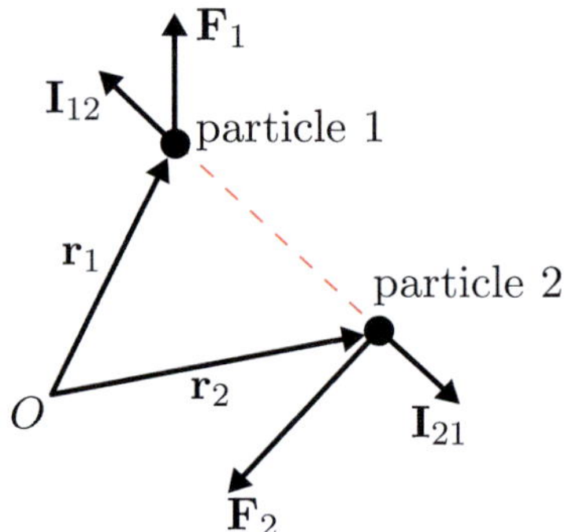

Figure 11   Two-particle system

Consider a two-particle system, as illustrated in Figure 11, to which Newton's third law applies. Particle 1, at position $\mathbf{r}_1$ relative to a fixed origin O, may be acted on by any number of external forces, but we will consider only the resultant of all these external forces, $\mathbf{F}_1$. The only other force on particle 1 is the internal force $\mathbf{I}_{12}$ exerted on it by particle 2. The total external force on particle 2, at position $\mathbf{r}_2$, is $\mathbf{F}_2$, and the only other force on particle 2 is the internal force $\mathbf{I}_{21}$ exerted on it by particle 1. Let the angular momentum of particle 1 relative to O be $\mathbf{L}_1$, and that of particle 2 relative to O be $\mathbf{L}_2$.

Applying the torque law for a particle (equation (6)) to particle 1 gives

$$\dot{\mathbf{L}}_1 = \mathbf{r}_1 \times (\mathbf{F}_1 + \mathbf{I}_{12}) = \mathbf{r}_1 \times \mathbf{F}_1 + \boldsymbol{\Gamma}_{12},$$

where $\boldsymbol{\Gamma}_{12}$ is the torque about O acting on particle 1 by the internal force from particle 2. Similarly, for particle 2 we have

$$\dot{\mathbf{L}}_2 = \mathbf{r}_2 \times (\mathbf{F}_2 + \mathbf{I}_{21}) = \mathbf{r}_2 \times \mathbf{F}_2 + \boldsymbol{\Gamma}_{21},$$

where $\boldsymbol{\Gamma}_{21}$ is the torque about O acting on particle 2 by the internal force from particle 1. Adding these two equations, we obtain

$$\dot{\mathbf{L}}_1 + \dot{\mathbf{L}}_2 = \mathbf{r}_1 \times \mathbf{F}_1 + \mathbf{r}_2 \times \mathbf{F}_2 + \boldsymbol{\Gamma}_{12} + \boldsymbol{\Gamma}_{21}.$$

Newton's third law implies that $\boldsymbol{\Gamma}_{12} + \boldsymbol{\Gamma}_{21} = \mathbf{0}$, so this reduces to

$$\dot{\mathbf{L}}_1 + \dot{\mathbf{L}}_2 = \mathbf{r}_1 \times \mathbf{F}_1 + \mathbf{r}_2 \times \mathbf{F}_2. \tag{12}$$

In equation (12), the right-hand side represents the total torque exerted on the two-particle system by the external forces, while the left-hand side gives the rate of change with time of the total angular momentum of the system. Thus equation (12) extends the torque law to a system of two

particles. The *torque law for a two-particle system* can then be stated as
follows. For a two-particle system, the rate of change of the total angular
momentum of the particles in the system is equal to the total torque
exerted on the system by external forces (where angular momentum and
torque are determined relative to the same fixed point).

Exercise 8

Figure 12 is a schematic representation of a type of children's playground
roundabout. Two children sit on seats at A and B, and the roundabout is
set in motion by pushing at C. It rotates horizontally about a fixed spindle
at O. The distances AO, BO and CO are a, b and c, respectively. The
mass of the child and seat at A is m_1, and the mass of the child and seat
at B is m_2. The positions of the seats are balanced so that $m_1 a = m_2 b$.
The mass of the rest of the roundabout is negligible.

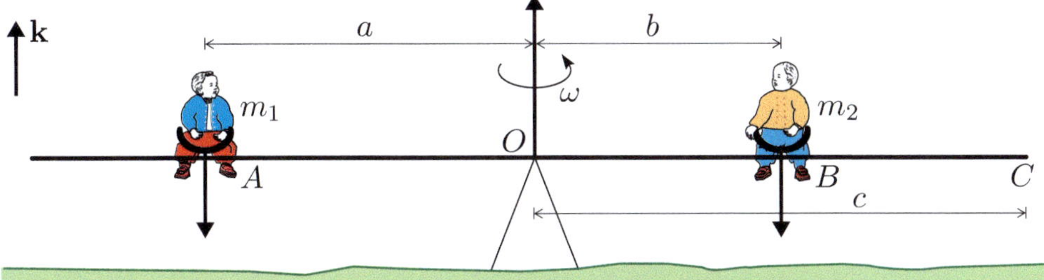

Figure 12   Two children on a playground roundabout

(a) Suppose that the roundabout is rotating anticlockwise at an angular
    speed ω. Model each 'child plus seat' as a particle, and show that the
    total angular momentum of the roundabout about O is $I\boldsymbol{\omega}$, where
    $I = m_1 a^2 + m_2 b^2$ and $\boldsymbol{\omega} = \omega\mathbf{k}$, with $\mathbf{k}$ being a unit vector pointing
    vertically upwards.

(b) Suppose that an adult pushes the roundabout so as to apply a force of
    constant magnitude F at C, in a horizontal direction at right angles
    to OC. Assuming that resistive forces are negligible, show that

$$I\dot{\omega} = cF.$$

   (Note that the roundabout is rotating anticlockwise, so $\dot{\theta} \geq 0$, thus
   $\omega = \dot{\theta}$ and $\dot{\omega} = \ddot{\theta}$.)

(c) If the distance AO is $1\,\mathrm{m}$, BO is $0.75\,\mathrm{m}$ and CO is $1.5\,\mathrm{m}$, and m_1 is
    $45\,\mathrm{kg}$, m_2 is $60\,\mathrm{kg}$ and F is $315\,\mathrm{N}$, then for how long will the adult
    need to push in order to get the roundabout rotating at 0.5 revolutions
    per second when the roundabout has started from rest?

The next example concerns a two-particle system executing a more
complicated motion.

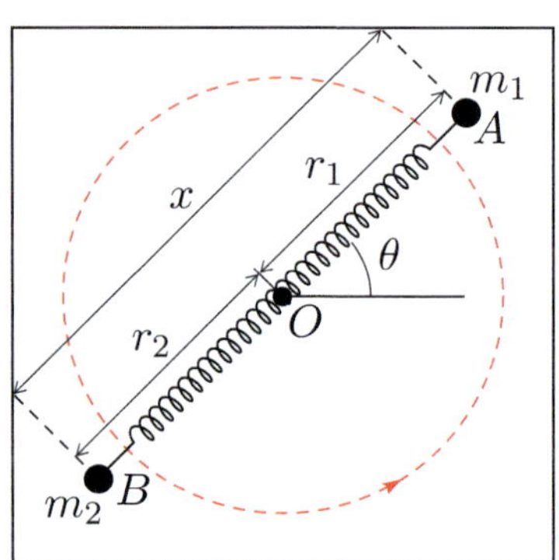

Figure 13   Two particles connected by a spring

Example 4

Consider two particles of masses m_1 and m_2, connected by a model spring and resting on a smooth horizontal table, as shown in Figure 13. The system is in motion: the two particles are rotating at the same angular velocity about the centre of mass of the system, and simultaneously they are oscillating as the spring extends and contracts. The total external force on the system is zero: the weight of each particle is balanced by the normal reaction from the table, and we assume that no other external forces are acting. The centre of mass is stationary initially. Therefore it remains stationary throughout the motion because the acceleration of the centre of mass is zero (since the total external force is zero). (This was shown in Unit 19.)

Take the origin O to be at the centre of mass, and suppose that particle A has polar coordinates (r_1, θ), and particle B has polar coordinates $(r_2, \theta - \pi)$. Let x be the distance between the two particles, and suppose that $\mathbf{k}$ is a unit vector pointing vertically upwards. Use the torque law for a two-particle system to show that the quantity $x^2\dot{\theta}$ is constant. Use this equation to describe the motion qualitatively.

Solution

The only external forces acting on the system are the weights of the particles and the normal reactions of the table balancing each weight. These normal reactions have the same points of action as the weights; consequently, not only is the total external force on the system zero, but the total external torque on the system is also zero. (The only other force acting on each particle is that from the spring, and this is an internal force.) It follows from the torque law that as the total external torque is zero, the angular momentum $\mathbf{L}$ of the system is a constant.

From equation (7), the angular momentum of particle A about O is $m_1 r_1^2 \dot{\theta}\mathbf{k}$, and that of particle B is $m_2 r_2^2 \dot{\theta}\mathbf{k}$. So the total angular momentum of the system is

$$\mathbf{L} = m_1 r_1^2 \dot{\theta}\mathbf{k} + m_2 r_2^2 \dot{\theta}\mathbf{k} = (m_1 r_1^2 + m_2 r_2^2)\dot{\theta}\mathbf{k}.$$

But $\mathbf{L}$ is a constant, as noted above, therefore

$$(m_1 r_1^2 + m_2 r_2^2)\dot{\theta} = c, \tag{13}$$

where c is a constant.

Because O is at the centre of mass of the two-particle system,

$$m_1 r_1 = m_2 r_2. \tag{14}$$

Now $x = r_1 + r_2$, so we can substitute for r_1 in equation (14) to obtain $m_1(x - r_2) = m_2 r_2$. Then solving for r_2 gives

$$r_2 = \frac{m_1 x}{m_1 + m_2}. \tag{15}$$

Substituting this into equation (14) gives r_1 in terms of x:

$$r_1 = \frac{m_2 x}{m_1 + m_2}. \tag{16}$$

Substituting from equations (15) and (16) into equation (13) yields

$$
\begin{aligned}
c &= \left(m_1 \frac{m_2^2 x^2}{(m_1 + m_2)^2} + m_2 \frac{m_1^2 x^2}{(m_1 + m_2)^2} \right) \dot{\theta} \\
&= \frac{m_1 m_2^2 + m_2 m_1^2}{(m_1 + m_2)^2} x^2 \dot{\theta} \\
&= \frac{m_1 m_2}{m_1 + m_2} x^2 \dot{\theta}.
\end{aligned}
$$

Since the masses m_1 and m_2 are constant, we can deduce that $x^2 \dot{\theta}$ must be constant.

As the system rotates, the spring extends and compresses as the particles oscillate in and out. The rotational motion of the particles is connected to these oscillations since $x^2 \dot{\theta}$ is a constant. When the spring is extended (x is larger) the angular speed $|\dot{\theta}|$ decreases, and vice versa.

Exercise 9

A binary star system consists of two stars of masses m_1 and m_2. Model each star as a particle, and assume that all the forces exerted on the system, other than the gravitational attraction between the two stars, can be ignored. Assume also that the distance d between the stars is constant, and that the common centre of mass of the two stars is fixed. Let the total angular momentum of the binary system have magnitude L. Express the period of rotation of the system in terms of m_1, m_2, d and L.

n-particle systems

You have seen above that the torque law for a two-particle system has the same form as that for a single particle. This is because the sum of the internal torques between the two particles is zero. The situation is analogous for a system of more than two particles. Therefore we can generalise the torque law to any number of particles.

Consider a system of n particles, which we will call particles $1, 2, 3, \ldots, n$, to which Newton's third law applies. Particle 1 may be acted on by various external forces, but we will denote the resultant of all the external forces on particle 1 by $\mathbf{F}_1$. In general, for $i = 1, 2, \ldots, n$, $\mathbf{F}_i$ denotes the resultant of all the external forces on particle i. As well as being subject to forces external to the system, particle 1 may be acted on by internal forces exerted by each of the other particles in the system: denote by $\mathbf{I}_{12}$ the force exerted on particle 1 by particle 2, denote by $\mathbf{I}_{13}$ the force exerted on particle 1 by particle 3, and so on. In general, let $\mathbf{I}_{ij}$ denote the force exerted on particle i by particle j (for i and j between 1 and n, with $i \neq j$). For $i = 1, 2, \ldots, n$, let the position of particle i be $\mathbf{r}_i$ (relative to a fixed origin O), and let the angular momentum of particle i relative to O be $\mathbf{L}_i$.

The torque law for a particle when applied to particle 1 gives

$$\dot{\mathbf{L}}_1 = \mathbf{r}_1 \times (\mathbf{F}_1 + \mathbf{I}_{12} + \mathbf{I}_{13} + \mathbf{I}_{14} + \cdots + \mathbf{I}_{1n}). \tag{17}$$

If $\boldsymbol{\Gamma}_{ij}$ is the torque exerted on particle i by the internal force from particle j, then $\boldsymbol{\Gamma}_{ij} = \mathbf{r}_i \times \mathbf{I}_{ij}$, and equation (17) becomes

$$\dot{\mathbf{L}}_1 = \mathbf{r}_1 \times \mathbf{F}_1 + \boldsymbol{\Gamma}_{12} + \boldsymbol{\Gamma}_{13} + \boldsymbol{\Gamma}_{14} + \cdots + \boldsymbol{\Gamma}_{1n}. \tag{18}$$

Similarly, the torque law for a particle when applied to particle 2 gives

$$\dot{\mathbf{L}}_2 = \mathbf{r}_2 \times (\mathbf{F}_2 + \mathbf{I}_{21} + \mathbf{I}_{23} + \mathbf{I}_{24} + \cdots + \mathbf{I}_{2n})$$
$$= \mathbf{r}_2 \times \mathbf{F}_2 + \boldsymbol{\Gamma}_{21} + \boldsymbol{\Gamma}_{23} + \boldsymbol{\Gamma}_{24} + \cdots + \boldsymbol{\Gamma}_{2n}. \tag{19}$$

For particle i, we get

$$\dot{\mathbf{L}}_i = \mathbf{r}_i \times (\mathbf{F}_i + \mathbf{I}_{i1} + \mathbf{I}_{i2} + \mathbf{I}_{i3} + \cdots + \mathbf{I}_{in})$$
$$= \mathbf{r}_i \times \mathbf{F}_i + \boldsymbol{\Gamma}_{i1} + \boldsymbol{\Gamma}_{i2} + \boldsymbol{\Gamma}_{i3} + \cdots + \boldsymbol{\Gamma}_{in}. \tag{20}$$

Note that the sum does not contain a term $\boldsymbol{\Gamma}_{ii}$.

Now, to obtain the total angular momentum of the system, we can sum versions of equation (20) for each of $i = 1, 2, \ldots, n$. When we do this, all the internal torques can be 'paired up': thus the term $\boldsymbol{\Gamma}_{12}$ from equation (18) can be paired with the term $\boldsymbol{\Gamma}_{21}$ from equation (19); the term $\boldsymbol{\Gamma}_{13}$ from equation (18) can be paired with the term $\boldsymbol{\Gamma}_{31}$ from the equivalent equation for particle 3; and generally, the term $\boldsymbol{\Gamma}_{ij}$ from the equation for particle i (where $i \neq j$) can be paired with the term $\boldsymbol{\Gamma}_{ji}$ from the equation for particle j. Each pair will sum to $\mathbf{0}$, since from Newton's third law, $\boldsymbol{\Gamma}_{ij} + \boldsymbol{\Gamma}_{ji} = \mathbf{0}$ (for $i \neq j$), so the total of all the internal torques for the n particles comprising the system will be $\mathbf{0}$. If $\mathbf{L}$ is the total angular momentum of the system (where $\mathbf{L} = \mathbf{L}_1 + \mathbf{L}_2 + \cdots + \mathbf{L}_n$), then from equation (20) (for $i = 1, 2, \ldots, n$) we have

$$\dot{\mathbf{L}} = \mathbf{r}_1 \times \mathbf{F}_1 + \mathbf{r}_2 \times \mathbf{F}_2 + \cdots + \mathbf{r}_n \times \mathbf{F}_n. \tag{21}$$

Here, $\mathbf{r}_1 \times \mathbf{F}_1$ is the torque exerted on particle 1 by the external forces acting on it, $\mathbf{r}_2 \times \mathbf{F}_2$ is the torque exerted on particle 2 by the external forces acting on it, and so on. Therefore the total external torque on the system is

$$\boldsymbol{\Gamma} = \mathbf{r}_1 \times \mathbf{F}_1 + \mathbf{r}_2 \times \mathbf{F}_2 + \cdots + \mathbf{r}_n \times \mathbf{F}_n.$$

So equation (21) can be written as

$$\dot{\mathbf{L}} = \boldsymbol{\Gamma}.$$

Thus we have shown that the rate of change of the total angular momentum of an n-particle system is equal to the total external torque acting on the system. This extends the torque law to an n-particle system, which we now refer to as just the torque law.

Torque law

Consider a system of n particles. Let $\mathbf{r}_i$ be the position vector (relative to a fixed origin O) of particle i, and let $\mathbf{F}_i$ be the total external force on particle i, for $i = 1, 2, \ldots, n$. The total external torque on the system about O is

$$\mathbf{\Gamma} = \sum_{i=1}^{n} \mathbf{r}_i \times \mathbf{F}_i. \tag{22}$$

The rate of change of the total angular momentum $\mathbf{L}$ of the system about O equals the total external torque acting on the system, that is,

$$\dot{\mathbf{L}} = \mathbf{\Gamma}. \tag{23}$$

In particular, when the total external torque about O is zero, the total angular momentum vector about O is conserved, that is, it is constant.

Exercise 10

Suppose that all the external forces on a system of particles are applied at the same fixed point X. What can you deduce about the angular momentum of the system?

3 Rigid-body rotation about a fixed axis

In this section we model a **rigid body** as an n-particle system where all the inter-particle distances remain constant. To simplify matters, we will consider rigid bodies that are rotating about a fixed axis. For such a body, the rotational motion can be expressed in a very convenient way by using the moment of inertia. This is done in Subsection 3.1.

Rigid bodies were introduced in Unit 2.

Moments of inertia were introduced in Unit 17.

Unit 17 showed how to use multiple integrals to calculate the moments of inertia of complicated shapes. In this unit we model systems using simple geometric shapes, and we will give a table of moments of inertia for these in Subsection 3.1. This table covers only moments of inertia about an axis through the centre of mass of the body. You may want to find moments of inertia about other axes, and this is covered in Subsection 3.2. The kinetic energy of a rotating rigid body can also be expressed in terms of its moment of inertia, and we consider that in Subsection 3.3.

3.1 Angular momentum and moments of inertia

In this section we confine our attention to motion in which a rigid body is rotating about a fixed axis. For simplicity, we choose a frame of reference in which the z-axis is the axis of rotation.

Now, in order to apply the torque law $\dot{\mathbf{L}} = \boldsymbol{\Gamma}$ to a rotating rigid body, we need to find an expression for the angular momentum of the body (relative to a point on the axis of rotation). Suppose that the body is composed of particles A, B, C, and so on. The particles that lie on the axis of rotation, like C and D in Figure 14(a), do not move. On the other hand, particles like A and B do move, but their distances from the z-axis remain constant, so they travel in circles centred on a point on the z-axis, and parallel to the (x, y)-plane. The angular velocity of the body (and of each of its constituent particles) is $\boldsymbol{\omega} = \dot{\theta}\mathbf{k}$, where θ is the angle through which the body is rotated about its axis of rotation, as shown in Figure 14(b).

We choose x-, y-, z-axes that are fixed in space, and an origin O that is fixed and on the axis of rotation. Here $\mathbf{k}$ is a unit vector in the z-direction.

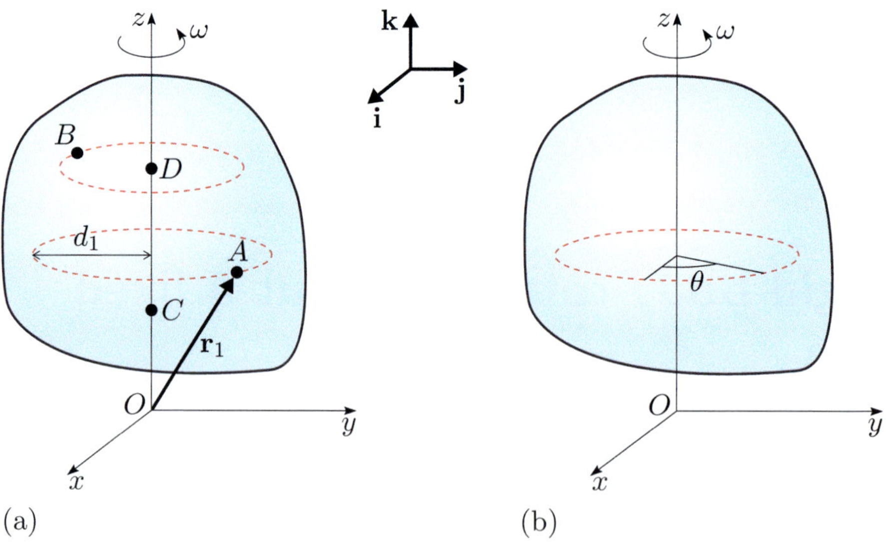

(a)

(b)

Figure 14   Rigid body rotating about the z-axis, showing (a) some individual particles, (b) the relationship with the angular displacement θ

First consider particle A, which has mass m_1 and position vector $\mathbf{r}_1$. From Unit 20, particle A has velocity $\dot{\mathbf{r}}_1 = \boldsymbol{\omega} \times \mathbf{r}_1$. Therefore its angular momentum is

$$
\begin{aligned}
\mathbf{L}_1 &= \mathbf{r}_1 \times m_1\dot{\mathbf{r}}_1 \\
&= \mathbf{r}_1 \times m_1(\boldsymbol{\omega} \times \mathbf{r}_1) \\
&= \mathbf{r}_1 \times m_1\dot{\theta}(\mathbf{k} \times \mathbf{r}_1).
\end{aligned} \tag{24}
$$

If $\mathbf{r}_1 = x_1\mathbf{i} + y_1\mathbf{j} + z_1\mathbf{k}$, we have

$$
\begin{aligned}
\mathbf{k} \times \mathbf{r}_1 &= \mathbf{k} \times (x_1\mathbf{i} + y_1\mathbf{j} + z_1\mathbf{k}) \\
&= x_1\mathbf{j} - y_1\mathbf{i}.
\end{aligned}
$$

Then, on substituting for $\mathbf{r}_1$ and $\mathbf{k} \times \mathbf{r}_1$ in equation (24), we obtain

$$
\begin{aligned}
\mathbf{L}_1 &= (x_1\mathbf{i} + y_1\mathbf{j} + z_1\mathbf{k}) \times m_1\dot{\theta}(x_1\mathbf{j} - y_1\mathbf{i}) \\
&= m_1\dot{\theta}(x_1^2\mathbf{k} + y_1^2\mathbf{k} - x_1z_1\mathbf{i} - y_1z_1\mathbf{j}).
\end{aligned} \tag{25}
$$

Let d_1 be the perpendicular distance of particle A from the axis of rotation. Then $d_1^2 = x_1^2 + y_1^2$, and equation (25) simplifies to

$$
\mathbf{L}_1 = m_1\dot{\theta}(-x_1z_1\mathbf{i} - y_1z_1\mathbf{j} + d_1^2\mathbf{k}).
$$

Suppose that the rigid body consists of n particles, where the ith particle (for $i = 1, 2, \ldots, n$) has mass m_i and position vector $\mathbf{r}_i = x_i\mathbf{i} + y_i\mathbf{j} + z_i\mathbf{k}$. The total angular momentum $\mathbf{L}$ of the body will be the sum of the angular momenta of all these particles, that is,

$$
\mathbf{L} = \dot{\theta}\sum_{i=1}^{n} m_i(-x_iz_i\mathbf{i} - y_iz_i\mathbf{j} + d_i^2\mathbf{k}), \tag{26}
$$

where d_i is the perpendicular distance of the ith particle from the z-axis. This angular momentum has non-zero components perpendicular to the axis of rotation (in the $\mathbf{i}$- and $\mathbf{j}$-directions). However, in many applications it is only the component in the direction of the axis of rotation that is needed (the situation in Exercise 11 below is an exception). With our choice of axes, this component is in the z-direction, and is

$$
L_z = \dot{\theta}\sum_{i=1}^{n} m_id_i^2.
$$

This result is very important. For any particular rigid body, the quantity $\sum_{i=1}^{n} m_id_i^2 = I$ is a constant, since both m_i and d_i are constant for all i. It depends on the distribution of mass about the axis of rotation, and as you saw in Unit 17, it is called the moment of inertia. Hence for a rigid body spinning about a fixed axis, the component of the angular momentum in the direction of the axis can be conveniently expressed in terms of its moment of inertia.

We have established this result with the z-axis as the axis of rotation. However, for any rigid body rotating about a fixed axis, we could choose axes such that the z-axis is the axis of rotation, so there is no loss of generality in our argument.

Angular momentum of a rigid body rotating about a fixed axis

Suppose that a rigid body is rotating about a fixed axis with angular velocity $\boldsymbol{\omega}$. Let $\mathbf{L}$ be the angular momentum of the body about a point O on the axis, and let L_{axis} be the component of $\mathbf{L}$ in the direction of the axis. Then

$$L_{\text{axis}} = I\dot{\theta}, \tag{27}$$

where I is the moment of inertia of the body about the axis of rotation.

Note that the *magnitude* of the component of the angular momentum in the axis of rotation can be written as $|L_{\text{axis}}| = I\omega$, where ω is the angular speed.

Exercise 11

Throughout this exercise, torque and angular momentum are measured about the origin O; also, $\mathbf{i}$, $\mathbf{j}$ and $\mathbf{k}$ are Cartesian unit vectors, with $\mathbf{i}$ and $\mathbf{j}$ in a horizontal plane and $\mathbf{k}$ pointing vertically upwards.

(a) Consider a rigid body modelled as a system of n particles lying on a curve C embedded in a vertical plane as shown in Figure 15. The curve C does not cross the z-axis or extend below the (x, y)-plane, and is moving anticlockwise with constant angular speed ω about the z-axis.

(i) Show that the angular momentum of the system can be expressed as

$$\mathbf{L} = I\omega\mathbf{k} - A\omega\mathbf{e}_r, \tag{28}$$

where I is the moment of inertia of the body about the z-axis, $A > 0$ is a constant, and $\mathbf{e}_r = \cos\theta\,\mathbf{i} + \sin\theta\,\mathbf{j}$ is the unit vector pointing horizontally from the z-axis within the vertical plane containing the rigid body, as shown in Figure 15. (You met this use of $\mathbf{e}_r$ in Unit 20.)

(*Hint*: Express the x- and y-coordinates of each of the particles in polar coordinates.)

(ii) Hence show that

$$\dot{\mathbf{L}} = -A\omega^2\mathbf{e}_\theta, \tag{29}$$

where the unit vector $\mathbf{e}_\theta = -\sin\theta\,\mathbf{i} + \cos\theta\,\mathbf{j}$ lies in the (x, y)-plane and is normal to $\mathbf{e}_r$, pointing in the direction of increasing θ. (You met this use of $\mathbf{e}_\theta$ in Unit 20.)

Deduce that the body must be subject to a non-zero torque in the tangential direction but pointing in the opposite direction to $\mathbf{e}_\theta$.

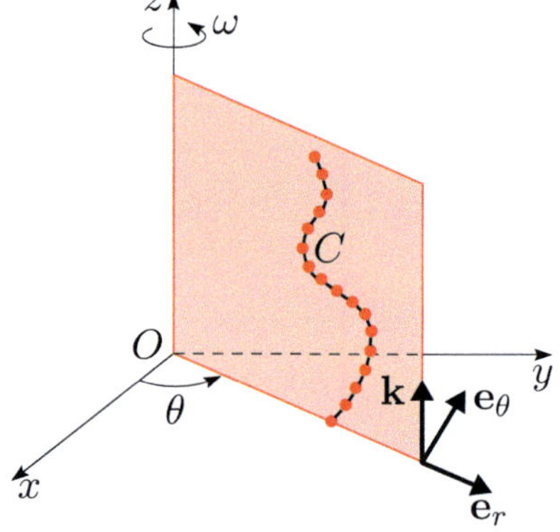

Figure 15   Rigid body modelled as n particles placed on a curve C embedded in a vertical plane rotating about the z-axis

(b) Consider a skater whose centre of mass is moving with constant angular speed ω in a circle parallel to the (x, y)-plane.

(i) By analysing the motion of an equivalent particle located at the skater's centre of mass, deduce the direction of the total external force $\mathbf{F}$ on the skater.

(ii) Suppose that the skater tries to follow the circle while remaining perfectly vertical. Use equation (29) to show that this is impossible. What must the skater do if she is to follow such a circle? (Model the skater as a rigid system of particles lying on a curve, as discussed in part (a). Assume that the only forces on the skater are her weight and the force exerted by the ice on her skates.)

Suppose that $\mathbf{k}$ is a unit vector along the axis of rotation of a rigid body, so $\boldsymbol{\omega} = \dot{\theta}\mathbf{k}$. If we look at just the $\mathbf{k}$-components, then the torque law and equation (27) yield

$$
\begin{aligned}
\boldsymbol{\Gamma} \cdot \mathbf{k} = \dot{\mathbf{L}} \cdot \mathbf{k} \\
= \frac{d}{dt}(\mathbf{L} \cdot \mathbf{k}) \\
= \frac{d}{dt}(I\dot{\theta}) \\
= I\ddot{\theta}.
\end{aligned}
$$

So in this situation the torque law leads to the following important and elegant result.

> ### Equation of rotational motion
>
> Suppose that a rigid body is rotating about a fixed axis, and that its angular displacement (from some fixed line normal to the axis) is θ. Then
>
> $$I\ddot{\theta} = \Gamma_{\text{axis}}, \tag{30}$$
>
> where Γ_{axis} is the component of the total external torque on the body in the direction of the axis of rotation.

The moments of inertia of a number of common regular geometric shapes are given in Table 1. In each case we assume that the rigid body is continuous and of uniform density, and we consider only objects of this type. The moments of inertia given in the table have been found by integration, as discussed in Unit 17. You may use these moments of inertia when doing the exercises in this unit.

The objects are therefore homogeneous rigid bodies, as defined in Unit 19.

Table 1   Moments of inertia of homogeneous rigid bodies

Object	Axis	Dimensions	Moment of inertia	Figure
Solid cylinder	Axis of cylinder	Radius R	$\frac{1}{2}MR^2$	16(a), axis AB
Solid cylinder	Normal to axis of cylinder	Radius R, length h	$\frac{1}{4}MR^2 + \frac{1}{12}Mh^2$	16(a), axis CD
Hollow cylinder	Axis of cylinder	Inner radius a, outer radius R	$\frac{1}{2}M(R^2 + a^2)$	16(b), axis AB
Hollow cylinder	Normal to axis of cylinder	Inner radius a, outer radius R, length h	$\frac{1}{4}M(R^2 + a^2) + \frac{1}{12}Mh^2$	16(b), axis CD
Solid rectangular cuboid	Normal to one pair of faces	Faces normal to axis have sides of lengths a, b	$\frac{1}{12}M(a^2 + b^2)$	16(c), axis AB
Thin straight rod	Normal to rod	Length h	$\frac{1}{12}Mh^2$	
Solid sphere	Through centre	Radius R	$\frac{2}{5}MR^2$	
Hollow sphere	Through centre	Inner radius a, outer radius R	$\frac{2}{5}M\dfrac{R^5 - a^5}{R^3 - a^3}$	
Thin spherical shell	Through centre	Radius R	$\frac{2}{3}MR^2$	

In each case, the mass of the object is M and the axis passes through its centre of mass G (see Figure 16).

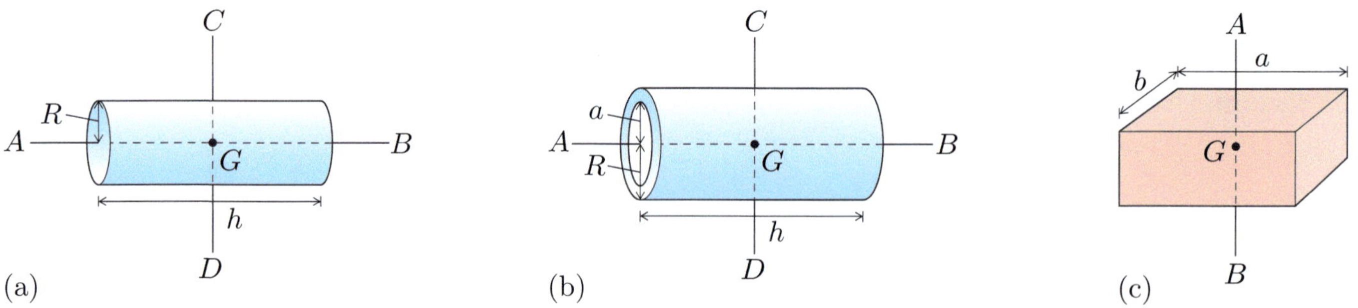

(a) (b) (c)

Figure 16   (a) Solid cylinder, (b) hollow cylinder, (c) solid rectangular cuboid

The moment of inertia of the 'thin straight rod' in Table 1 is calculated using the assumption that all its particles lie on a straight line. While this

will never be absolutely true for a three-dimensional object, we sometimes
use such limiting cases as convenient models. Similarly, a 'thin spherical
shell' is an object whose particles are confined to the surface of a sphere.
The moment of inertia of a thin spherical shell is, in fact, the limit of that
for a hollow sphere as $a \to R$, but that limit is not obvious, so this case is
given separately in the table.

The next example illustrates the use of Table 1 to find the moment of
inertia of a 'thin disc', which is another example of a two-dimensional
model of a three-dimensional object.

Example 5

Consider a uniform solid circular disc of mass M and radius R, and
negligible thickness. What is its moment of inertia about an axis normal to
the disc and through its centre?

Solution

The distribution of the mass of the disc relative to the specified axis is the
same as for a solid cylinder rotating about its own axis. From Table 1, the
moment of inertia in this case is $\frac{1}{2}MR^2$.

Exercise 12

Show that the moment of inertia about the axis of a thin cylindrical shell
of mass M and radius R is MR^2.

To find the moment of inertia of a compound object formed from several
simple shapes joined together, we find the moment of inertia of each
component part separately, and then sum these. We use this approach in
the next example.

Example 6

A playground roundabout can be modelled as a uniform solid circular disc
of radius $1.2\,\text{m}$ and mass $240\,\text{kg}$. The disc is horizontal and rotates about a
vertical axis through its centre, making 0.5 revolutions per second. A man
of mass $80\,\text{kg}$ is standing stationary with respect to the disc at a position
$0.2\,\text{m}$ from the centre O of the disc.

Suppose that the man moves towards the edge of the disc, stopping when
he is $1\,\text{m}$ away from the centre. Assume that the external forces on the
roundabout are such that the component of the total external torque
about O in the direction of the axis of rotation is zero. Assume also that
the man can be modelled as a particle. At what rate will the disc be
rotating when he is in his new position?

Solution

Since Γ_{axis}, the component of the total external torque about O in the direction of the axis of rotation, is zero, from equations (27) and (30) it follows that L_{axis}, the angular momentum about the axis of rotation through O, is conserved. To proceed, we need to know the moment of inertia of the combined 'man-plus-roundabout' system for each of the two positions of the man.

Now, the moment of inertia of the roundabout about the axis through O (in kg m^2) is

$$\tfrac{1}{2}MR^2 = \tfrac{1}{2} \times 240(1.2)^2 = 172.8.$$

If we model the man as a particle of mass m located at a distance r from O, then his moment of inertia is mr^2. So when he is $0.2\,\text{m}$ from the centre of the disc, the moment of inertia of the combined system is

$$I_1 = 172.8 + 80(0.2)^2 = 176.$$

With the man $1\,\text{m}$ from the centre of the disc, the moment of inertia of the combined system is

$$I_2 = 172.8 + 80(1)^2 = 252.8.$$

When the man is nearer to the centre of the disc, the angular speed ω_1 is 0.5 revolutions per second, that is, $\omega_1 = \pi$ (in rad s^{-1}). Suppose that the angular speed is ω_2 after the man has moved closer to the edge of the disc. Conservation of angular momentum gives

$$I_1\omega_1 = I_2\omega_2,$$

hence

$$\omega_2 = \frac{176}{252.8}\pi \simeq 2.19.$$

Therefore after the man has moved closer to the edge of the roundabout, the roundabout rotates more slowly, at about $2.19\,\text{rad s}^{-1}$, that is, at about 0.35 revolutions per second.

The following example and exercise consolidate ideas presented in this subsection.

Example 7

A bucket used to draw water from a well has mass m and is attached to a light inextensible rope that is wound round a heavy wheel (see Figure 17). The wheel is a uniform solid disc, with centre O, radius R and mass M. The distance from O to the surface of the water in the well is h. Suppose that the bucket is released from rest at a point X, level with O, and allowed to fall down the well, causing the wheel to rotate. Model the bucket as a particle and the rope as a model string, and assume that the force supporting the wheel acts at the point O and that there is no friction there.

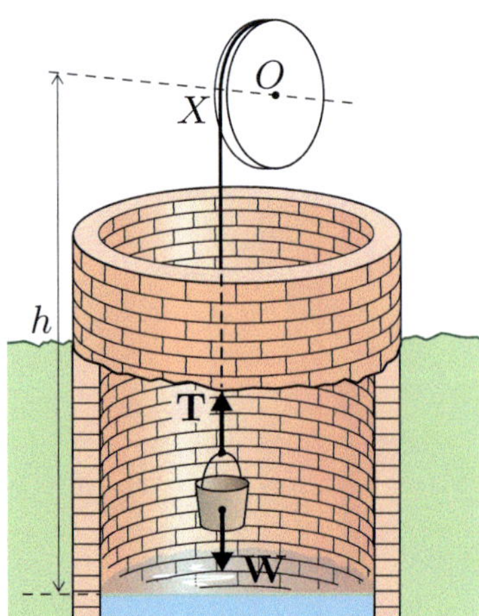

Figure 17   Bucket attached to a wheel centred at O, drawing water from a well

(a) If $\mathbf{T}$ denotes the tension in the rope, and z denotes the distance that the bucket has travelled down the well in time t, write down the equation of motion for the bucket in terms of m, $|\mathbf{T}|$ and z.

(b) What is the torque acting on the wheel about its centre?

(c) Suppose that the wheel has rotated through an angle θ while the bucket has been falling. What is the relationship between z and θ?

(d) Write down the equation of rotational motion for the wheel.

(e) Find an expression for the acceleration $\ddot{z}$ of the bucket by eliminating $|\mathbf{T}|$ and θ from the equations of motion obtained in parts (a) and (d), and by using the result from part (c).

(f) Hence find the time taken for the bucket to descend the well.

Solution

(a) Choose Cartesian unit vectors with $\mathbf{k}$ pointing vertically downwards, $\mathbf{j}$ in the direction of the axis of rotation of the wheel, and $\mathbf{i}$ horizontal, as shown in Figure 18. The origin O is at the centre of the wheel.

The forces acting on the bucket are the tension force in the string, $-|\mathbf{T}|\mathbf{k}$, and the force due to gravity, $\mathbf{W} = mg\mathbf{k}$. So by Newton's second law,

$$m\ddot{z}\mathbf{k} = mg\mathbf{k} - |\mathbf{T}|\mathbf{k}.$$

Resolving in the $\mathbf{k}$-direction gives the equation of motion:

$$m\ddot{z} = mg - |\mathbf{T}|. \tag{31}$$

(b) The force exerted on the wheel by the rope is $|\mathbf{T}|\mathbf{k}$, acting at the point X whose position relative to O is $-R\mathbf{i}$ (see Figure 18). Therefore the torque acting on the wheel about its centre is
$(-R\mathbf{i}) \times |\mathbf{T}|\mathbf{k} = R|\mathbf{T}|\mathbf{j}.$

(c) If the bucket has fallen a distance z, then the quantity of rope that has unwound from the wheel as the bucket falls must also be of length z. As the bucket falls, the wheel turns through an angle θ anticlockwise, so the part of the perimeter of the wheel that has moved past the point X in Figure 19 has length $R\theta$. Hence $z = R\theta$.

(d) The wheel is a disc turning about an axis through its centre and normal to its plane. It has moment of inertia $I = \frac{1}{2}MR^2$ (from Example 5) about the axis of rotation. Then from equation (30), the equation of rotational motion of the wheel (in the $\mathbf{j}$-direction) is

$$I\ddot{\theta} = \Gamma_{\text{axis}},$$

and from part (b), $\Gamma_{\text{axis}} = R|\mathbf{T}|$. So we obtain

$$\tfrac{1}{2}MR^2\ddot{\theta} = R|\mathbf{T}|,$$

or equivalently,

$$MR\ddot{\theta} = 2|\mathbf{T}|. \tag{32}$$

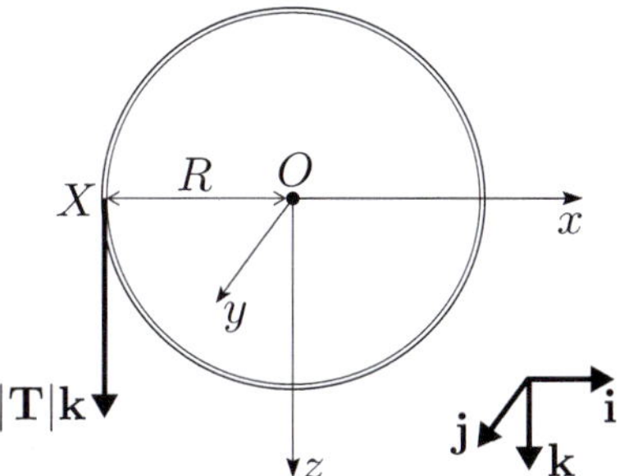

Figure 18 Force on the wheel

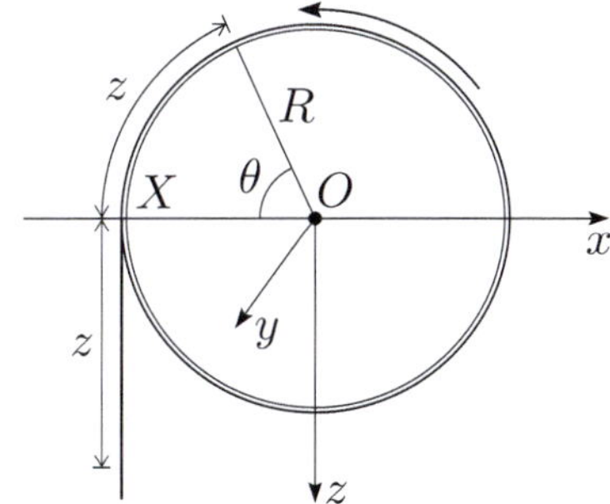

Figure 19 Rotation of the wheel

(e) From part (c) we have $z = R\theta$, and since R is constant, $\ddot{z} = R\ddot{\theta}$. On substituting in equation (32), we find

$$M\ddot{z} = 2|\mathbf{T}|. \tag{33}$$

Now from equation (31), $|\mathbf{T}| = mg - m\ddot{z}$. So substituting for $|\mathbf{T}|$ in equation (33) yields

$$M\ddot{z} = 2(mg - m\ddot{z}).$$

Hence

$$\ddot{z} = \frac{2mg}{M + 2m}. \tag{34}$$

(f) The expression for $\ddot{z}$ in equation (34) is a constant, so using the constant acceleration formula $z = z_0 + v_0 t + \frac{1}{2}a_0 t^2$ with $z = h$, $z_0 = 0$, $v_0 = 0$ and a_0 given by the right-hand side of equation (34), we have

$$h = \frac{mg}{M + 2m}\, t^2.$$

Rearranging this gives the time for the bucket to reach the surface of the water as $\sqrt{h(M + 2m)/mg}$.

Exercise 13

A roundabout of mass $250\,\text{kg}$ is modelled as a uniform solid disc of radius $1.2\,\text{m}$, which turns in a horizontal plane about a vertical axis through its centre O. It is being pushed with a force $\mathbf{F}$, using a handle at X as in Figure 20. This force, which has constant magnitude $100\,\text{N}$, is applied at a distance of $1.5\,\text{m}$ from O, and is horizontal and normal to OX. A force $\mathbf{R}$ resists the rotational motion. It has magnitude $c\omega$, where c is a constant and ω is the angular speed of the roundabout; its point of action is $0.1\,\text{m}$ from O, and its direction is opposite to the velocity of that point on the roundabout. Assume throughout that $\dot{\theta} \geq 0$ (anticlockwise rotation), so $\omega = \dot{\theta}$.

(a) Obtain a differential equation in terms of ω for the rotational motion of the roundabout.

(b) If the pushing starts at time $t = 0$ with the roundabout at rest, find ω as a function of t.

(c) If $c = 10$, what is the maximum possible angular speed that the roundabout could reach according to this model? Do you think that this could be achieved in practice?

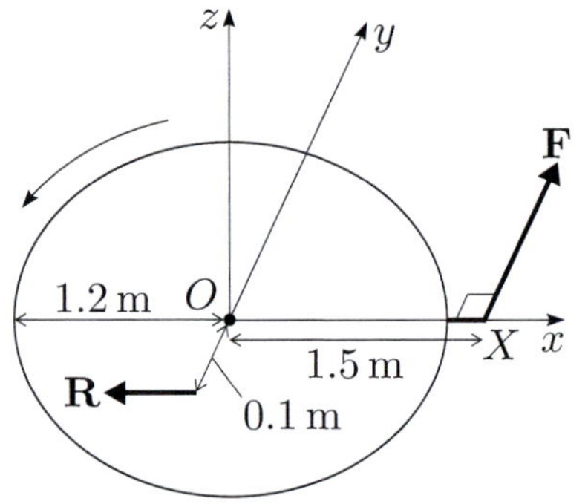

Figure 20 Roundabout rotating about O and pushed by force $\mathbf{F}$ with resistive force $\mathbf{R}$

3.2 Parallel axes theorem

The moments of inertia of various objects about their centres of mass were given in Table 1. To find the moments of inertia about some other axis, we can use the *parallel axes theorem*.

Parallel axes theorem

Suppose that I_{AB} is the moment of inertia of a rigid body of mass M about a line AB (see Figure 21(a)). Let EF be a line through the centre of mass of the body that is parallel to AB, let the distance between the lines AB and EF be d, and let I_{EF} be the moment of inertia of the body about EF. Then

$$I_{AB} = I_{EF} + Md^2. \tag{35}$$

To see this, we follow the argument set out below.

Figure 21(a) shows a rigid body of mass M, with the centre of mass O taken as the origin. Suppose that we want to find the moment of inertia of the body about the line AB. As shown in Figure 21(b), choose as the z-axis a line EF through O that is parallel to AB. Let the perpendicular distance between these two parallel lines be d. Choose as the x-axis a line through O that intersects the line AB and is normal to AB.

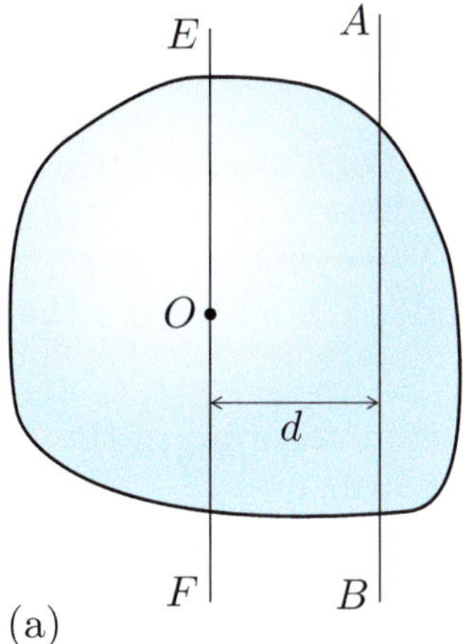
(a)

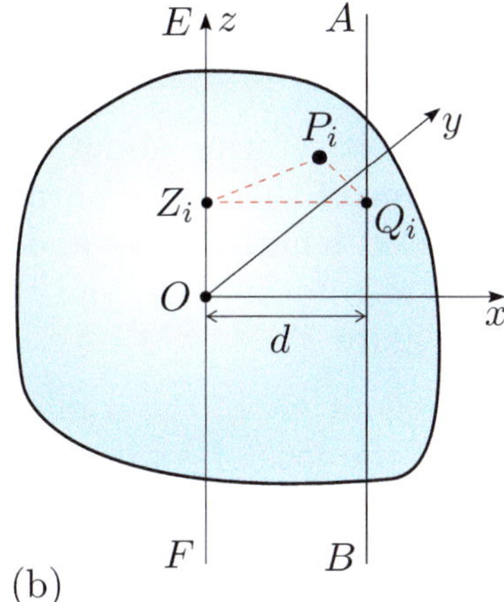
(b)

Figure 21   Rigid body, showing (a) two axes of rotation, EF and AB, and (b) the coordinate axes and a particular particle P_i

Suppose that the rigid body is composed of n particles P_i, where for $i = 1, 2, \ldots, n$ the ith particle has mass m_i and position (x_i, y_i, z_i). The distance of particle P_i from the z-axis is d_i ($P_i Z_i$ in Figures 21(b) and 22), where $d_i^2 = x_i^2 + y_i^2$. Also, the particle's distance from the line AB is s_i ($P_i Q_i$ in Figures 21(b) and 22), where

$$\begin{aligned}
s_i^2 &= (d - x_i)^2 + y_i^2 \\
&= d^2 - 2x_i d + x_i^2 + y_i^2 \\
&= d^2 - 2x_i d + d_i^2.
\end{aligned}$$

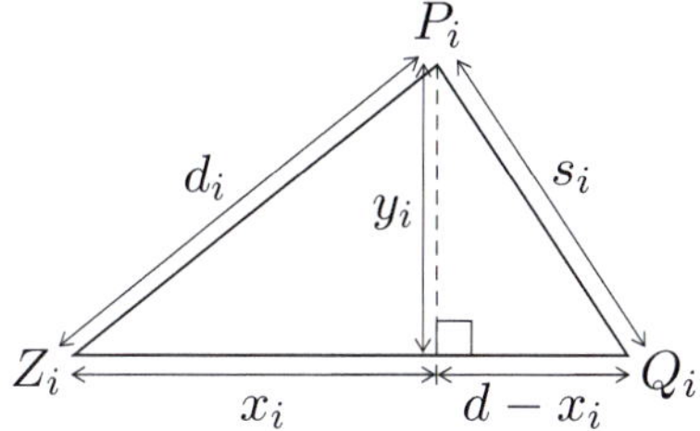

Figure 22   Part of a section through the rigid body in Figure 21(b), parallel to the (x, y)-plane

By definition (see Unit 17), the moment of inertia of the rigid body about the line AB is

$$
\begin{aligned}
I_{AB} &= \sum_{i=1}^{n} m_i s_i^2 \\
&= \sum_{i=1}^{n} m_i (d^2 - 2x_i d + d_i^2) \\
&= d^2 \sum_{i=1}^{n} m_i - 2d \sum_{i=1}^{n} m_i x_i + \sum_{i=1}^{n} m_i d_i^2.
\end{aligned}
\tag{36}
$$

Now $\sum_{i=1}^{n} m_i = M$ and $\sum_{i=1}^{n} m_i d_i^2 = I_{EF}$, where I_{EF} is the moment of inertia of the object about the z-axis (the line EF). Also, from Unit 19, Subsection 2.1, $\left(\sum_{i=1}^{n} m_i x_i\right)/M$ is the x-coordinate of the centre of mass of the body. Since the origin was chosen at the centre of mass, we have $\sum_{i=1}^{n} m_i x_i = 0$. Hence equation (36) becomes

$$
I_{AB} = I_{EF} + Md^2,
$$

giving the parallel axes theorem.

___

Example 8

An ice skater is rotating about a fixed axis AB at an angular speed of $8\,\mathrm{rad\,s^{-1}}$. One of the skater's arms is modelled as a uniform solid cylinder of mass 5.5 kg, length 0.66 m and diameter 0.08 m. The cylinder is normal to the axis of rotation AB, and the end of the cylinder is 0.09 m from the axis (see Figure 23).

Find the magnitude of the component of the angular momentum in the direction of the axis of rotation for this model of the arm.

Solution

From Table 1, the moment of inertia of the cylinder about an axis EF through its centre of mass and normal to the cylinder is

$$
I_{EF} = \tfrac{1}{4}MR^2 + \tfrac{1}{12}Mh^2 = \tfrac{1}{4} \times 5.5(0.04)^2 + \tfrac{1}{12} \times 5.5(0.66)^2 \simeq 0.202
$$

(in $\mathrm{kg\,m^2}$). The axis of rotation AB is $0.33 + 0.09 = 0.42$ (in m) from the centre of mass of the cylinder, so from the parallel axes theorem, the moment of inertia of the cylinder about the axis of rotation AB is

$$
I_{AB} \simeq 0.202 + 5.5(0.42)^2 \simeq 1.172
$$

(in $\mathrm{kg\,m^2}$). As the skater's angular speed is $\omega = 8$ (in $\mathrm{rad\,s^{-1}}$), it follows from equation (27) that for this model of the arm, the component of angular momentum in the direction of the axis of rotation has magnitude

$$
|L_{\mathrm{axis}}| = I_{AB}\,\omega \simeq 1.172 \times 8 = 9.376
$$

(in $\mathrm{kg\,m^2\,s^{-1}}$).

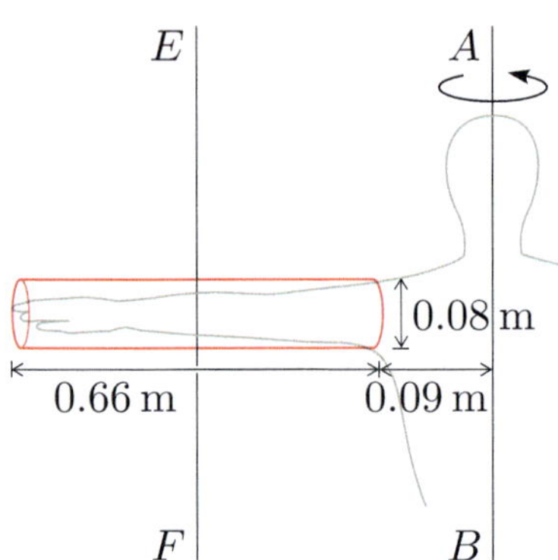

Figure 23 Skater with outstretched arm

Note that we could treat other parts of the body in a similar fashion and, in principle, sum up the moments of inertia of all the individual body parts to obtain the moment of inertia of the whole body.

Exercise 14

The mace used by the leader of a troupe of drum majorettes can be modelled as a sphere of radius r attached to the end of a uniform cylindrical rod of length d and radius R. The mass of the sphere is M, and the mass of the rod is m.

Find the moment of inertia of the mace about an axis AB through the end of the rod and normal to it (as shown in Figure 24).

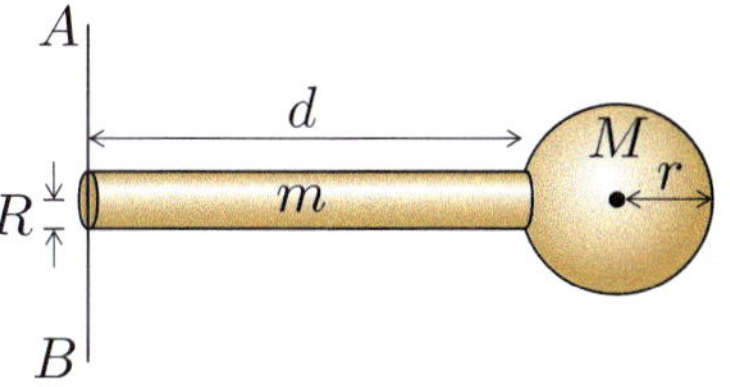

Figure 24   Mace rotating about AB

3.3 Kinetic energy of a rotating rigid body

You saw in Unit 17 how the kinetic energy of a rotating particle can be expressed in terms of its moment of inertia. Recall that the kinetic energy of a particle with mass m moving at speed v is given by $T = \frac{1}{2}mv^2$. If the particle is moving in a circle of radius d (the distance from the axis of rotation) with angular speed ω, then $v = d\omega$, so we have $T = \frac{1}{2}md^2\omega^2 = \frac{1}{2}I\omega^2$, where $I = md^2$ is the moment of inertia of the particle.

The kinetic energy of a rotating rigid body can be expressed in a similar way. Consider a rigid body rotating about a fixed axis with angular speed ω. The kinetic energy of the body is the sum of the kinetic energies of all its constituent particles. If the ith particle is a distance d_i from the axis of rotation, then its speed is $d_i\omega$ and its kinetic energy is $\frac{1}{2}m_i(d_i\omega)^2$. Hence the total kinetic energy of the body is

$$\sum_{i=1}^{n} \tfrac{1}{2}m_i(d_i\omega)^2 = \tfrac{1}{2}\left(\sum_{i=1}^{n} m_i d_i^2\right)\omega^2 = \tfrac{1}{2}I\omega^2,$$

where I is the body's moment of inertia about the axis of rotation.

Kinetic energy of a rigid body rotating about a fixed axis

Suppose that a rigid body is rotating with angular speed ω about a fixed axis. Let I be the moment of inertia of the body about the axis of rotation. Then the kinetic energy of the body is

$$T = \tfrac{1}{2}I\omega^2. \tag{37}$$

Note that as long as rotation is about a fixed axis, the expression $\frac{1}{2}I\omega^2$ gives the *total* kinetic energy of the body.

Example 9

A planet is modelled as a uniform solid sphere of radius 6400 km and mass 6.0×10^{24} kg. It is turning on its axis once every 24 hours.

If the axis of rotation is fixed, what is the kinetic energy of the planet?

Solution

The planet's kinetic energy is $\frac{1}{2}I\omega^2$, where

$$\omega = \frac{2\pi}{24 \times 60^2}$$

(in $\mathrm{rad\,s^{-1}}$), and from Table 1,

$$I = \tfrac{2}{5}MR^2 = \tfrac{2}{5} \times 6 \times 10^{24} \times (6.4 \times 10^6)^2$$

(in $\mathrm{kg\,m^2}$). Therefore the kinetic energy of the planet (in joules) is

$$\tfrac{1}{2} \times \tfrac{2}{5} \times 6 \times 10^{24} \times (6.4 \times 10^6)^2 \left(\frac{2\pi}{24 \times 60^2} \right)^2 \simeq 2.6 \times 10^{29}.$$

Exercise 15

A drum majorette's baton is modelled as a uniform cylindrical rod with spheres of equal mass at either end. The rod has mass 0.1 kg, length 0.8 m and diameter 0.04 m. Each sphere has diameter 0.1 m and mass 0.25 kg. The baton is rotated at 1 revolution per second about a fixed axis that is normal to the axis of the cylinder and through the centre of mass (see Figure 25).

Determine the kinetic energy of the baton.

Exercise 16

(a) In the situation considered in Example 6, determine the total kinetic energy of the system (man plus roundabout) when the man is stationary at 0.2 m from the centre of the roundabout, and then when he is stationary at 1 m from the centre. (Use the results found in the solution to Example 6, as needed.)

(b) In the reverse of the situation in Example 6, the man starts 1 m from the centre of the roundabout and moves inwards until he is 0.2 m from the centre. What happens to the kinetic energy of the system when he does this?

In a system such as that dealt with in Example 6 and Exercise 16, the angular momentum is constant because the total external torque is zero (both quantities are measured about the centre of the roundabout and refer to components in the direction of the axis of rotation). However, the internal forces can change the kinetic energy of the system. This means that the kinetic energy is *not* necessarily constant, as you saw

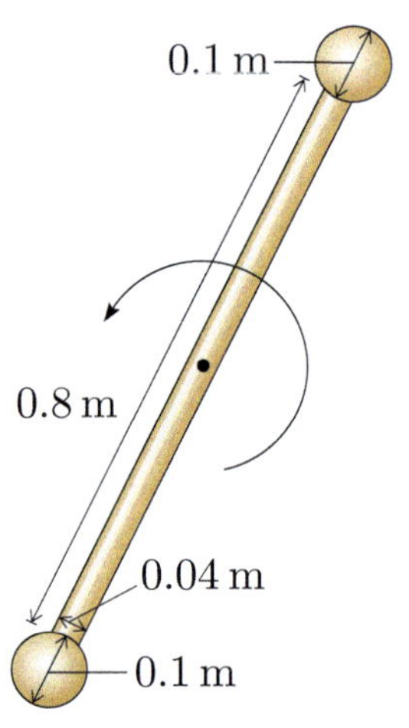

Figure 25   Drum majorette's baton rotating about an axis through its centre of mass

in Exercise 16(b), where the effort put in by the man in moving across the roundabout increased the total kinetic energy of that system.

Exercise 17

A rigid body of mass M is rotating about a fixed axis with angular speed ω. Suppose that the centre of mass is following a circle of radius R at speed v, and that I_G is the moment of inertia of the body about an axis parallel to the axis of rotation and through the centre of mass.

Show that the kinetic energy of the body is

$$\tfrac{1}{2}I_G\omega^2 + \tfrac{1}{2}Mv^2.$$

Exercise 18

A diver begins a dive at rest and vertical in the handstand position. She starts to rotate from this position with negligible angular speed. She lets go of the diving board when her body makes an angle α with the vertical (where $0 < \alpha \leq \tfrac{\pi}{2}$), and she then starts the 'in-flight' part of the dive. Let ω be the angular speed of the diver at the moment of letting go of the board. Assume that while she is in contact with the board at O, the diver is a rigid body able to rotate about an axis through O, normal to the plane of Figure 26.

Use the principle of conservation of mechanical energy to obtain ω in terms of α and the following parameters: l, the distance from the diver's hands (in the handstand position) to her centre of mass; m, the mass of the diver; I_G, the moment of inertia of the diver about an axis through her centre of mass when her body is straight.

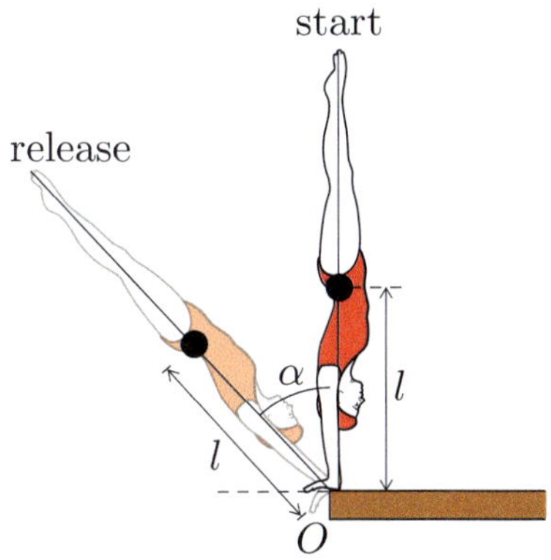

Figure 26   Diver beginning a dive in the handstand position

Exercise 19

Consider the diver from Exercise 18 when she is part way through her dive (as shown in Figure 27). Suppose that she has rotated through an angle θ ($\theta < \alpha$) in time t, and in that position the force exerted on her hands by the board has component vectors $R_1\mathbf{i}$ and $R_2\mathbf{j}$, as shown in the figure.

(a) With the origin and axes shown in Figure 27, let $x\mathbf{i} + y\mathbf{j}$ be the position of the diver's centre of mass. Express x and y in terms of θ. By differentiating, show that

$$\ddot{x} = l\dot{\theta}^2\sin\theta - l\ddot{\theta}\cos\theta, \quad \ddot{y} = -l\dot{\theta}^2\cos\theta - l\ddot{\theta}\sin\theta. \tag{38}$$

(b) Let I_O be the diver's moment of inertia about an axis through O. Give the equation of rotational motion for the diver in terms of θ, I_O, m and l.

(c) Obtain another equation for the rotational motion by applying the principle of conservation of mechanical energy to the diver. Verify that differentiation of this equation leads to the equation of rotational motion that you found in part (b).

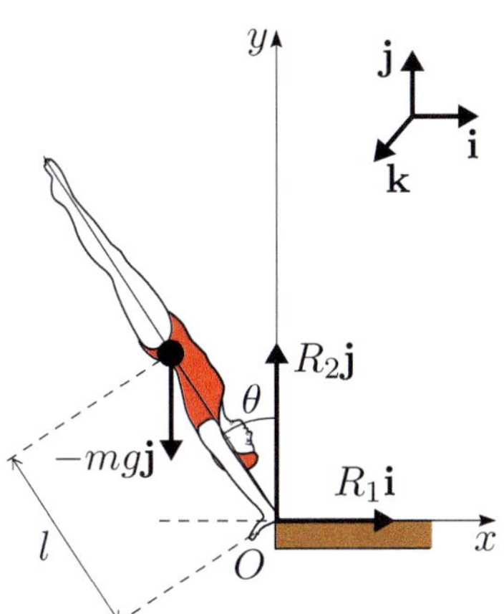

Figure 27   Diver part way through her dive

(d) (i) Write down Newton's second law for the motion of the diver's centre of mass, and obtain two equations of motion.

   (ii) Use equations (38) to substitute for $\ddot{x}$ and $\ddot{y}$ in these equations of motion.

(e) (i) Use the results of parts (b) and (c) to substitute for $\dot{\theta}^2$ and $\ddot{\theta}$ in your equation for R_1 in part (d)(ii). Hence show that

$$R_1 = mg\,\frac{ml^2}{I_O}\sin\theta\,(2 - 3\cos\theta). \tag{39}$$

   (ii) Deduce that for angles θ between 0 and $\pi/2$, the horizontal component R_1 of the force exerted by the diving board is zero when $\theta = 0$ or when $\theta = \arccos\frac{2}{3}$.

(f) Estimate I_O by modelling the diver as a uniform thin straight rod of length $2l$. Then estimate, as a multiple of the magnitude of her weight, the magnitude of the horizontal force that the diver would need to exert on the board to be able to remain in contact with it until $\theta = \frac{\pi}{2}$.

(g) At what point is the diver likely to lose contact with the board? Assume that there is nothing on the diving board on which the diver can grip, so she cannot pull on the board but can only push on it.

4 Rotation about a moving axis

The motion of an extended body may be much more complicated than that considered in Section 3, where we confined our attention to a *rigid* body rotating about a *fixed* axis. In this section we look at a more general situation where the axis of rotation is moving. The general case is considered in Subsection 4.1, and here we derive the result used in Section 1 that the motion can be decomposed into the motion of the centre of mass together with the rotation about the centre of mass. In Subsection 4.2, we consider the motion of a *rigid* body when the body is rotating about an axis whose *direction is fixed*, though the centre of mass may be moving. In Subsection 4.3, we look at the motion of rolling objects, such as cans rolling down slopes.

We will not consider the motion of a rigid body when the direction of the axis of rotation varies.

4.1 Torque law relative to the centre of mass

In Unit 19 we showed how Newton's second law of motion can be extended to an n-particle system. You saw there that the centre of mass of the system moves as if it is a single particle with the same mass as the whole system, and with all the external forces applied to this particle. The motion of the centre of mass is often referred to as the **linear motion** of the system, to distinguish it from the rotational motion of the system.

That is not to suggest that the centre of mass will always travel in a straight line!

It is convenient to analyse complicated motion in two parts: the linear motion, and the motion relative to the centre of mass. Now, to apply the torque law as stated in Section 2, we need to work relative to a *fixed* origin. However, as you will now see, we can still use the torque law when these quantities are calculated relative to a point that is *moving*, as long as *that point is at the centre of mass.*

To demonstrate this, we need some notation. Throughout this subsection, we will consider an extended body modelled as a system of n particles, with total mass M and centre of mass at a position $\mathbf{R}$ relative to some fixed point O. The ith particle of the system (for $i = 1, 2, \ldots, n$) has mass m_i and position relative to O given by

$$\mathbf{r}_i = \mathbf{R} + \mathbf{r}_i^{\text{rel}}, \tag{40}$$

where $\mathbf{r}_i^{\text{rel}}$ denotes the position of the particle relative to the centre of mass. In the following discussion it is also useful to obtain a relationship between the velocities of the particles relative to the different origins by differentiating equation (40) to give

$$\dot{\mathbf{r}}_i = \dot{\mathbf{R}} + \dot{\mathbf{r}}_i^{\text{rel}}. \tag{41}$$

There is an additional equation involving $\mathbf{r}_i^{\text{rel}}$ that results from the fact that $\mathbf{R}$ is the centre of mass of the system. From Unit 19, we have

$$M\mathbf{R} = \sum_{i=1}^{n} m_i \mathbf{r}_i = \sum_{i=1}^{n} m_i \left(\mathbf{R} + \mathbf{r}_i^{\text{rel}}\right)$$

$$= \sum_{i=1}^{n} m_i \mathbf{R} + \sum_{i=1}^{n} m_i \mathbf{r}_i^{\text{rel}}$$

$$= M\mathbf{R} + \sum_{i=1}^{n} m_i \mathbf{r}_i^{\text{rel}}.$$

This is where we use the fact that $\mathbf{R}$ is the centre of mass of the system.

Thus, as one would expect,

$$\sum_{i=1}^{n} m_i \mathbf{r}_i^{\text{rel}} = \mathbf{0}. \tag{42}$$

This takes us to the following important definitions.

This equation states that if position vectors are taken relative to the centre of mass, then the centre of mass is at the origin.

Torque and angular momentum relative to the centre of mass

If $\mathbf{r}_i^{\text{rel}}$ denotes the position of the ith particle relative to the centre of mass, and $\mathbf{F}_i$ denotes the total external force on the ith particle, then the total external torque on the system relative to the centre of mass is defined by

$$\boldsymbol{\Gamma}^{\text{rel}} = \sum_{i=1}^{n} \mathbf{r}_i^{\text{rel}} \times \mathbf{F}_i, \tag{43}$$

and the total angular momentum relative to the centre of mass is defined as

$$\mathbf{L}^{\text{rel}} = \sum_{i=1}^{n} \mathbf{r}_i^{\text{rel}} \times m_i \dot{\mathbf{r}}_i^{\text{rel}}. \tag{44}$$

We aim to relate $\boldsymbol{\Gamma}^{\mathrm{rel}}$ and $\mathbf{L}^{\mathrm{rel}}$ to the corresponding quantities calculated relative to the fixed origin O, and start by looking at the total angular momentum relative to O, which is defined as

$$\mathbf{L} = \sum_{i=1}^{n} \mathbf{r}_i \times m_i \dot{\mathbf{r}}_i.$$

Substituting for $\mathbf{r}_i$ and $\dot{\mathbf{r}}_i$, using equations (40) and (41), gives

$$\mathbf{L} = \sum_{i=1}^{n} (\mathbf{R} + \mathbf{r}_i^{\mathrm{rel}}) \times m_i (\dot{\mathbf{R}} + \dot{\mathbf{r}}_i^{\mathrm{rel}}).$$

Expanding the brackets gives

$$\mathbf{L} = \sum_{i=1}^{n} (\mathbf{R} \times m_i \dot{\mathbf{R}} + \mathbf{R} \times m_i \dot{\mathbf{r}}_i^{\mathrm{rel}} + \mathbf{r}_i^{\mathrm{rel}} \times m_i \dot{\mathbf{R}} + \mathbf{r}_i^{\mathrm{rel}} \times m_i \dot{\mathbf{r}}_i^{\mathrm{rel}}).$$

The last term can be recognised as the total angular momentum relative to the centre of mass (equation (44)), so we have

$$\mathbf{L} = \sum_{i=1}^{n} (\mathbf{R} \times m_i \dot{\mathbf{R}} + \mathbf{R} \times m_i \dot{\mathbf{r}}_i^{\mathrm{rel}} + \mathbf{r}_i^{\mathrm{rel}} \times m_i \dot{\mathbf{R}}) + \mathbf{L}^{\mathrm{rel}}.$$

As $\mathbf{R}$ is independent of i, this expression can be written as

$$\mathbf{L} = \mathbf{R} \times \left(\sum_{i=1}^{n} m_i \right) \dot{\mathbf{R}} + \mathbf{R} \times \left(\sum_{i=1}^{n} m_i \dot{\mathbf{r}}_i^{\mathrm{rel}} \right) + \left(\sum_{i=1}^{n} m_i \mathbf{r}_i^{\mathrm{rel}} \right) \times \dot{\mathbf{R}} + \mathbf{L}^{\mathrm{rel}}.$$

The first bracketed term above is the total mass M of the system. The third bracketed term is zero by equation (42), and the second bracketed term is zero by differentiating equation (42). So we arrive at the result

$$\mathbf{L} = \mathbf{R} \times M\dot{\mathbf{R}} + \mathbf{L}^{\mathrm{rel}}.$$

The next exercise asks you to derive a similar relationship that holds between the torque relative to the centre of mass and the torque relative to O.

Exercise 20

Starting from the definition of $\boldsymbol{\Gamma}$, the total external torque about O, and using equations (40) and (43), show that

$$\boldsymbol{\Gamma} = \mathbf{R} \times \mathbf{F} + \boldsymbol{\Gamma}^{\mathrm{rel}},$$

where $\mathbf{F} = \sum_{i=1}^{n} \mathbf{F}_i$ is the total external force on the system.

The two results derived above are worth re-stating formally.

Decomposition theorems

Let $\mathbf{R}$ be the position vector of the centre of mass of a body relative to some fixed point O, let M be the total mass of the body, and let $\mathbf{F}$ be the total external force on the body. If the total angular momentum of the body relative to the centre of mass is $\mathbf{L}^{\mathrm{rel}}$, then the total angular momentum of the body relative to O is given by

$$\mathbf{L} = \mathbf{R} \times M\dot{\mathbf{R}} + \mathbf{L}^{\mathrm{rel}}. \tag{45}$$

If $\boldsymbol{\Gamma}^{\mathrm{rel}}$ is the total external torque on the body relative to the centre of mass, then the total external torque on the body about O is given by

$$\boldsymbol{\Gamma} = \mathbf{R} \times \mathbf{F} + \boldsymbol{\Gamma}^{\mathrm{rel}}. \tag{46}$$

This states that both the total angular momentum and the total external torque of an n-particle system can be decomposed into the corresponding quantity for an equivalent particle located at the centre of mass plus the corresponding quantity for the rotational motion relative to the centre of mass.

Now we move on to derive the central result of this subsection, and one of the key results of the unit, which is the *torque law relative to the centre of mass*. To derive this result, we start by differentiating equation (45) to obtain

$$\dot{\mathbf{L}} = \frac{d}{dt}\left(\mathbf{R} \times M\dot{\mathbf{R}}\right) + \dot{\mathbf{L}}^{\mathrm{rel}}.$$

Using the torque law relative to a fixed origin O gives $\dot{\mathbf{L}} = \boldsymbol{\Gamma}$, so

$$\boldsymbol{\Gamma} = \frac{d}{dt}\left(\mathbf{R} \times M\dot{\mathbf{R}}\right) + \dot{\mathbf{L}}^{\mathrm{rel}}.$$

Using the product rule for differentiating the cross product gives

$$\boldsymbol{\Gamma} = \dot{\mathbf{R}} \times M\dot{\mathbf{R}} + \mathbf{R} \times M\ddot{\mathbf{R}} + \dot{\mathbf{L}}^{\mathrm{rel}}.$$

See Unit 20.

The first term on the right-hand side is zero, since $\dot{\mathbf{R}} \times \dot{\mathbf{R}} = \mathbf{0}$, so

$$\boldsymbol{\Gamma} - \mathbf{R} \times M\ddot{\mathbf{R}} = \dot{\mathbf{L}}^{\mathrm{rel}}.$$

By Newton's second law for n-particle systems, we have $\mathbf{F} = M\ddot{\mathbf{R}}$, hence

$$\boldsymbol{\Gamma} - \mathbf{R} \times \mathbf{F} = \dot{\mathbf{L}}^{\mathrm{rel}}.$$

The result $\mathbf{F} = M\ddot{\mathbf{R}}$ was established in Unit 19, Subsection 2.1.

Now we use equation (46) to obtain the desired relationship between the torque and angular momentum relative to the centre of mass:

$$\boldsymbol{\Gamma}^{\mathrm{rel}} = \dot{\mathbf{L}}^{\mathrm{rel}}.$$

Torque law relative to the centre of mass

The total external torque on an extended body relative to its centre of mass is equal to the rate of change of the total angular momentum relative to the centre of mass. So if $\boldsymbol{\Gamma}^{\mathrm{rel}}$ is the total external torque on the body relative to the centre of mass, and $\mathbf{L}^{\mathrm{rel}}$ is the total angular momentum of the body relative to the centre of mass, as defined in equations (43) and (44), then

$$\boldsymbol{\Gamma}^{\mathrm{rel}} = \dot{\mathbf{L}}^{\mathrm{rel}}. \tag{47}$$

Remember that we are modelling the extended body as a system of n particles.

This result shows that we can extend the torque law of Section 2 by considering torques and angular momentum *relative to the centre of mass*. Therefore to model rotational motion, we can work in a frame of reference where the centre of mass is taken as the origin and thought of as fixed for the purpose of treating the relative motion. We can then deal with motion of the centre of mass separately; this can be done by applying Newton's second law to an equivalent particle located at the centre of mass, that is, $\mathbf{F} = M\ddot{\mathbf{R}}$.

We can use the torque law relative to the centre of mass to justify an assumption that we made in Section 1, that the angular momentum relative to the centre of mass is conserved for projectiles in flight.

Conservation of angular momentum: special case

Suppose that each particle in a system of n particles is subject to an external force of the form $cm_i\mathbf{k}$, where c is a constant, m_i is the mass of the ith particle, and $\mathbf{k}$ is a fixed vector, and that there are no other external forces on the system. Then $\mathbf{\Gamma}^{\mathrm{rel}} = \mathbf{0}$, so $\mathbf{L}^{\mathrm{rel}}$ is constant.

The next exercise asks you to establish this result.

Exercise 21

(a) Use equation (43) and the torque law relative to the centre of mass to establish the boxed result above.

(b) Show that the boxed result applies to a body (such as a diver or gymnast) in flight and subject only to gravity.

Exercise 22

Suppose that all the external forces acting on a system of particles are directed towards the origin. Show that $\mathbf{\Gamma}^{\mathrm{rel}} = -\mathbf{R} \times \mathbf{F}$.

We end this subsection with another decomposition theorem, this time for kinetic energy. This theorem is valuable if we wish to tackle a problem about rotational motion by using conservation of mechanical energy (rather than by using equations of motion). In terms of the vectors defined at the start of this subsection, the square of the speed of the ith particle is $\dot{\mathbf{r}}_i \cdot \dot{\mathbf{r}}_i$, so the total kinetic energy of the system is

Recall that $|\dot{\mathbf{r}}_i|^2 = \dot{\mathbf{r}}_i \cdot \dot{\mathbf{r}}_i$.

$$T = \sum_{i=1}^{n} \tfrac{1}{2}m_i\dot{\mathbf{r}}_i \cdot \dot{\mathbf{r}}_i.$$

Using equation (41) gives

$$T = \sum_{i=1}^{n} \tfrac{1}{2}m_i\left(\dot{\mathbf{R}} + \dot{\mathbf{r}}_i^{\mathrm{rel}}\right) \cdot \left(\dot{\mathbf{R}} + \dot{\mathbf{r}}_i^{\mathrm{rel}}\right).$$

Expanding the brackets gives

$$T = \sum_{i=1}^{n} \tfrac{1}{2} m_i \left(\dot{\mathbf{R}} \cdot \dot{\mathbf{R}} + 2\dot{\mathbf{R}} \cdot \dot{\mathbf{r}}_i^{\text{rel}} + \dot{\mathbf{r}}_i^{\text{rel}} \cdot \dot{\mathbf{r}}_i^{\text{rel}} \right).$$

Here we have used $\mathbf{a} \cdot \mathbf{b} = \mathbf{b} \cdot \mathbf{a}$ in order to collect terms.

Using the fact that $\dot{\mathbf{R}}$ is independent of i allows us to rearrange to

$$T = \tfrac{1}{2} \left(\sum_{i=1}^{n} m_i \right) \dot{\mathbf{R}} \cdot \dot{\mathbf{R}} + \dot{\mathbf{R}} \cdot \left(\sum_{i=1}^{n} m_i \dot{\mathbf{r}}_i^{\text{rel}} \right) + \tfrac{1}{2} \sum_{i=1}^{n} m_i \dot{\mathbf{r}}_i^{\text{rel}} \cdot \dot{\mathbf{r}}_i^{\text{rel}}.$$

The first bracketed term in this equation is the total mass M of the system. By differentiating equation (42) with respect to t, we can show that the second bracketed term is zero. So the equation reduces to

$$T = \tfrac{1}{2} M \dot{\mathbf{R}} \cdot \dot{\mathbf{R}} + \tfrac{1}{2} \sum_{i=1}^{n} m_i \dot{\mathbf{r}}_i^{\text{rel}} \cdot \dot{\mathbf{r}}_i^{\text{rel}}. \tag{48}$$

This result can be stated in words as follows.

Kinetic energy decomposition theorem

The kinetic energy of an extended body is equal to the kinetic energy of an equivalent particle that has the velocity of the body's centre of mass, plus the sum of the kinetic energies due to the motion, relative to the centre of mass, of all the body's constituent particles.

The term 'equivalent particle' is used to mean a particle of the same mass as the body located at its centre of mass.

4.2   Rigid body rotating with fixed orientation

Consider a rigid body that is in motion, rotating about an axis that may itself move but remains pointing in the same direction. A cylindrical can rolling down a slope, with its axis pointing in the same horizontal direction throughout, provides an example of such motion. In this situation, the centre of mass of the rigid body may be moving, but the motion *relative to the centre of mass* is of the type considered in Section 3. The position of each particle in the rigid body, relative to the centre of mass, is constrained in the same way that the position of each particle was in our discussion in Subsection 3.1. This means that arguments similar to those in Section 3 can be used to deduce expressions, in terms of the moment of inertia, for the angular momentum and kinetic energy of the rigid body, relative to its centre of mass.

To illustrate these points, take a rigid body of mass M that is rotating about an axis of fixed orientation through its centre of mass with angular velocity $\boldsymbol{\omega} = \dot{\theta}\mathbf{k}$, where $\mathbf{k}$ is a fixed unit vector and θ is the usual anticlockwise angular displacement relative to the axis. Let I be the body's moment of inertia about that axis. Then the $\mathbf{k}$-component of the angular momentum of the body relative to the centre of mass is $I\dot{\theta}$, while the kinetic energy of the body relative to the centre of mass is $\tfrac{1}{2} I \omega^2$, where $\omega = |\boldsymbol{\omega}| = |\dot{\theta}|$ is the angular speed. Combining these results with the torque law relative to the centre of mass (equation (47)) and the kinetic energy decomposition theorem (equation (48)) leads to the following.

Rigid body rotating with fixed orientation

A rigid body of mass M is rotating about an axis of fixed orientation through its centre of mass, with angular velocity $\boldsymbol{\omega} = \dot{\theta}\mathbf{k}$, where $\mathbf{k}$ is a fixed unit vector and θ is the anticlockwise angular displacement relative to the axis of rotation. Let I be the moment of inertia of the body about the axis of rotation.

The $\mathbf{k}$-component $L_{\text{axis}}^{\text{rel}}$ $(= \mathbf{L}^{\text{rel}} \cdot \mathbf{k})$ of the angular momentum of the body relative to the centre of mass is given by

$$L_{\text{axis}}^{\text{rel}} = I\dot{\theta}. \tag{49}$$

The **equation of relative rotational motion** of the body is

$$\Gamma_{\text{axis}}^{\text{rel}} = I\ddot{\theta}, \tag{50}$$

where $\Gamma_{\text{axis}}^{\text{rel}}$ $(= \boldsymbol{\Gamma}^{\text{rel}} \cdot \mathbf{k})$ is the $\mathbf{k}$-component of the total external torque relative to the centre of mass.

The kinetic energy T of the body is the sum of the kinetic energy of an equivalent particle at the centre of mass and the rotational kinetic energy relative to the centre of mass, that is,

$$T = \tfrac{1}{2}M|\dot{\mathbf{R}}|^2 + \tfrac{1}{2}I\omega^2, \tag{51}$$

where $\mathbf{R}$ is the position vector of the centre of mass and ω $(= |\dot{\theta}|)$ is the angular speed.

Exercise 23

You considered an identical baton in Exercise 15. Use any results from the solution to that exercise that you find useful.

A drum majorette's baton is modelled as a uniform cylindrical rod with spheres of equal mass at each end. The rod has mass $0.1\,\text{kg}$, length $0.8\,\text{m}$ and diameter $0.04\,\text{m}$. Each sphere has diameter $0.1\,\text{m}$ and mass $0.25\,\text{kg}$. The baton has been thrown upwards and is rotating at 1 revolution per second about a horizontal axis through its centre of mass and normal to the axis of the cylinder. Its centre of mass, which was initially at O, is moving vertically upwards at $5\,\text{m\,s}^{-1}$.

(a) Find the kinetic energy of the baton at the instant described.

(b) Find the magnitude of the $\mathbf{k}$-component of the angular momentum of the baton, relative to O, where $\mathbf{k}$ is a unit vector in the direction of the axis of rotation.

Recall the Highland sport of tossing the caber. The process of tossing the caber can be divided into five phases, as shown in Figure 7. In the following exercise, we consider an early part of the toss, that is, the transition between phases (i) and (ii).

Exercise 24

(a) During a caber-tossing competition, a competitor runs forward holding the caber (carrying the end X shown in Figure 28) and stops suddenly. While the competitor is running, the caber is vertical, and the whole caber has the same forward speed v. When the competitor stops, a large force is exerted at X for a very short time, with the effect that the end X of the caber becomes stationary. Model the caber as a uniform thin rod of length $2l$ and mass m, and assume that the motion of the caber is confined to two dimensions, in the (x, y)-plane in Figure 28.

   (i)   What will be the angular speed of the caber just after the competitor stops?

   (ii) What will be the kinetic energy of the caber just after the competitor stops?

(b) Suppose that after stopping, the competitor holds the end X of the caber stationary while the caber falls forward under gravity. Assuming that resistive forces are negligible, estimate the angular speed of the caber when it makes an angle θ with the vertical.

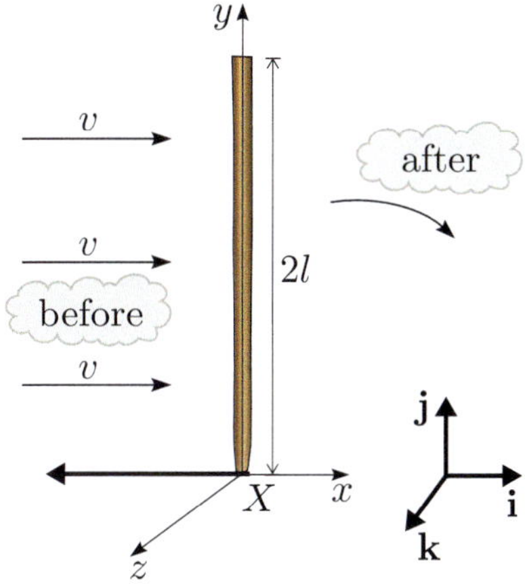

Figure 28   The instant of change between phases (i) and (ii) of Figure 7

We now consider a much later part of the motion of the caber: the start of phase (v) (see Figure 7) or the moment at which it strikes the ground (see Figure 29). The aim of the toss is to ensure that the caber finishes lying on the ground with the end X, which was originally being held by the competitor, now furthest away from him. Even if the caber strikes the ground as shown in the figure, before it has rotated sufficiently for X to have moved to the right of Y, it may maintain sufficient rotation after impact for X to swing past Y, and for the caber to fall with X pointing away from the competitor. How might we model the effect of the caber hitting the ground on its rotational motion?

When an object hits the ground, there is an **impact**, during which the object is subject to forces of great magnitude over a short period of time. These forces drastically change the motion of the object. To model the impact, we assume that after hitting the ground, the end Y of the caber is stationary, and that during impact, all the forces on the caber are acting at the point Y. Take an origin at the point Y, and consider the torque and angular momentum about that point. Since we are assuming that all the external forces act at Y during the impact, the torque about Y is zero. Then by the torque law, the *angular momentum about Y is conserved during the impact.*

You are asked to develop this model further in the following exercise.

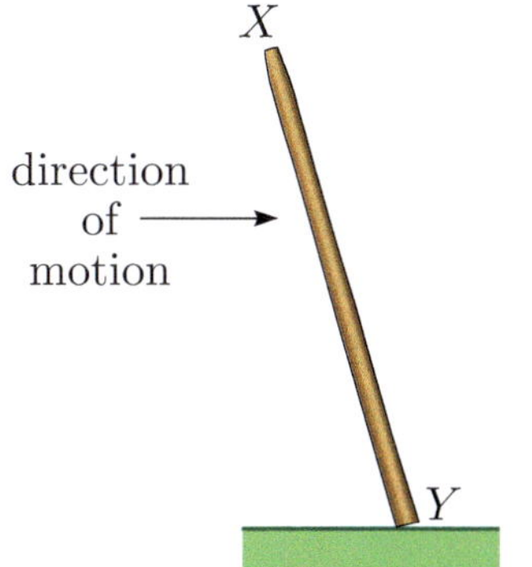

Figure 29   Instant where the caber strikes the ground at the start of phase (v)

We are assuming that during the impact, the force due to gravity is negligible compared with the forces acting at Y.

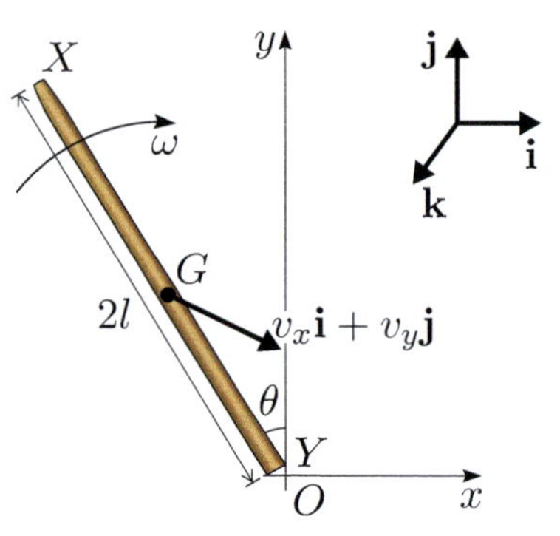

Figure 30   Coordinate set-up at the start of phase (v)

Exercise 25

A caber is in flight, and its end Y is about to strike the ground at O at an angle θ from the vertical. Use the axes shown in Figure 30, and assume that the motion of the caber is confined to the (x, y)-plane throughout. Just before the end Y comes in contact with the ground, the centre of mass G of the caber has velocity $v_x\mathbf{i} + v_y\mathbf{j}$, and the angular speed of the caber about G is ω, rotating clockwise. Model the caber as a uniform thin straight rod of length $2l$ and mass m.

(a) Use a decomposition theorem to find the **k**-component of the angular momentum of the caber about O just before it hits the ground.

(b) Let ω_1 be the angular speed of rotation of the caber about O just after it hits the ground. Show that

$$\omega_1 = \tfrac{1}{4}\omega + \frac{3}{4l}(v_x \cos\theta + v_y \sin\theta). \tag{52}$$

(c) To rotate past the vertical, the caber must have sufficient kinetic energy after impact that it reaches the vertical with non-zero kinetic energy. Use this fact to show that in order for the caber to rotate past the vertical, we must have that

$$\omega_1^2 > \frac{3g(1 - \cos\theta)}{2l}. \tag{53}$$

The values of θ, v_x, v_y and ω, as defined in Exercise 25, when the caber strikes the ground will be determined by the way that the competitor launches the caber. If we know these values and the length $2l$ of the caber, then we can calculate ω_1 from equation (52). Then condition (53) enables us to determine whether or not the caber will pass the vertical.

4.3 Rolling objects

It is interesting to compare how solid and hollow cylindrical cans roll down an inclined plane – a situation mentioned in the Introduction. We are now in a position to model that situation quantitatively. To do so, we will assume that there is no loss of mechanical energy when a cylinder rolls down a slope. This is an important assumption that will be fully justified later on.

Consider a cylinder of mass M and radius R rolling down a plane inclined at an angle α to the horizontal (Figure 31(a)). We will first look at the behaviour of a uniform *solid* cylinder. Its moment of inertia about an axis through its centre of mass is $\tfrac{1}{2}MR^2$ (from Table 1). The cylinder starts from rest at the origin O, and we want to find how long it will take to reach the point A, where the distance OA is l.

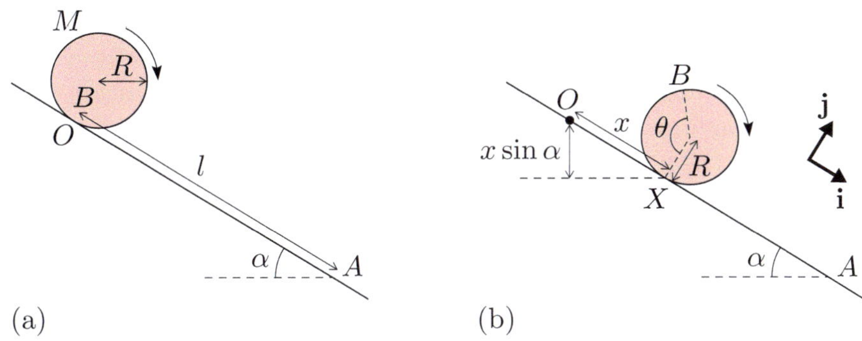

(a)   (b)

Figure 31   Cylinder rolling down an inclined plane: (a) initial position,
(b) after it has rolled a distance x along the plane

Take the situation where the cylinder has rolled as far as X (Figure 31(b)),
and choose Cartesian unit vectors $\mathbf{i}$ and $\mathbf{j}$ as shown in the figure. Let
$OX = x$, and suppose that in rolling from O to X the cylinder has turned
through an angle θ, without any slipping having occurred. Note that since
the cylinder is rotating clockwise, θ becomes increasingly negative. The
distance OX must be equal to the length of the circumference of the
cylinder from B to X, where B is the point of contact between the
cylinder and the slope at the outset. Thus taking 0 as the initial value of θ,
so that its subsequent values are negative, we have

Note that, as usual, θ is measured positive in an anticlockwise sense.

$$x = -R\theta. \tag{54}$$

We refer to this equation as the **rolling condition**.

The rolling condition is often referred to as the no slip condition.

When the cylinder is at the point shown in Figure 31(b), its centre of mass
has position $x\mathbf{i} + R\mathbf{j}$. Hence the velocity of the centre of mass is $\dot{x}\mathbf{i}$, since R
is constant. The cylinder is rotating clockwise at an angular speed $|\dot{\theta}|$.
Therefore by equation (51), the cylinder has kinetic energy

$$T = \tfrac{1}{2}M\dot{x}^2 + \tfrac{1}{2}\left(\tfrac{1}{2}MR^2\right)\dot{\theta}^2. \tag{55}$$

From equation (54) we have $R|\dot{\theta}| = |\dot{x}|$, so

$$T = \tfrac{1}{2}M\dot{x}^2 + \tfrac{1}{4}M\dot{x}^2 = \tfrac{3}{4}M\dot{x}^2.$$

Because the cylinder starts from rest at O, its initial kinetic energy is zero.
In moving from O to X, the centre of mass of the cylinder descends a
vertical distance $x\sin\alpha$. So the potential energy of the cylinder is reduced
by $Mgx\sin\alpha$. Then, if mechanical energy is conserved, we have

$$\tfrac{3}{4}M\dot{x}^2 = Mgx\sin\alpha. \tag{56}$$

By differentiating each side of this equation with respect to t, we obtain

$$\tfrac{3}{2}M\dot{x}\ddot{x} = Mg\dot{x}\sin\alpha. \tag{57}$$

On dividing by $M\dot{x}$ and rearranging, we obtain

The cylinder is not stationary, so $\dot{x}$ is not zero.

$$\ddot{x} = \tfrac{2}{3}g\sin\alpha. \tag{58}$$

You saw this in Unit 3.

Now, if the cylinder were simply *sliding* down the slope (without friction) rather than rolling down, it would have acceleration $g \sin \alpha$. But from equation (58) we see that the rolling cylinder has a lower acceleration than if it were to slide down the slope. This is because in the case of the rolling object, some of its potential energy has been converted into kinetic energy of rotation, while for a sliding object all the kinetic energy is associated with the linear motion of the centre of mass.

Exercise 26

(a) Adapt the foregoing argument to obtain an expression for the acceleration of a hollow cylinder of mass M and radius R when, starting from rest, it is rolling down the same slope as considered above. (Model the hollow cylinder as a thin cylindrical shell, all of whose mass is at a distance R from its axis.)

(b) In a 'race' over a distance of $2\,\mathrm{m}$ down a slope angled at $\frac{\pi}{6}$ to the horizontal, how much faster will a solid cylinder travel than a hollow one? Do the masses of the cylinders matter?

We assume that air resistance is negligible.

We now consider the forces on a cylinder as it rolls down a slope without slipping. Since the cylinder starts from rest, it is gaining angular momentum about its centre of mass as it rolls, so there must be some force supplying a torque relative to the centre of mass. This cannot be the weight $\mathbf{W}$ – you saw in Exercise 21 that the force of gravity on a body gives a zero torque relative to the centre of mass. Another force on the cylinder is that from the slope: the normal reaction $\mathbf{N}$ of the slope acts through the centre of mass, so again it gives a zero torque about the centre of mass. That leaves only a force between the slope and the cylinder in the direction of the slope, which is supplied by friction. Therefore we must include friction $\mathbf{F}$ in our model if the model is to have any hope of predicting the motion that is observed.

If there were no friction between the slope and the cylinder, the cylinder would simply slide down the slope without rolling.

Exercise 27

(a) Use Newton's second law (for the motion of an equivalent particle at the centre of mass) and the equation of relative rotational motion (equation (50)) to find an expression for the acceleration of a solid cylinder of mass M and radius R when it is rolling (without slipping) down a slope of angle α to the horizontal, as in Figure 32.

(b) The condition for the cylinder to roll without slipping is $|\mathbf{F}| \leq \mu|\mathbf{N}|$, where μ is the coefficient of static friction. Obtain a condition relating α and μ that must hold if only rolling is to occur.

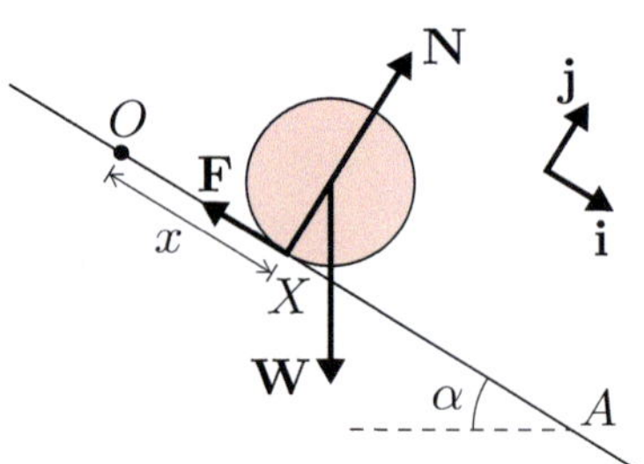

Figure 32   Forces on a rolling cylinder

In a system where there is friction, you might reasonably expect there to be a loss of mechanical energy. However, this is not necessarily the case. A cornering car needs a sideways frictional force between each tyre and the road to avoid skidding, as you saw in Unit 20, but as long as the car does not skid sideways, there is no loss of mechanical energy due to this force. The point of contact between the tyre and the road does not move in the direction of the frictional force, thus no mechanical energy is lost to friction. The situation is similar for the rolling cylinder, though this is perhaps more difficult to see. The point on the cylinder that is in contact with the slope at any instant does not move relative to the slope (if it did, the cylinder would skid and the rolling condition would not hold). Hence there is no loss of mechanical energy due to the frictional force. The truth of this assertion was demonstrated in Exercise 27(a), where you saw that the equation for the acceleration of the centre of mass obtained using Newton's second law, and making no assumption about the mechanical energy, is the same as equation (58) obtained earlier in this subsection using the assumption that mechanical energy is conserved.

That is, the frictional force $\mathbf{F}$ is normal to the velocity $\dot{\mathbf{r}}$, so $\mathbf{F} \cdot \dot{\mathbf{r}} = 0$. Therefore the work done by $\mathbf{F}$ is zero (see Unit 16).

Learning outcomes

After studying this unit, you should be able to:

- model the motion of an extended body as the motion of an equivalent particle at the centre of mass combined with the motion of the body relative to its centre of mass

- find the moments of inertia of rigid bodies of common geometrical shapes about axes of symmetry by reference to Table 1, and use the parallel axes theorem to find the moments of inertia of such bodies about other axes

- determine the angular momentum and kinetic energy of a rigid body rotating about an axis whose direction is fixed, using the appropriate decomposition theorem if necessary

- for an extended body subjected to an impact at a particular point, use conservation of angular momentum about that point to relate the motions of the body before and after the impact

- give the equation of rotational motion for a rigid body rotating about a fixed axis, or the equation of relative rotational motion for a rigid body rotating about an axis whose direction is fixed

- use the rolling condition to relate the translational and rotational motions of a body rolling across a plane surface without slipping.

Solutions to exercises

Solution to Exercise 1

(a) To determine the time of flight t, we need only consider the vertical motion of the centre of mass. The centre of mass starts with an upward velocity component (in $\mathrm{m\,s^{-1}}$) of $4\cos\frac{\pi}{6} = 2\sqrt{3}$, but is subject to a constant acceleration of magnitude g (taken to be $9.81\,\mathrm{m\,s^{-2}}$) downwards, and it descends from an initial height (above ground level) of $3 - 1.2\cos\frac{\pi}{3} = 2.4$ to a final height of 1.2 (both in m). Hence, from the constant acceleration equation,

$$1.2 = 2.4 + 2\sqrt{3}\,t - \tfrac{1}{2} \times 9.81 t^2,$$

or

$$\tfrac{1}{2} \times 9.81 t^2 - 2\sqrt{3}\,t - 1.2 = 0.$$

Solving for t, and rejecting the negative root, we obtain $t = 0.96$. So the gymnast is in flight for about $0.96\,\mathrm{s}$.

(b) The gymnast rotates through $\frac{5\pi}{3}$ (from Example 2(b)) in approximately $0.96\,\mathrm{s}$, and therefore has an angular speed of $5\pi/(3 \times 0.96) \simeq 5.45$ (in $\mathrm{rad\,s^{-1}}$). Suppose that his moment of inertia about an axis through his centre of mass is I_T in the tuck position, and I_E when fully extended. Just after leaving the bar, he is fully extended and has angular speed $\frac{10}{3}\,\mathrm{rad\,s^{-1}}$ (from Example 2(a)).

Now, angular momentum is given by $I\omega$, where I is the moment of inertia and ω is the angular speed. Since angular momentum is conserved, we have

$$I_\mathrm{E}\tfrac{10}{3} = I_\mathrm{T}5.45.$$

Hence $I_\mathrm{T}/I_\mathrm{E} = 10/(3 \times 5.45) \simeq 0.61$. So adopting a tuck position must reduce the moment of inertia by at least 40%, if it is to allow the gymnast to complete the dismount. (In practice, the reduction in the moment of inertia would need to be greater than this, since the gymnast will need time to get into and out of the tuck position.)

Solution to Exercise 2

The pairs of forces shown in parts (a), (b) and (c) of Figure 8 satisfy equation (2). Those in part (d) do not.

Solution to Exercise 3

The pair of forces that satisfy equation (2) but do not satisfy Newton's third law is in Figure 8(c).

(The forces in Figures 8(a) and 8(b) satisfy Newton's third law: they are opposite in direction and act in the same straight line. Figure 8(d) does not show forces satisfying Newton's third law: although they act in the same straight line, they are *not* in opposite directions.)

Solution to Exercise 4

Differentiating the given expression for $\mathbf{r}$, we have

$$\dot{\mathbf{r}} = -6\sin(2t)\mathbf{i} + 8\cos(2t)\mathbf{j}.$$

Then at $t = 0$, the expressions for $\mathbf{r}$ and $\dot{\mathbf{r}}$ reduce to

$$\mathbf{r} = 3\mathbf{i} + 5\mathbf{k},$$
$$\dot{\mathbf{r}} = 8\mathbf{j}.$$

So the angular momentum of the particle about O at $t = 0$ is

$$\mathbf{r} \times m\dot{\mathbf{r}} = (3\mathbf{i} + 5\mathbf{k}) \times 20(8\mathbf{j}) = 160(3\mathbf{k} - 5\mathbf{i}).$$

Solution to Exercise 5

(a) The angular speed ω of the particle is v/R, and the angular velocity $\boldsymbol{\omega}$ is $(v/R)\mathbf{k}$, using the right-hand grip rule (see Unit 20) for the direction.

(b) By definition, $\mathbf{L} = \mathbf{r} \times m\dot{\mathbf{r}}$. Now, using $\boldsymbol{\omega}$ as defined in part (a), we have from the hint that

$$\mathbf{r} = R\big(\cos(\omega t)\mathbf{i} + \sin(\omega t)\mathbf{j}\big),$$

and therefore (since ω is constant)

$$\dot{\mathbf{r}} = R\omega\big(-\sin(\omega t)\mathbf{i} + \cos(\omega t)\mathbf{j}\big).$$

Then, substituting for $\mathbf{r}$ and $\dot{\mathbf{r}}$ in $\mathbf{L} = \mathbf{r} \times m\dot{\mathbf{r}}$, we find

$$\begin{aligned}
\mathbf{L} &= R\big(\cos(\omega t)\mathbf{i} + \sin(\omega t)\mathbf{j}\big) \times mR\omega\big(-\sin(\omega t)\mathbf{i} + \cos(\omega t)\mathbf{j}\big) \\
&= mR^2\omega\big(\cos^2(\omega t)\mathbf{k} + \sin^2(\omega t)\mathbf{k}\big) \\
&= mR^2\omega\mathbf{k}.
\end{aligned}$$

(c) For a single particle, the moment of inertia is mr^2 (from Unit 17), where r is the distance of the particle from the relevant axis. In this case $r = R$, so $I = mR^2$.

(d) With $\boldsymbol{\omega} = \omega\mathbf{k}$ and using the result from part (c), we have

$$I\boldsymbol{\omega} = mR^2\boldsymbol{\omega} = mR^2\omega\mathbf{k},$$

which is equal to $\mathbf{L}$ as given by the solution to part (b).

Solution to Exercise 6

(a) As $\mathbf{r} = x\mathbf{i} + y\mathbf{j}$, it follows that

$$\mathbf{r} \times \dot{\mathbf{r}} = (x\mathbf{i} + y\mathbf{j}) \times (\dot{x}\mathbf{i} + \dot{y}\mathbf{j}) = x\dot{y}\mathbf{k} - y\dot{x}\mathbf{k} = (x\dot{y} - y\dot{x})\mathbf{k}.$$

(b) The angular momentum of the particle about O is given by

$$\mathbf{L} = \mathbf{r} \times m\dot{\mathbf{r}} = m(\mathbf{r} \times \dot{\mathbf{r}}) = m(x\dot{y} - y\dot{x})\mathbf{k},$$

from part (a). But

$$x = r\cos\theta, \quad y = r\sin\theta.$$

Hence, using the product and chain rules,

$$\dot{x} = \dot{r}\cos\theta + r\,\frac{d}{dt}(\cos\theta) = \dot{r}\cos\theta - r\dot{\theta}\sin\theta,$$

$$\dot{y} = \dot{r}\sin\theta + r\,\frac{d}{dt}(\sin\theta) = \dot{r}\sin\theta + r\dot{\theta}\cos\theta.$$

Then

$$\begin{aligned}
x\dot{y} - y\dot{x} &= r\cos\theta(\dot{r}\sin\theta + r\dot{\theta}\cos\theta) - r\sin\theta(\dot{r}\cos\theta - r\dot{\theta}\sin\theta)\\
&= r\dot{r}\cos\theta\sin\theta + r^2\dot{\theta}\cos^2\theta - r\dot{r}\cos\theta\sin\theta + r^2\dot{\theta}\sin^2\theta\\
&= r^2\dot{\theta}.
\end{aligned}$$

Hence we have

$$\begin{aligned}
\mathbf{L} &= m(x\dot{y} - y\dot{x})\mathbf{k}\\
&= mr^2\dot{\theta}\mathbf{k},
\end{aligned}$$

as required.

Solution to Exercise 7

In Example 3(b)(i), we found

$$\dot{\mathbf{r}} = -R\omega\sin(\omega t)\mathbf{i} + R\omega\cos(\omega t)\mathbf{j}.$$

Differentiating this gives the acceleration

$$\ddot{\mathbf{r}} = -R\omega^2\cos(\omega t)\mathbf{i} - R\omega^2\sin(\omega t)\mathbf{j}.$$

The total force on the particle is $\mathbf{F} = m\ddot{\mathbf{r}}$. So, substituting for $\ddot{\mathbf{r}}$ as given above, we have

$$\mathbf{F} = -mR\omega^2\big(\cos(\omega t)\mathbf{i} + \sin(\omega t)\mathbf{j}\big).$$

Now, $\boldsymbol{\Gamma} = \mathbf{r} \times \mathbf{F}$, which, on substituting for $\mathbf{r}$ from equation (8) and for $\mathbf{F}$ as given above, becomes

$$\begin{aligned}
\boldsymbol{\Gamma} &= \big(R\cos(\omega t)\mathbf{i} + R\sin(\omega t)\mathbf{j} + h\mathbf{k}\big) \times (-mR\omega^2)\big(\cos(\omega t)\mathbf{i} + \sin(\omega t)\mathbf{j}\big)\\
&= -mR^2\omega^2\big(\cos(\omega t)\sin(\omega t)\mathbf{k} - \cos(\omega t)\sin(\omega t)\mathbf{k}\big)\\
&\quad - mhR\omega^2\big(\cos(\omega t)\mathbf{j} - \sin(\omega t)\mathbf{i}\big)\\
&= mhR\omega^2\big(\sin(\omega t)\mathbf{i} - \cos(\omega t)\mathbf{j}\big).
\end{aligned}$$

Comparing this with equation (11), we see that we have obtained the same expression for $\boldsymbol{\Gamma}$.

Solution to Exercise 8

(a) From the solution to Exercise 5(b), the angular momentum, about O, of the particle at A is $m_1 a^2\boldsymbol{\omega}$, while the angular momentum, about O, of the particle at B is $m_2 b^2\boldsymbol{\omega}$. So the total angular momentum of the two-particle system modelling the roundabout is

$$m_1 a^2\boldsymbol{\omega} + m_2 b^2\boldsymbol{\omega} = I\boldsymbol{\omega},$$

where $I = m_1 a^2 + m_2 b^2$.

(b) Since the force at C is applied in a horizontal direction at right angles to OC, it exerts a torque about O of $cF\mathbf{k}$. There are other external forces on the system that need to be taken into account: the force at O exerted by the fixed spindle, and the weight of each of the two particles. The force from the spindle has zero torque about O. Each weight exerts a non-zero torque about O, but the condition $m_1 a = m_2 b$ ensures that the torques exerted by the two weights are equal and opposite, so their sum is zero. Hence the total external torque on the two-particle system is $cF\mathbf{k}$.

Then, as the total angular momentum of the system is $I\boldsymbol{\omega}$ (from part (a)), the torque law applied to the system consisting of the two particles at A and B gives

$$\frac{d}{dt}(I\boldsymbol{\omega}) = cF\mathbf{k}.$$

Because I is a constant and $\boldsymbol{\omega} = \omega\mathbf{k}$, this equation can be rewritten as

$$I\dot{\omega}\mathbf{k} = cF\mathbf{k},$$

which simplifies to

$$I\dot{\omega} = cF.$$

(c) From the previous equation,

$$\dot{\omega} = \frac{cF}{I}.$$

Now $c = 1.5$ and $F = 315$, and I can be calculated from

$$I = m_1 a^2 + m_2 b^2 = (45 \times 1^2) + (60 \times 0.75^2) = 78.75.$$

So, on substituting, we have (in $\mathrm{rad\,s^{-2}}$)

$$\dot{\omega} = \frac{1.5 \times 315}{78.75} = 6.$$

Rotations at 0.5 revolutions per second require an angular speed of $\pi\,\mathrm{rad\,s^{-1}}$. A force of magnitude $315\,\mathrm{N}$ increases the angular speed from 0 to $6\,\mathrm{rad\,s^{-1}}$ in one second, so to reach an angular speed of $\pi\,\mathrm{rad\,s^{-1}}$ starting from rest will take $\frac{\pi}{6}$ seconds, which is about $0.5\,\mathrm{s}$.

Solution to Exercise 9

The only force assumed to be acting on each star is the gravitational force due to the other star, which is an internal force. Hence the total external force acting on the two-particle system is zero, and consequently so is the external torque. It follows from the torque law that the angular momentum of the system must be conserved, so its magnitude L is constant.

Suppose that the distances of the stars from their common centre of mass O are d_1 and d_2, and that the angular velocity of each star about O is $\boldsymbol{\omega} = \omega\mathbf{k}$. (Since the two stars are rotating as though tied together by a rigid rod, their angular velocities must be the same.) Then from equation (7), the angular momentum of star 1 is $m_1 d_1^2 \omega\mathbf{k}$, and that of star 2 is $m_2 d_2^2 \omega\mathbf{k}$.

The total angular momentum of the system is

$$\mathbf{L} = m_1 d_1^2 \omega \mathbf{k} + m_2 d_2^2 \omega \mathbf{k} = (m_1 d_1^2 + m_2 d_2^2)\omega \mathbf{k}.$$

By an argument similar to that used in Example 4, this must be equal to

$$\frac{m_1 m_2}{m_1 + m_2} d^2 \omega \mathbf{k}.$$

Therefore

$$L = \frac{m_1 m_2}{m_1 + m_2} d^2 \omega,$$

and this can be rearranged to give

$$\omega = \frac{m_1 + m_2}{m_1 m_2 d^2} L.$$

Thus the period of rotation of the system is

$$\frac{2\pi}{\omega} = \frac{2\pi m_1 m_2 d^2}{(m_1 + m_2)L}.$$

Solution to Exercise 10

Apply the torque law about the point X. Since all the external forces are applied at X, the total external torque about X is zero. Hence the rate of change of angular momentum about X is zero, that is, the angular momentum of the system *about* X is constant.

Solution to Exercise 11

(a) (i) Start from equation (26), which can be expressed as

$$\mathbf{L} = I\dot{\theta}\mathbf{k} - \dot{\theta}\sum_{i=1}^{n} m_i z_i(x_i\mathbf{i} + y_i\mathbf{j}),$$

where the definition $I = \sum_{i=1}^{n} m_i d_i^2$ has been used. Now write $x_i = d_i \cos\theta$ and $y_i = d_i \sin\theta$, since d_i is the perpendicular distance of the ith particle from the z-axis. Note also that $\dot{\theta} = \omega$, since the body is rotating anticlockwise with angular speed ω. Hence

$$\mathbf{L} = I\omega\mathbf{k} - \omega\sum_{i=1}^{n} m_i d_i z_i(\cos\theta\,\mathbf{i} + \sin\theta\,\mathbf{j})$$

$$= I\omega\mathbf{k} - \omega\sum_{i=1}^{n} m_i d_i z_i \mathbf{e}_r$$

$$= I\omega\mathbf{k} - A\omega\mathbf{e}_r,$$

where

$$A = \sum_{i=1}^{n} m_i d_i z_i,$$

which is constant since m_i, d_i and z_i are constants. Moreover, we must have $A > 0$ since for all $1 \le i \le n$ we have $m_i > 0$, $d_i > 0$ and $z_i > 0$, except possibly for the bottom-most particle where z_i may be zero.

(ii)   None of I, ω, $\mathbf{k}$ or A varies with time, but the vector $\mathbf{e}_r$ does (because its direction is changing). Hence differentiating equation (28) with respect to time yields $\dot{\mathbf{L}} = -A\omega\dot{\mathbf{e}}_r$. But from Unit 20 we have $\dot{\mathbf{e}}_r = \dot{\theta}\mathbf{e}_\theta = \omega\mathbf{e}_\theta$ (using the chain rule). Thus $\dot{\mathbf{L}} = -A\omega^2\mathbf{e}_\theta$, as required.

By the torque law, the total torque on the body is $\boldsymbol{\Gamma} = \dot{\mathbf{L}}$. Since A and ω^2 are positive constants, we see from equation (29) that $\dot{\mathbf{L}}$, and therefore $\boldsymbol{\Gamma}$, is a negative multiple of $\mathbf{e}_\theta$. Thus in order to keep the body rotating with constant angular speed, it must be subjected to a non-zero torque in a direction tangential to the motion but pointing in the opposite direction to $\mathbf{e}_\theta$.

(b) (i)   Since the skater's centre of mass is following a circle at constant angular speed, its acceleration is towards the z-axis, that is, in the direction of $-\mathbf{e}_r$. Thus by Newton's second law, the total external force $\mathbf{F}$ on the skater must be in this direction.

(ii)   Let $\mathbf{W}$ be the skater's weight, and let the force exerted on her skates by the ice have component forces $\mathbf{N}$ vertically upwards and $\mathbf{R}$ horizontally. So $\mathbf{F} = \mathbf{W} + \mathbf{N} + \mathbf{R}$. From part (b)(i), the vertical component of $\mathbf{F}$ is zero, so $\mathbf{W} + \mathbf{N} = \mathbf{0}$. If the skater is perfectly vertical, then $\mathbf{N}$ and $\mathbf{W}$ have exactly the same line of action, so the sum of their torques must also be zero. Since $\mathbf{F}$ must be in the direction of $-\mathbf{e}_r$, the component of $\mathbf{R}$ normal to $\mathbf{e}_r$ must be zero. Therefore if O is the centre of the circle followed by the skater's feet, the line of action of $\mathbf{R}$ passes through O, and hence $\mathbf{R}$ has zero torque about O. This means that if the skater is vertical, the total torque about O will be zero. But part (a) showed that a system of this kind must be subject to a non-zero torque in the tangential direction! So it is impossible for the skater to follow a circle while remaining vertical.

To follow a circle, the skater needs to adopt a position in which there is a non-zero torque in the tangential direction. To do this, she must lean towards the centre of the circle. This will mean that $\mathbf{W}$ and $\mathbf{N}$ do not have the same line of action, so they provide a non-zero torque. However, note that if the skater leaned in the opposite direction, away from the centre of the circle while following the circle, there would still be a non-zero torque but in the opposite tangential direction to that predicted in part (a), so this motion would be impossible.

Solution to Exercise 12

The required moment of inertia can be obtained from that of a hollow cylinder (inner radius a, outer radius R) about its axis, as given in Table 1, by taking the limit as $a \to R$, in which case

$$I = \tfrac{1}{2}M(R^2 + a^2) \to MR^2.$$

Solution to Exercise 13

(a) With unit vectors $\mathbf{i}$, $\mathbf{j}$, $\mathbf{k}$ in the directions of the x-, y-, z-axes shown in Figure 20, we have $\mathbf{F} = 100\mathbf{j}$ and $\mathbf{R} = -c\omega\mathbf{i}$. So the total torque about O is

$$\mathbf{\Gamma} = 1.5\mathbf{i} \times \mathbf{F} + (-0.1\mathbf{j}) \times \mathbf{R} = 150\mathbf{k} - 0.1c\omega\mathbf{k}.$$

Thus the z-component of the total torque on the roundabout is $150 - 0.1c\omega$.

The moment of inertia of the roundabout (from Example 5) is $\frac{1}{2}(250)(1.2)^2 = 180$, so from equation (30), the equation of rotational motion for the roundabout is

$$180\dot{\omega} = 150 - 0.1c\omega.$$

(b) Employing methods from Unit 1 (either separation of variables or the integrating factor method can be used in this case), we can obtain the general solution of the differential equation derived in part (a), which is

$$\omega = Ae^{-ct/1800} + \frac{1500}{c},$$

where A is a constant. Since the roundabout is at rest at the outset, we have $\omega = 0$ at $t = 0$, so

$$0 = A + \frac{1500}{c},$$

thus

$$A = -\frac{1500}{c}.$$

The particular solution satisfying this initial condition is therefore

$$\omega = \frac{1500}{c}\left(1 - e^{-ct/1800}\right).$$

(c) If pushing continues indefinitely, the final expression for ω implies that the rotational speed will increase to almost $1500/c$ as the exponential term becomes negligible, and it will then become steady. For $c = 10$, this maximum angular speed is $150\,\mathrm{rad\,s^{-1}}$. To follow the roundabout at that angular speed, the pusher would need to be travelling at a speed of $1.5(150) = 225\,\mathrm{m\,s^{-1}}$. This is impossible for a human pusher to achieve.

Solution to Exercise 14

The moment of inertia I_{M} of the mace about AB is the sum of the moment of inertia I_{S} of the sphere about AB and the moment of inertia I_{R} of the rod about AB. Using Table 1, in conjunction with the parallel axes theorem, we have

$$I_{\mathrm{S}} = \tfrac{2}{5}Mr^2 + M(d+r)^2,$$

$$I_{\mathrm{R}} = \tfrac{1}{4}mR^2 + \tfrac{1}{12}md^2 + m\left(\tfrac{1}{2}d\right)^2 = \tfrac{1}{4}mR^2 + \tfrac{1}{3}md^2,$$

so

$$I_{\mathrm{M}} = I_{\mathrm{S}} + I_{\mathrm{R}} = \tfrac{2}{5}Mr^2 + M(d+r)^2 + \tfrac{1}{4}mR^2 + \tfrac{1}{3}md^2.$$

Solution to Exercise 15

The kinetic energy of the baton is given by $\frac{1}{2}I\omega^2$, where I is the moment of inertia of the baton about the axis of rotation, and ω is the angular speed. Now, the moment of inertia of the baton is the sum of the moments of inertia of the cylindrical rod and the two spheres.

The cylinder is rotating about its centre of mass, and its moment of inertia (from Table 1) is

$$\tfrac{1}{4}(0.1)(0.02)^2 + \tfrac{1}{12}(0.1)(0.8)^2 \simeq 0.005\,34$$

(in $\mathrm{kg\,m^2}$).

Each sphere has its centre of mass $0.45\,\mathrm{m}$ from the centre of the baton. To find the moment of inertia of a sphere about the axis of rotation, we find the moment of inertia about its centre of mass from Table 1 and use the parallel axes theorem, to obtain

$$\tfrac{2}{5}(0.25)(0.05)^2 + 0.25(0.45)^2 \simeq 0.0509$$

(in $\mathrm{kg\,m^2}$).

So the moment of inertia of the baton about the axis of rotation is

$$I = 0.005\,34 + 2(0.0509) \simeq 0.1071$$

(in $\mathrm{kg\,m^2}$). The angular speed ω of the baton is $2\pi\,\mathrm{rad\,s^{-1}}$. Hence the kinetic energy of the baton (in joules) is

$$\tfrac{1}{2}I\omega^2 = \tfrac{1}{2}(0.1071)(2\pi)^2$$
$$\simeq 2.11.$$

Solution to Exercise 16

(a) As you saw in Example 6, when the man is $0.2\,\mathrm{m}$ from the centre of the roundabout, the angular speed of the roundabout is $\omega_1 = \pi$ (in $\mathrm{rad\,s^{-1}}$). The moment of inertia of the combined system under these circumstances is $I_1 = 176$ (in $\mathrm{kg\,m^2}$). So the kinetic energy of the system (in joules) is

$$\tfrac{1}{2}I_1\omega_1^2 = \tfrac{1}{2} \times 176\pi^2$$
$$\simeq 869.$$

Similarly, when the man is $1\,\mathrm{m}$ from the centre, the angular speed is $\omega_2 \simeq 2.19$ (in $\mathrm{rad\,s^{-1}}$). The moment of inertia of the combined system is $I_2 = 252.8$ (in $\mathrm{kg\,m^2}$). So the kinetic energy of the system (in joules) is

$$\tfrac{1}{2}I_2\omega_2^2 \simeq \tfrac{1}{2} \times 252.8(2.19)^2$$
$$\simeq 606.$$

(b) When the man moves inwards, the kinetic energy of the system *increases*, from about 606 joules to about 869 joules.

Solution to Exercise 17

The kinetic energy of the body is given by $\frac{1}{2}I\omega^2$, where I is the moment of inertia of the body about the axis of rotation. By the parallel axes theorem,

$$I = I_G + MR^2.$$

The centre of mass is following a circle of radius R at angular speed ω, so $R\omega = v$, and hence the kinetic energy of the body is

$$\begin{aligned}
\tfrac{1}{2}I\omega^2 &= \tfrac{1}{2}(I_G + MR^2)\omega^2 \\
&= \tfrac{1}{2}I_G\omega^2 + \tfrac{1}{2}M(R\omega)^2 \\
&= \tfrac{1}{2}I_G\omega^2 + \tfrac{1}{2}Mv^2.
\end{aligned}$$

Solution to Exercise 18

The diver's centre of mass starts at a height l above the board. In the position at which the diver lets go of the board, the vertical displacement (above the board) of her centre of mass is $l\cos\alpha$, so the potential energy has decreased by

$$mgl(1 - \cos\alpha).$$

Let I_O be the moment of inertia of the diver about O. Then the kinetic energy at the time of letting go of the board is $\frac{1}{2}I_O\omega^2$.

The kinetic energy at the outset is zero, as the diver is at rest, so by the conservation of mechanical energy, we have

$$\tfrac{1}{2}I_O\omega^2 = mgl(1 - \cos\alpha).$$

The parallel axes theorem gives

$$I_O = I_G + ml^2.$$

Hence

$$\omega = \sqrt{\frac{2mgl(1 - \cos\alpha)}{I_G + ml^2}}.$$

Solution to Exercise 19

(a) From Figure 27, we have

$$x = -l\sin\theta, \quad y = l\cos\theta.$$

Differentiating with respect to t, the chain rule is used to obtain

$$\dot{x} = -l\dot{\theta}\cos\theta, \quad \dot{y} = -l\dot{\theta}\sin\theta.$$

Differentiating again with respect to t (using both the product and chain rules) gives

$$\begin{aligned}
\ddot{x} &= l\dot{\theta}^2\sin\theta - l\ddot{\theta}\cos\theta, \\
\ddot{y} &= -l\dot{\theta}^2\cos\theta - l\ddot{\theta}\sin\theta.
\end{aligned}$$

(b) The diver is rotating about a fixed axis through O. The component of the total external torque about O in the direction of the axis of rotation (the $\mathbf{k}$-direction in Figure 27) is $mgl\sin\theta$ since $\boldsymbol{\Gamma} = (-l\sin\theta\,\mathbf{i} + l\cos\theta\,\mathbf{j}) \times (-mg\,\mathbf{j}) = mgl\sin\theta\,\mathbf{k}$. So from equation (30), the equation of rotational motion about O for the diver is

$$I_O\ddot{\theta} = mgl\sin\theta.$$

(c) As in Exercise 18, the principle of conservation of mechanical energy gives

$$\tfrac{1}{2}I_O\dot{\theta}^2 = mgl(1 - \cos\theta).$$

Differentiating this equation with respect to t and using the chain rule gives

$$\tfrac{1}{2}I_O(2\dot{\theta}\ddot{\theta}) = mgl\dot{\theta}\sin\theta,$$

which simplifies to (since $\dot{\theta} \neq 0$)

$$I_O\ddot{\theta} = mgl\sin\theta,$$

the equation of rotational motion derived in part (b).

(d) (i) By applying Newton's second law, we obtain

$$R_1\mathbf{i} + R_2\mathbf{j} - mg\mathbf{j} = m\ddot{\mathbf{r}} = m(\ddot{x}\mathbf{i} + \ddot{y}\mathbf{j}).$$

Resolving in the x- and y-directions gives

$$R_1 = m\ddot{x},$$
$$R_2 - mg = m\ddot{y}.$$

(ii) Substituting from equation (38) for $\ddot{x}$ and $\ddot{y}$ in the results obtained in part (d)(i) gives

$$R_1 = ml(\dot{\theta}^2\sin\theta - \ddot{\theta}\cos\theta),$$
$$R_2 = mg + ml(-\dot{\theta}^2\cos\theta - \ddot{\theta}\sin\theta).$$

(e) (i) From parts (b) and (c), we have

$$I_O\ddot{\theta} = mgl\sin\theta,$$
$$\tfrac{1}{2}I_O\dot{\theta}^2 = mgl(1 - \cos\theta).$$

Then, from the results of part (d)(ii), we obtain

$$R_1 = ml(\dot{\theta}^2\sin\theta - \ddot{\theta}\cos\theta)$$
$$= ml\left(\frac{2mgl}{I_O}(1 - \cos\theta)\sin\theta - \frac{mgl}{I_O}\sin\theta\cos\theta\right)$$
$$= mg\,\frac{ml^2}{I_O}\sin\theta\,(2 - 3\cos\theta).$$

(ii) From this equation we have $R_1 = 0$ when $\sin\theta = 0$, with the only solution in the given range $0 \le \theta \le \frac{\pi}{2}$ being $\theta = 0$, or when $2 - 3\cos\theta = 0$, that is, when $3\cos\theta = 2$, with the only solution in the given range being $\theta = \arccos\frac{2}{3}$ (which is 0.8411, or about 48°).

(f) From Table 1, the moment of inertia of a thin rod of length $2l$ and mass m about an axis through its centre of mass is $\frac{1}{12}m(2l)^2 = \frac{1}{3}ml^2$. So for this model of the diver, using the parallel axes theorem,
$$I_O = \tfrac{1}{3}ml^2 + ml^2 = \tfrac{4}{3}ml^2.$$

Then from equation (39), we obtain
$$R_1 = \tfrac{3}{4}mg\sin\theta\,(2 - 3\cos\theta).$$

At $\theta = \frac{\pi}{2}$, the right-hand side of this equation is $\frac{3}{2}mg$. Therefore if the diver were still in contact with the board when $\theta = \frac{\pi}{2}$, she would need to be exerting a horizontal force on the board equal in magnitude to R_1 (and opposite in direction), that is, about 50% greater than the magnitude of her weight.

(g) Equation (39) shows that $R_1 < 0$ for small positive values of θ, since $\sin\theta \geq 0$ for $0 < \theta < \pi$, and for $\theta \simeq 0$, $\cos\theta \simeq 1$ so $2 - 3\cos\theta \simeq -1$. The board is 'pushing' (horizontally) on the diver's hands, and by Newton's third law, this corresponds to the diver pushing on the board. However, for $\theta > \arccos\frac{2}{3}$, we have $R_1 > 0$. We are assuming that the diver cannot pull on the board to provide a force in this direction, so the diver is unlikely to be able to rotate past $\arccos\frac{2}{3}$. The diver is therefore likely to lose contact with the board when θ is approximately $\arccos\frac{2}{3}$ (i.e. about 48°).

Solution to Exercise 20

Start from the definition of the total external torque relative to the fixed origin O, namely
$$\boldsymbol{\Gamma} = \sum_{i=1}^{n} \mathbf{r}_i \times \mathbf{F}_i.$$

Now use equation (40) to substitute for $\mathbf{r}_i$, to yield
$$\boldsymbol{\Gamma} = \sum_{i=1}^{n} (\mathbf{R} + \mathbf{r}_i^{\text{rel}}) \times \mathbf{F}_i.$$

Expanding the bracket gives
$$\boldsymbol{\Gamma} = \sum_{i=1}^{n} \mathbf{R} \times \mathbf{F}_i + \sum_{i=1}^{n} \mathbf{r}_i^{\text{rel}} \times \mathbf{F}_i.$$

The first term on the right-hand side simplifies because $\mathbf{R}$ is independent of i, while the second term is by definition the total external torque relative to the centre of mass (equation (43)), so
$$\boldsymbol{\Gamma} = \mathbf{R} \times \left(\sum_{i=1}^{n} \mathbf{F}_i \right) + \boldsymbol{\Gamma}^{\text{rel}} = \mathbf{R} \times \mathbf{F} + \boldsymbol{\Gamma}^{\text{rel}},$$

as required.

Solution to Exercise 21

(a) The total external force on the ith particle is $\mathbf{F}_i = cm_i\mathbf{k}$. Now, from equation (43),

$$\boldsymbol{\Gamma}^{\text{rel}} = \sum_{i=1}^{n} \mathbf{r}_i^{\text{rel}} \times \mathbf{F}_i = \sum_{i=1}^{n} \mathbf{r}_i^{\text{rel}} \times cm_i\mathbf{k} = \left(\sum_{i=1}^{n} m_i\mathbf{r}_i^{\text{rel}} \right) \times c\mathbf{k}.$$

But from equation (42), $\sum_{i=1}^{n} m_i\mathbf{r}_i^{\text{rel}} = \mathbf{0}$, so

$$\boldsymbol{\Gamma}^{\text{rel}} = \mathbf{0} \times c\mathbf{k} = \mathbf{0}.$$

Then from the torque law relative to the centre of mass (equation (47)), $\mathbf{L}^{\text{rel}}$ is constant.

(b) If we model the body in flight as a system of particles, then each particle of mass m_i is subject only to a force $\mathbf{F}_i = m_i g\mathbf{k}$, where $\mathbf{k}$ is a unit vector pointing vertically downwards. So the only external force on each particle has the form $cm_i\mathbf{k}$, where $c = g$ is a constant, hence the boxed result applies, that is, the angular momentum of the body relative to the centre of mass is conserved.

Solution to Exercise 22

If the external force on the ith particle is acting towards the origin, it must act along the line of the position vector $\mathbf{r}_i$ of the particle, so $\mathbf{F}_i = c_i\mathbf{r}_i$, for some scalar c_i. The external torque on this particle about the origin is

$$\boldsymbol{\Gamma}_i = \mathbf{r}_i \times \mathbf{F}_i = \mathbf{r}_i \times c_i\mathbf{r}_i = \mathbf{0}.$$

Then, summing over all particles, $\boldsymbol{\Gamma} = \mathbf{0}$. Hence from the torque decomposition theorem (equation (46)),

$$\mathbf{R} \times \mathbf{F} + \boldsymbol{\Gamma}^{\text{rel}} = \boldsymbol{\Gamma} = \mathbf{0}.$$

Thus $\boldsymbol{\Gamma}^{\text{rel}} = -\mathbf{R} \times \mathbf{F}$, as required.

Solution to Exercise 23

(a) The model of the baton is identical to that used in Exercise 15, and the angular speed (about the centre of mass) is the same as there. So the kinetic energy due to the rotation of the baton is $2.11\,\text{J}$. The total mass of the baton is $0.6\,\text{kg}$, and the speed of the centre of mass is $5\,\text{m s}^{-1}$. Hence the kinetic energy (in joules) of an equivalent particle at the centre of mass is

$$\tfrac{1}{2}(0.6)5^2 = 7.5.$$

Therefore by equation (51), the total kinetic energy of the baton (in joules) is $7.5 + 2.11 = 9.61$.

(b) An equivalent particle at the centre of mass of the baton would be moving vertically upwards. Since the centre of mass is vertically above O, this means that the position vector $\mathbf{R}$ of the centre of mass and the velocity $\dot{\mathbf{R}}$ of the centre of mass have the same direction.

Hence $\mathbf{R} \times \dot{\mathbf{R}} = \mathbf{0}$. So the angular momentum of an equivalent particle at the centre of mass is

$$\mathbf{R} \times M\dot{\mathbf{R}} = M(\mathbf{R} \times \dot{\mathbf{R}}) = \mathbf{0}.$$

Therefore by the angular momentum decomposition theorem (equation (45)), the angular momentum of the baton about O is equal to its angular momentum relative to the centre of mass. By equation (49), the $\mathbf{k}$-component of this is $I\omega$, where $I = 0.1071$ (from Exercise 15) and $\omega = 2\pi$, so

$$I\omega = 0.1071(2\pi) \simeq 0.6729.$$

So the angular momentum of the baton, relative to O, has a $\mathbf{k}$-component with magnitude approximately $0.67\,\mathrm{kg\,m^2\,s^{-1}}$.

Solution to Exercise 24

(a) (i)   During the brief period while the competitor is in the act of stopping, the large force exerted at X has zero torque about X. We would be justified in regarding the effect of the torque exerted by the weight of the caber during this short period as negligible, but since the caber is vertically above X, the line of action of its weight passes through X, thus will have zero torque about X anyway. Hence while the competitor is in the act of stopping, angular momentum about X is constant. Thus the angular momentum of the caber about X is the same just before and just after the competitor stops.

Before the competitor stops, the whole caber is moving forward in the $\mathbf{i}$-direction at speed v. We now use the angular momentum decomposition theorem, equation (45), to find the angular momentum of the caber about X. The angular momentum relative to the centre of mass is zero, since the caber is stationary relative to the centre of mass. The angular momentum (about X) of an equivalent particle of mass m and velocity $v\mathbf{i}$ at the centre of mass is

$$l\mathbf{j} \times mv\mathbf{i} = -mvl\mathbf{k}.$$

Therefore the total angular momentum of the caber about X is $-mvl\mathbf{k}$.

Just after the competitor stops, the caber will be rotating about an axis through X (in the negative $\mathbf{k}$-direction). Suppose that its angular speed is ω, that is, its angular velocity is $\boldsymbol{\omega} = -\omega\mathbf{k}$, and its moment of inertia about the z-axis through X is I. Now, angular momentum about X is the same just before and just after the competitor stops, so equating $\mathbf{k}$-components gives

$$-mvl = -I\omega,$$

thus

$$\omega = mvl/I.$$

Using Table 1 and the parallel axes theorem, we have

$$I = ml^2 + \tfrac{1}{12}m(2l)^2 = \tfrac{4}{3}ml^2.$$

Hence on substituting this into $\omega = mvl/I$, we obtain $\omega = \tfrac{3}{4}v/l$.

So just after the competitor stops, the caber will be rotating forwards (clockwise) at an angular speed of $\tfrac{3}{4}v/l$.

(ii) The kinetic energy of the caber just after the competitor stops is

$$\tfrac{1}{2}I\omega^2 = \tfrac{1}{2}(\tfrac{4}{3}ml^2)\left(\frac{3v}{4l}\right)^2 = \tfrac{3}{8}mv^2.$$

(b) Suppose that the caber has angular speed ω_1 about X when its angle with the vertical is θ. Since we assume that resistive forces are negligible, we can apply the conservation of mechanical energy, which implies that

$$\tfrac{1}{2}I\omega^2 + mgl = \tfrac{1}{2}I\omega_1^2 + mgl\cos\theta.$$

Rearranging gives

$$\omega_1^2 = \omega^2 + \frac{2mgl}{I}(1 - \cos\theta),$$

which, on substituting for $\omega = 3v/(4l)$ and I from part (a)(i), yields

$$\omega_1^2 = \frac{9v^2}{16l^2} + \frac{3g}{2l}(1 - \cos\theta).$$

So when the caber makes an angle θ with the vertical, its angular speed is

$$\omega_1 = \sqrt{\frac{9v^2}{16l^2} + \frac{3g}{2l}(1 - \cos\theta)}.$$

Solution to Exercise 25

(a) From the angular momentum decomposition theorem, the angular momentum $\mathbf{L}$ of the caber about O just prior to impact is the sum of the angular momentum $\mathbf{L}_G$ of an equivalent particle at the centre of mass, and the angular momentum $\mathbf{L}^{\mathrm{rel}}$ relative to the centre of mass. Now

$$\mathbf{L}_G = \mathbf{R} \times m\dot{\mathbf{R}},$$

where $\mathbf{R} = (-l\sin\theta)\mathbf{i} + (l\cos\theta)\mathbf{j}$ and $\dot{\mathbf{R}} = v_x\mathbf{i} + v_y\mathbf{j}$. So

$$\begin{aligned}
\mathbf{L}_G &= \big((-l\sin\theta)\mathbf{i} + (l\cos\theta)\mathbf{j}\big) \times m(v_x\mathbf{i} + v_y\mathbf{j}) \\
&= ml\big(-(v_y\sin\theta)\mathbf{k} - (v_x\cos\theta)\mathbf{k}\big) \\
&= -ml(v_x\cos\theta + v_y\sin\theta)\mathbf{k}.
\end{aligned}$$

From Table 1, the moment of inertia of the caber (modelled as a thin rod of length $2l$) about an axis through its centre of mass is $I = \tfrac{1}{12}m(2l)^2 = \tfrac{1}{3}ml^2$. Now, since the angular speed is ω and the caber is rotating clockwise, the angular velocity is $\boldsymbol{\omega} = -\omega\mathbf{k}$ (i.e. $\dot\theta = -\omega$). So using equation (49), the $\mathbf{k}$-component of the angular momentum of the caber relative to the centre of mass is $L^{\mathrm{rel}}_{\mathrm{axis}} = I(-\omega) = -\tfrac{1}{3}ml^2\omega$.

The **k**-component of the angular momentum **L** just before the caber hits the ground is the sum of the **k**-component of $\mathbf{L}^{\text{rel}}$ and the **k**-component of $\mathbf{L}_G$, and thus is

$$-ml\left(\tfrac{1}{3}l\omega + v_x \cos\theta + v_y \sin\theta\right).$$

(b) From the parallel axes theorem, the moment of inertia of the caber about an axis through O is

$$\tfrac{1}{3}ml^2 + ml^2 = \tfrac{4}{3}ml^2.$$

Just after impact we have $\boldsymbol{\omega} = -\omega_1\mathbf{k}$ since the caber is continuing to rotate clockwise but with angular speed ω_1. Assuming that the angular momentum about O is unchanged by the impact, we have, on equating **k**-components of the angular momentum just before impact (from part (a)) and just after impact,

$$-ml\left(\tfrac{1}{3}l\omega + v_x \cos\theta + v_y \sin\theta\right) = -\tfrac{4}{3}ml^2\omega_1.$$

Hence

$$\omega_1 = \frac{3}{4l}\left(\tfrac{1}{3}l\omega + v_x \cos\theta + v_y \sin\theta\right)$$

$$= \tfrac{1}{4}\omega + \frac{3}{4l}(v_x \cos\theta + v_y \sin\theta).$$

(c) After impact, the caber is rotating about its end O at an angular speed ω_1, so its kinetic energy is

$$\tfrac{1}{2}I\omega_1^2 = \tfrac{1}{2}\left(\tfrac{4}{3}ml^2\right)\omega_1^2 = \tfrac{2}{3}ml^2\omega_1^2.$$

To reach the vertical, its gain in potential energy (which equals loss in kinetic energy) must be

$$mgl(1 - \cos\theta).$$

To pass the vertical, the caber must reach the vertical while still retaining some kinetic energy, so we need

$$\tfrac{2}{3}ml^2\omega_1^2 - mgl(1 - \cos\theta) > 0,$$

that is (as $m > 0$ and $l > 0$),

$$\omega_1^2 > \frac{3g(1 - \cos\theta)}{2l}.$$

Solution to Exercise 26

(a) For a cylindrical shell with all its mass at a distance R from its axis, the moment of inertia about the centre of mass is MR^2. Then the total kinetic energy of the cylinder is

$$T = \tfrac{1}{2}M\dot{x}^2 + \tfrac{1}{2}(MR^2)\dot{\theta}^2 = M\dot{x}^2$$

(compare with equation (55)).

The assumption of conservation of mechanical energy then gives

$$M\dot{x}^2 = Mgx \sin\alpha$$

This moment of inertia was obtained in Exercise 12.

(compare with equation (56)). Differentiating with respect to time, we have

$$2M\dot{x}\ddot{x} = Mg\dot{x}\sin\alpha$$

(compare with equation (57)), so

$$\ddot{x} = \tfrac{1}{2}g\sin\alpha$$

(compare with equation (58)).

(We see that a hollow cylinder has a smaller acceleration than a solid one. A higher proportion of the potential energy lost goes into rotational kinetic energy in the case of a hollow cylinder.)

(b) From the final expressions for $\ddot{x}$ for solid and hollow cylinders, we can see that the acceleration is independent of the mass in each case. With $\alpha = \frac{\pi}{6}$, $\sin\alpha = \frac{1}{2}$, and substituting into the expressions for $\ddot{x}$, we obtain

$$\ddot{x}_{\text{solid}} = \tfrac{1}{3}g, \quad \ddot{x}_{\text{hollow}} = \tfrac{1}{4}g.$$

To find the time τ to move $2\,\text{m}$ down the slope at constant acceleration a_0, starting from rest, we can use the formula $x = x_0 + v_0 t + \frac{1}{2}a_0 t^2$ from Unit 3 (with $x_0 = v_0 = 0$) to obtain $\frac{1}{2}a_0\tau^2 = 2$. So $\tau = 2/\sqrt{a_0}$, and substituting for a_0 from the above expressions for $\ddot{x}_{\text{solid}}$ and $\ddot{x}_{\text{hollow}}$, the times (in seconds) are

$$\tau_{\text{solid}} = 1.11, \quad \tau_{\text{hollow}} = 1.28.$$

Therefore the solid cylinder is $0.17\,\text{s}$ faster.

Solution to Exercise 27

(a) The centre of mass has acceleration $\ddot{x}\mathbf{i}$, where x is displacement down the slope. So Newton's second law applied to an equivalent particle at the centre of mass gives

$$M\ddot{x}\mathbf{i} = \mathbf{W} + \mathbf{N} + \mathbf{F}$$
$$= Mg(\sin\alpha\,\mathbf{i} - \cos\alpha\,\mathbf{j}) + |\mathbf{N}|\mathbf{j} - |\mathbf{F}|\mathbf{i}.$$

Resolving in the $\mathbf{i}$- and $\mathbf{j}$-directions, we obtain

$$M\ddot{x} = Mg\sin\alpha - |\mathbf{F}|,$$
$$0 = -Mg\cos\alpha + |\mathbf{N}|.$$

The rotational acceleration of the cylinder about its centre of mass is $\ddot{\theta}$, where $x = -R\theta$ (from equation (54)), so $\ddot{x} = -R\ddot{\theta}$. The moment of inertia of a solid cylinder of mass M about an axis in the $\mathbf{k}$-direction through its centre of mass is $I = \frac{1}{2}MR^2$ (see Table 1). The total external torque relative to the centre of mass is given by

$$\boldsymbol{\Gamma}^{\text{rel}} = (-R\mathbf{j})) \times (-|\mathbf{F}|\mathbf{i}) = -R|\mathbf{F}|\mathbf{k},$$

so

$$\Gamma^{\text{rel}}_{\text{axis}} = -R|\mathbf{F}|.$$

Hence the equation of relative rotational motion, $\Gamma_{\text{axis}}^{\text{rel}} = I\ddot{\theta}$, implies that

$$-R|\mathbf{F}| = \tfrac{1}{2}MR^2\ddot{\theta} = -\tfrac{1}{2}MR\ddot{x}.$$

So we have

$$|\mathbf{F}| = \tfrac{1}{2}M\ddot{x}.$$

Substituting this into $M\ddot{x} = Mg\sin\alpha - |\mathbf{F}|$ gives

$$Mg\sin\alpha - \tfrac{1}{2}M\ddot{x} = M\ddot{x},$$

hence

$$g\sin\alpha = \tfrac{3}{2}\ddot{x},$$

which can be rearranged as

$$\ddot{x} = \tfrac{2}{3}g\sin\alpha.$$

This is the same as equation (58), which was obtained under the assumption of conservation of mechanical energy.

(b) From part (a) we have

$$|\mathbf{N}| = Mg\cos\alpha$$

and

$$|\mathbf{F}| = \tfrac{1}{2}M\ddot{x} = \tfrac{1}{2}M\left(\tfrac{2}{3}g\sin\alpha\right) = \tfrac{1}{3}Mg\sin\alpha.$$

So

$$\frac{|\mathbf{F}|}{|\mathbf{N}|} = \frac{Mg\sin\alpha}{3Mg\cos\alpha} = \tfrac{1}{3}\tan\alpha,$$

or equivalently,

$$|\mathbf{F}| = \tfrac{1}{3}\tan\alpha\,|\mathbf{N}|.$$

The condition that the cylinder rolls without slipping is $|\mathbf{F}| \le \mu|\mathbf{N}|$, that is, $\tfrac{1}{3}\tan\alpha\,|\mathbf{N}| \le \mu|\mathbf{N}|$, so

$$\tan\alpha \le 3\mu.$$

Index